Synthesis Lectures on Mechanical Engineering

This series publishes short books in mechanical engineering (ME), the engineering branch that combines engineering, physics and mathematics principles with materials science to design, analyze, manufacture, and maintain mechanical systems. It involves the production and usage of heat and mechanical power for the design, production and operation of machines and tools. This series publishes within all areas of ME and follows the ASME technical division categories.

Kingsley Ukoba · Tien-Chien Jen

Shaping Tomorrow: Thin Films and 3D Printing in the Fourth Industrial Revolution 1

Fundamentals

 Springer

Kingsley Ukoba (ID)
Department of Mechanical Engineering
Science
University of Johannesburg
Johannesburg, Gauteng, South Africa

Tien-Chien Jen (ID)
Department of Mechanical Engineering
Science
University of Johannesburg
Johannesburg, Gauteng, South Africa

ISSN 2573-3168   ISSN 2573-3176 (electronic)
Synthesis Lectures on Mechanical Engineering
ISBN 978-3-031-84123-1   ISBN 978-3-031-84124-8 (eBook)
https://doi.org/10.1007/978-3-031-84124-8

This Springer imprint is published by the registered company Springer Nature Switzerland AG
The registered company address is: Gewerbestrasse 11, 6330 Cham, Switzerland

If disposing of this product, please recycle the paper.

Competing Interests The authors have no competing interests to declare that are relevant to the content of this manuscript.

Contents

Industry 4.0 and Key Drivers

1

1.1 Introduction

In modern industry, the advent of Industry 4.0 marks a significant paradigm shift, characterized by the integration of digital technologies, automation, data exchange, and advanced manufacturing techniques. This chapter investigates the core principles of Industry 4.0 and identifies the key drivers propelling its evolution.

In recent years, the landscape of industrial production has undergone a profound transformation, ushering in a new era known as Industry 4.0. This paradigm shift represents the culmination of technological advancements in digitalization, connectivity, automation, and data analytics, converging to redefine the way goods are manufactured, distributed, and consumed. At the heart of Industry 4.0 lies the vision of interconnected, intelligent, and autonomous manufacturing systems, capable of driving unprecedented levels of efficiency, productivity, and innovation. Industry 4.0, often called the Fourth Industrial Revolution, builds upon the foundation laid by its predecessors—mechanization, mass production, and automation—to propel manufacturing into the digital age. Unlike previous industrial revolutions, which focused primarily on enhancing physical processes and labor efficiency, Industry 4.0 harnesses the power of digital technologies to create smart factories that are responsive, adaptable, and data-driven.

1.2 The Evolution of Industry 4.0

Industry 4.0, also known as the Fourth Industrial Revolution, represents a significant leap forward in the evolution of industrial production. Its origins can be traced back to Germany, where it was first introduced as part of the government's High-Tech Strategy 2020 initiative at the Hannover Messe trade fair in 2011. Since then, Industry 4.0 has

© The Author(s), under exclusive license to Springer Nature Switzerland AG 2025
K. Ukoba and T.-C. Jen, *Shaping Tomorrow: Thin Films and 3D Printing in the Fourth Industrial Revolution 1*, Synthesis Lectures on Mechanical Engineering,
https://doi.org/10.1007/978-3-031-84124-8_1

gained momentum worldwide, reshaping the manufacturing landscape and revolutioniz-ing the way goods are produced, distributed, and consumed. The concept of Industry 4.0 builds upon the foundation laid by its predecessors, marking a continuation of the ongo-ing transformation of industrial processes. The concept of Industry 4.0 traces its origins to Germany, where it was introduced as part of the government's High-Tech Strategy 2020 initiative in 2011. Coined at the Hannover Messe trade fair, Industry 4.0 signifies integrating cyber-physical systems, the Internet of Things (IoT), cloud computing, and artificial intelligence (AI) into industrial processes. Since its inception, Industry 4.0 has gained widespread traction worldwide, shaping the agendas of governments, businesses, and academia alike.

The industrial revolution has seen a lot of innovation and generations as shown in Fig. 1.1. The First Industrial Revolution, which began in the late eighteenth century, saw the mechanization of production through water and steam power, leading to the rise of factories and mass production. The Second Industrial Revolution, characterized by the advent of electricity and assembly lines in the late nineteenth and early twentieth cen-turies, further accelerated industrialization and ushered in an era of mass consumption. The Third Industrial Revolution, often referred to as the Digital Revolution, emerged in the latter half of the twentieth century with the widespread adoption of computers, automation, and digital technologies. This period saw the automation of manufactur-ing processes, the introduction of computer-aided design (CAD) and computer-aided manufacturing (CAM), and the rise of information technology (IT) systems for enter-prise resource planning (ERP) and supply chain management. Industry 4.0 represents the culmination of these technological advancements, fusing digitalization, connectivity, automation, and data analytics to create smart, interconnected manufacturing systems. Unlike its predecessors, which focused primarily on enhancing physical processes and labor efficiency, Industry 4.0 emphasizes the integration of cyber-physical systems (CPS) with the Internet of Things (IoT), cloud computing, and artificial intelligence (AI). At its core, Industry 4.0 seeks to create "smart factories" that are capable of autonomously monitoring, analyzing, and optimizing production processes in real-time. Through the use of IoT sensors, devices, and machines communicate and share data seamlessly, enabling intelligent decision-making and adaptive responses to changing conditions. This inter-connectedness extends beyond the factory floor to encompass the entire value chain, facilitating collaboration, transparency, and agility across suppliers, manufacturers, and customers. The evolution of Industry 4.0 represents a transformative shift in the way goods are manufactured, distributed, and consumed. By embracing digitalization, connectivity, automation, and data analytics, organizations can unlock new opportunities for efficiency, agility, and competitiveness in an increasingly interconnected and dynamic global econ-omy. As Industry 4.0 continues to evolve, the synergy between technology, innovation, and human ingenuity will shape the future of industry in unprecedented ways. Industry 4.0 represents a transformative era in manufacturing, driven by digitalization, connectiv-ity, automation, and advanced technologies. By embracing the key drivers of Industry 4.0,

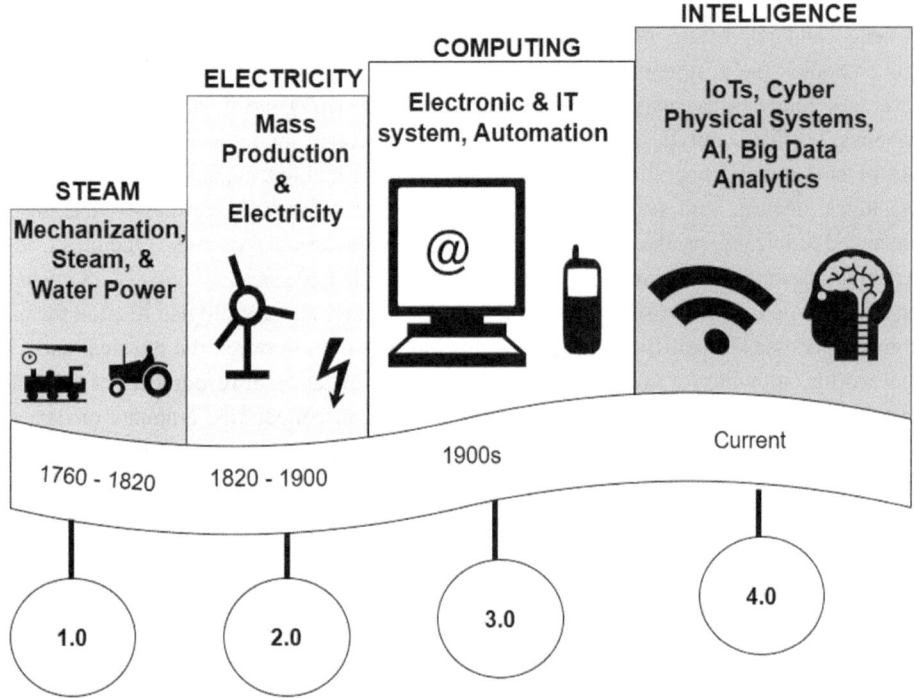

Fig. 1.1 Schematic of industrial revolution

organizations can unlock new opportunities for growth, competitiveness, and sustainability in an increasingly interconnected and dynamic global economy. As the Fourth Industrial Revolution continues to unfold, the synergy between technology, innovation, and human ingenuity will shape the future of industry in unprecedented ways.

1.3   Understanding Industry 4.0

Industry 4.0, also known as the Fourth Industrial Revolution, represents a convergence of cyber-physical systems, the Internet of Things (IoT), cloud computing, and artificial intelligence (AI) to revolutionize industrial processes. At its core, Industry 4.0 seeks to create "smart factories" that are interconnected, data-driven, and capable of autonomous decision-making.

In the vast landscape of technological advancements, few concepts have captured the imagination and intrigue of industries worldwide like Industry 4.0. It represents a paradigm shift in manufacturing and production processes, promising unprecedented levels of automation, connectivity, and data-driven decision-making. To truly grasp the essence of Industry 4.0, one must examine its core principles, understand its key components, and appreciate its implications across various sectors. At its heart, Industry 4.0 embodies the fusion of digital technologies with traditional industrial practices. It leverages innovations such as the Internet of Things (IoT), artificial intelligence (AI), robotics, big data analytics, and cloud computing to create interconnected systems capable of autonomous operation and intelligent decision-making. This integration enables machines, devices, and systems to communicate and collaborate seamlessly, leading to enhanced efficiency, productivity, and agility in manufacturing processes. One of the fundamental aspects of Industry 4.0 is the concept of cyber-physical systems (CPS), where physical components are seamlessly integrated with digital technologies to monitor and control processes in real-time. These systems blur the lines between the physical and virtual worlds, allowing for precise coordination and optimization of resources. For example, in a smart factory, sensors embedded in machinery can collect vast amounts of data on performance and conditions, which are then analyzed to predict maintenance needs or optimize production schedules.

The Internet of Things (IoT) plays a pivotal role in enabling connectivity within Industry 4.0 ecosystems. By interconnecting devices, sensors, and machines, IoT facilitates the seamless exchange of data across the manufacturing value chain. This interconnectedness enables unprecedented levels of visibility and transparency, allowing stakeholders to monitor operations remotely and make informed decisions in real-time. For instance, IoT-enabled sensors in supply chain logistics can track the movement of goods from raw material suppliers to the end customer, providing insights into inventory levels, delivery times, and potential bottlenecks. Artificial intelligence (AI) is another cornerstone of Industry 4.0, empowering machines to analyze data, learn from patterns, and make autonomous decisions. Machine learning algorithms can optimize production processes by identifying inefficiencies, predicting equipment failures, and recommending corrective actions. Moreover, AI-powered predictive maintenance can reduce downtime and extend the lifespan of machinery, leading to significant cost savings for manufacturers. Robotics plays a central role in automating repetitive tasks and enhancing precision in manufacturing environments. Collaborative robots, or cobots, work alongside human operators to perform tasks that require dexterity and flexibility. These robots can be programmed to adapt to changing production requirements swiftly, improving efficiency and ensuring consistent quality. Furthermore, advancements in autonomous mobile robots (AMRs) enable the seamless movement of materials within factory floors, optimizing logistics and reducing manual handling.

Big data analytics is instrumental in extracting actionable insights from the vast amounts of data generated by Industry 4.0 systems. By analyzing production data, market trends, and customer preferences, manufacturers can optimize processes, tailor products to specific customer needs, and identify new business opportunities. Predictive analytics can anticipate demand fluctuations, enabling proactive inventory management and resource allocation. Additionally, data-driven decision-making enhances agility and responsiveness to dynamic market conditions, allowing companies to stay ahead of the competition. Cloud computing infrastructure provides the scalability and flexibility needed to support Industry 4.0 applications effectively. By leveraging cloud services, manufacturers can access computing resources on-demand, scale their operations seamlessly, and deploy new applications rapidly. Cloud-based platforms also facilitate collaboration and knowledge sharing across geographically dispersed teams, enabling seamless integration of processes and data across the organization. Security and privacy are paramount concerns in the era of Industry 4.0, where interconnected systems are vulnerable to cyber threats and data breaches. Robust cybersecurity measures, including encryption, authentication, and intrusion detection, are essential to safeguarding sensitive information and ensuring the integrity of industrial networks. Moreover, compliance with data protection regulations such as GDPR is crucial to maintaining trust and credibility with customers and partners.

1.4   Harbinger of Industry 4.0

In the annals of industrial revolutions, few phenomena have sparked as much anticipation and transformation as Industry 4.0. Often hailed as the fourth industrial revolution, it represents a seismic shift in the way we conceive, design, and produce goods and services. Serving as the harbinger of this paradigm shift, Industry 4.0 embodies a convergence of digital technologies, automation, and data-driven insights, reshaping industries and societies worldwide. At its essence, Industry 4.0 signifies the fusion of physical and digital realms, blurring the lines between the tangible world of machinery and the intangible realm of data and algorithms. It marks the dawn of a new era where cyber-physical systems orchestrate manufacturing processes with unprecedented efficiency, agility, and intelligence. But what makes Industry 4.0 a harbinger of change, and what implications does it hold for the future of industry and beyond?

The advent of Industry 4.0 is characterized by several key technological advancements, each serving as a harbinger of profound transformation across various domains. Chief among these advancements is the Internet of Things (IoT), which connects an ever-expanding network of sensors, devices, and machinery, enabling real-time monitoring, control, and optimization of industrial processes. This interconnectedness lays the groundwork for smart factories, where machines communicate seamlessly and adapt autonomously to changing conditions. Another harbinger of Industry 4.0 is artificial intelligence (AI), which imbues machines with the ability to learn, reason, and make decisions autonomously. AI algorithms analyze vast datasets to uncover patterns, predict outcomes, and optimize processes, revolutionizing everything from predictive maintenance to quality control. Machine learning algorithms, in particular, enable adaptive and self-learning systems capable of continuously improving performance over time. Robotics also stands as a harbinger of Industry 4.0, ushering in an era of collaborative automation where humans and machines work hand in hand to achieve unprecedented levels of productivity and flexibility. Collaborative robots, or cobots, assist human workers in tasks that require precision, dexterity, or repetitive motion, enhancing efficiency and safety on the factory floor. Meanwhile, autonomous mobile robots (AMRs) navigate warehouse environments autonomously, optimizing material handling and logistics operations.

Big data analytics serves as another harbinger of Industry 4.0, empowering organizations to derive actionable insights from the vast volumes of data generated by interconnected systems. By harnessing the power of advanced analytics techniques such as predictive modeling and prescriptive analytics, companies can optimize processes, anticipate market trends, and make data-driven decisions with confidence. The ability to extract value from data lies at the heart of Industry 4.0's transformative potential. Cloud computing infrastructure emerges as a crucial harbinger of Industry 4.0, providing the scalable computing power and storage capacity needed to support data-intensive applications and services. Cloud platforms enable organizations to deploy and scale Industry 4.0 solutions rapidly, without the need for substantial upfront investments in hardware or infrastructure. This accessibility democratizes access to advanced technologies, leveling the playing field for organizations of all sizes. Cybersecurity also emerges as a critical harbinger of Industry 4.0, as interconnected systems become increasingly vulnerable to cyber threats and attacks. Robust cybersecurity measures are essential to safeguarding industrial networks, protecting sensitive data, and ensuring the integrity of cyber-physical systems. As Industry 4.0 adoption accelerates, organizations must prioritize cybersecurity to mitigate risks and build trust in digital ecosystems.

The implications of Industry 4.0 extend far beyond the realm of manufacturing, permeating into diverse sectors such as healthcare, transportation, energy, and beyond. In healthcare, for example, Industry 4.0 technologies enable remote patient monitoring, personalized treatment plans, and predictive analytics for early disease detection. In transportation, autonomous vehicles and smart infrastructure revolutionize mobility, reducing congestion and emissions while enhancing safety and efficiency. In energy, smart grids and renewable technologies optimize power generation and distribution, paving the way for a more sustainable future. Industry 4.0 stands as a harbinger of unprecedented change, heralding a new era of interconnectedness, intelligence, and automation. By harnessing the power of digital technologies, organizations can unlock new levels of efficiency, productivity, and innovation. However, realizing the full potential of Industry 4.0 requires a holistic approach that addresses technological, organizational, and societal challenges. Only by embracing change and fostering collaboration can industries thrive in the age of Industry 4.0. As the harbinger of a new industrial revolution, Industry 4.0 holds the promise of a brighter, more connected future for industries and societies worldwide.

1.5   Implications of Industry 4.0

The implications of Industry 4.0 extend far beyond the realm of manufacturing, influencing various sectors such as healthcare, transportation, and energy. In healthcare, for example, smart devices and wearable sensors enable remote patient monitoring and personalized treatment plans. In transportation, autonomous vehicles and smart infrastructure promise to revolutionize mobility and reduce traffic congestion. In energy, smart grids and renewable technologies pave the way for sustainable and efficient power generation and distribution.

The emergence of Industry 4.0 heralds a profound transformation across various sectors, promising to reshape economies, redefine business models, and revolutionize the way we work and live. The implications of this technological paradigm extend far beyond the confines of manufacturing, permeating into diverse industries and impacting every aspect of society. At the core of Industry 4.0 lies the convergence of digital technologies with traditional industrial practices, ushering in an era of unprecedented connectivity, automation, and data-driven decision-making. This convergence enables the creation of cyber-physical systems (CPS) where physical components are seamlessly integrated with digital technologies to monitor, control, and optimize processes in real-time.

One of the primary implications of Industry 4.0 is the transformation of manufacturing processes, leading to enhanced efficiency, productivity, and agility. By leveraging innovations such as the Internet of Things (IoT), artificial intelligence (AI), robotics, and big data analytics, manufacturers can streamline operations, reduce downtime, and improve product quality. This transformation extends beyond the factory floor, influencing supply chain management, logistics, and customer interactions. The advent of Industry 4.0 also brings significant implications for the workforce, necessitating a shift in skills and capabilities. Automation and robotics are increasingly augmenting human labor, automating repetitive tasks, and enhancing precision. As a result, there is a growing demand for workers with expertise in data analytics, cybersecurity, AI, and robotics. Additionally, Industry 4.0 necessitates a culture of lifelong learning and continuous upskilling to adapt to evolving technological landscapes. Furthermore, Industry 4.0 has profound implications for business models and organizational structures. Traditional hierarchical structures are giving way to more agile, networked organizations capable of rapid adaptation and innovation. Digital platforms and ecosystems are enabling new forms of collaboration and value creation, blurring the boundaries between industries and fostering innovation ecosystems.

In terms of sustainability and environmental impact, Industry 4.0 offers both opportunities and challenges. On the one hand, smart technologies and data analytics can optimize resource usage, reduce waste, and minimize environmental footprint. For example, predictive maintenance can extend the lifespan of machinery, reducing the need for replacements and conserving resources. On the other hand, the proliferation of connected devices and digital infrastructure poses challenges in terms of energy consumption, electronic waste, and cybersecurity risks. From a societal perspective, Industry 4.0 has implications for job creation, economic development, and social inclusion. While automation may lead to job displacement in some sectors, it also creates opportunities for new types of employment and entrepreneurship. Moreover, Industry 4.0 has the potential to drive economic growth, spur innovation, and improve living standards, particularly in developing countries.

In terms of governance and policy, Industry 4.0 poses challenges related to data privacy, cybersecurity, intellectual property rights, and ethical considerations. As industries become more interconnected and reliant on digital technologies, there is a need for robust regulatory frameworks to ensure accountability, transparency, and trust. Policymakers must strike a balance between promoting innovation and safeguarding public interests, fostering an environment conducive to responsible innovation and technological development. The implications of Industry 4.0 are far-reaching and multifaceted, touching upon every aspect of society, economy, and governance. While offering tremendous opportunities for innovation, growth, and efficiency, Industry 4.0 also presents challenges related to workforce adaptation, environmental sustainability, and ethical considerations. Embracing these implications requires a holistic approach that fosters collaboration, invests in education and training, and prioritizes ethical and inclusive technological development. Only by navigating these challenges effectively can societies harness the full potential of Industry 4.0 to create a more prosperous, sustainable, and inclusive future.

Industry 4.0 represents a transformative force that is reshaping the global industrial landscape. By harnessing the power of digital technologies, connectivity, and data analytics, companies can unlock new levels of efficiency, productivity, and innovation. However, realizing the full potential of Industry 4.0 requires a holistic approach that addresses technological, organizational, and societal challenges. Only by embracing change and fostering collaboration can industries thrive in the age of Industry 4.0.

Industry 4.0 technologies have the potential to significantly enhance various aspects of manufacturing and supply chain management. Implementing predictive maintenance systems based on Industry 4.0 principles can lead to a reduction in downtime by up to 30% and decrease maintenance costs by 25–35% (Armbrust et al., 2010). Furthermore, the adoption of Industry 4.0 solutions in supply chain management can result in inventory reduction by 20–25% and lead time reduction by 15–20%. Integration of smart energy management systems in Industry 4.0 environments can achieve energy savings of 10–15% (Ho et al., 2022). Moreover, the implementation of AI-driven quality control systems can reduce defect rates by 20–25% and improve product quality by 15–20%. Effective cybersecurity measures are crucial in Industry 4.0 environments, as they can reduce the risk of cyber attacks targeting these systems by 20–30%. Organizations leveraging cloud computing for Industry 4.0 initiatives can achieve a substantial return on investment (ROI) of 300–400% over a five-year period. Additionally, Industry 4.0 adoption can lead to the creation of new job roles related to data analytics, cybersecurity, and robotics, while also necessitating workforce reskilling in 80–85% of existing job roles. Companies embracing Industry 4.0 technologies can experience significant revenue growth of 10–15% annually compared to non-adopters. By 2025, it is projected that a large percentage of manufacturing companies globally will have implemented Industry 4.0 initiatives to some extent.

1.5.1 Global Impact of Industry 4.0

The Fourth Industrial Revolution (4IR) is expected to bring significant economic transformations across various industries. The World Economic Forum predicts that by 2025, the 4IR could unlock up to $3.7 trillion in economic value (Paschen et al., 2019). Moreover, McKinsey Global Institute projects that by 2030, the economic impact of 4IR technologies could range between $14 and $33 trillion annually (Williamson et al., 2021). However, this revolution is not without its challenges. The World Economic Forum's Future of Jobs Report suggests that by 2022, 42% of core job skills will undergo significant changes due to automation and 4IR technologies (Wei et al., 2022). Oxford Economics estimates that by 2030, automation could displace up to 20 million manufacturing jobs globally (Wei et al., 2022).

The integration of digital technologies into daily life and business operations has been facilitated by the widespread access to the internet and mobile connectivity. As of 2022, over 60% of the global population had internet access, and there were approximately 111 mobile cellular subscriptions per 100 inhabitants globally (Mertanen et al., 2021). This connectivity has paved the way for the adoption of digital transformation initiatives, with worldwide spending expected to exceed $2.8 trillion by 2022 (Nawaz et al., 2023).

Artificial Intelligence (AI) and the Internet of Things (IoT) are key drivers of the 4IR. The global AI market is projected to reach $327.5 billion by 2028 (Perminova, 2021), while the IoT market is expected to grow from $250.72 billion in 2020 to $1.39 trillion by 2026, with a compound annual growth rate of 29.4% (Fouad, 2022). Additionally, the 3D printing market is anticipated to reach $63.46 billion by 2028, driven by advancements in materials and applications (Sethi et al., 2022). The COVID-19 pandemic has further accelerated digital transformation efforts, with remote work and digital collaboration tools becoming ubiquitous (Autor, 2015). Educational Technology (EdTech) platforms have experienced significant growth, with the global online education market projected to reach $585.48 billion by 2027, fueled by the demand for remote learning solutions (Arrieta et al., 2020). The Fourth Industrial Revolution is reshaping economies and industries globally, with technologies like AI, IoT, and 3D printing playing pivotal roles. The digital transformation accelerated by the pandemic underscores the importance of adapting to technological advancements for economic growth and sustainability. These detailed statistics underscore the multifaceted impact of the Fourth Industrial Revolution on economies, societies, and industries worldwide.

1.6   Key Drivers of Industry 4.0

The proliferation of digital technologies and interconnected devices forms the backbone of Industry 4.0. Through IoT sensors, devices, and machines communicate and share data in real-time, enabling seamless integration and coordination across the manufacturing ecosystem. Industry 4.0 harnesses the power of big data and advanced analytics to derive actionable insights from vast amounts of manufacturing data. Predictive analytics, machine learning algorithms, and data visualization tools empower decision-makers to optimize processes, improve quality, and drive innovation. Automation plays a pivotal role in Industry 4.0, enabling repetitive tasks to be performed by machines with precision and efficiency. Collaborative robots (cobots), autonomous vehicles, and automated guided vehicles (AGVs) streamline production processes, enhance safety, and reduce labor costs. Artificial Intelligence and Machine Learning technologies enable machines to learn from data, adapt to changing conditions, and make intelligent decisions autonomously. From predictive maintenance and quality control to adaptive manufacturing and supply chain optimization, AI augments human capabilities and drives operational excellence. Additive manufacturing, including 3D printing, revolutionizes traditional manufacturing processes by enabling the fabrication of complex geometries, customized components, and on-demand production. By eliminating the need for tooling and reducing material waste, additive manufacturing accelerates prototyping, reduces lead times, and fosters innovation. With increased connectivity comes the need for robust cybersecurity measures to protect sensitive data and critical infrastructure from cyber threats. Industry 4.0 emphasizes the implementation of cybersecurity protocols, encryption techniques, and secure communication channels to safeguard digital assets and ensure data privacy.

1.7   Technological Enablers of Industry 4.0

The advent of Industry 4.0 is propelled by a convergence of transformative technologies reshaping the landscape of manufacturing and industry. These technological enablers serve as the foundation for the digital revolution, fostering unprecedented levels of connectivity, efficiency, and innovation across various sectors. Here are some key components driving this paradigm shift. Figure 1.2 shows the major 4IR technologies.

1.7.1   Internet of Things (IoT)

The Internet of Things (IoT) stands as a cornerstone of Industry 4.0, connecting physical devices and machinery to the Internet, enabling data exchange and automation on a vast scale. Through embedded sensors, actuators, and connectivity modules, IoT systems facilitate real-time monitoring, predictive maintenance, and seamless integration of

Fig. 1.2 Schematic of key technology enablers of industry 4.0

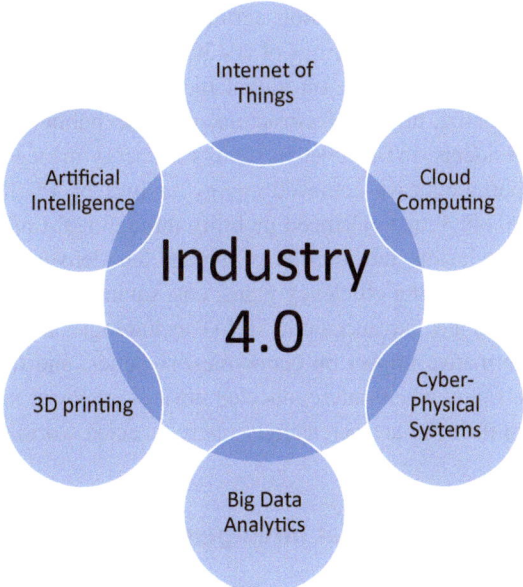

machines and processes. The Internet of Things (IoT) has achieved remarkable milestones across various industries, driving significant advancements in connectivity, automation, and data-driven decision-making. With an estimated 25 billion connected devices globally as of 2022, IoT has become ubiquitous, permeating almost every aspect of modern life. In manufacturing, IoT has revolutionized traditional production processes, leading to the emergence of smart factories equipped with interconnected sensors, robots, and machines. These smart manufacturing systems have witnessed productivity gains of up to 25% and cost reductions of 10–20% through optimized asset utilization, predictive maintenance, and real-time monitoring of production processes.

In healthcare, IoT-powered medical devices and wearables have enabled remote patient monitoring, personalized treatment plans, and early disease detection. Remote monitoring solutions have shown a 30% reduction in hospital readmissions and a 20% decrease in healthcare costs, while wearable health trackers have empowered individuals to take proactive measures towards improving their well-being. The transportation sector has benefited immensely from IoT applications, with connected vehicles, smart infrastructure, and logistics optimization leading to enhanced safety, efficiency, and sustainability. IoT-enabled fleet management solutions have achieved fuel savings of up to 15% and reduced accident rates by 25%, while smart traffic management systems have alleviated congestion and reduced travel times by 10–20% in urban areas. In agriculture, IoT has enabled precision farming techniques such as soil monitoring, crop analytics, and automated irrigation, leading to higher crop yields, reduced resource consumption, and improved environmental sustainability. Precision agriculture solutions have demonstrated yield increases of 10–20% and water savings of 20–30%, contributing to food security and mitigating the impact of climate change on global food production. Smart cities leverage IoT technologies to optimize urban infrastructure, enhance public services, and improve quality of life for residents. IoT-enabled systems for energy management, waste disposal, and public transportation have achieved energy savings of 15–30%, reduced waste collection costs by 20–40%, and enhanced mobility through real-time traffic monitoring and optimization. In conclusion, the Internet of Things has delivered significant achievements across industries, driving efficiency gains, cost savings, and innovation at unprecedented scales. With continued advancements in IoT technologies and adoption, the potential for further transformative impact on economies, societies, and the environment remains vast, promising a future where interconnected devices empower individuals, businesses, and governments to thrive in an increasingly interconnected world.

1.7.2   Artificial Intelligence (AI)

Artificial Intelligence (AI) plays a pivotal role in Industry 4.0 by imbuing machines with cognitive capabilities, enabling them to perceive, reason, and make decisions autonomously. From machine learning algorithms optimizing production workflows to neural networks enhancing quality control and predictive analytics, AI-driven systems empower organizations to extract actionable insights and drive efficiency gains. The subset of artificial intelligence are shown in Fig. 1.3.

Machine Learning:

Fig. 1.3 Schematic of classes of artificial intelligence

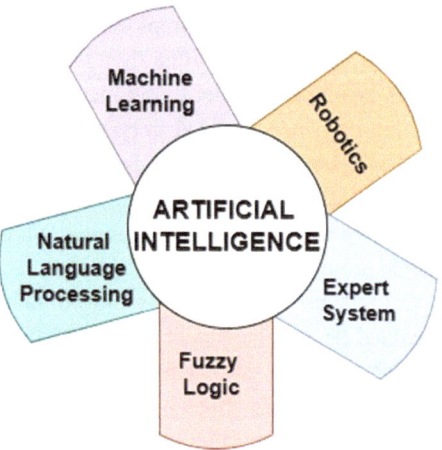

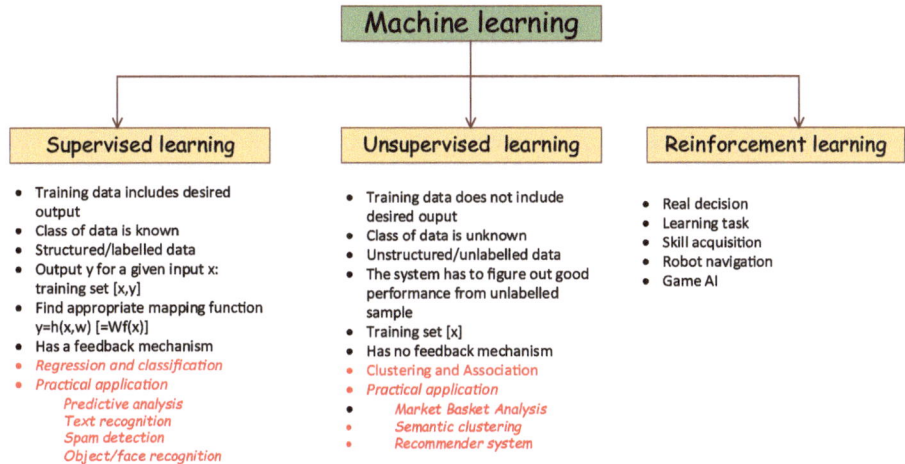

Fig. 1.4 Classification of machine learning

Machine learning (ML) is a subset of artificial intelligence (AI) that focuses on the development of algorithms and statistical models that enable computers to learn and improve from experience without being explicitly programmed. Machine learning algorithms analyze data, identify patterns, and make decisions or predictions based on the observed patterns. ML has become a critical tool in various industries, driving innovation, efficiency, and automation across diverse domains.

Classifications of Machine Learning:

Machine learning is classified into three distinct categories using the signal and feedback approach. This is shown in Fig. 1.4 to be reinforcement, supervised and unsupervised.

1. *Supervised Learning*: In supervised learning, the algorithm learns from labeled data, where each input is associated with a corresponding output label. The algorithm aims to learn a mapping function that can accurately predict the output for new, unseen inputs. Examples of supervised learning algorithms include linear regression, logistic regression, decision trees, support vector machines (SVM), and neural networks.
2. *Unsupervised Learning*: In unsupervised learning, the algorithm learns from unlabeled data, where the input data is not accompanied by output labels. The algorithm explores the structure of the data to identify patterns, clusters, or associations among the data points. Examples of unsupervised learning algorithms include clustering algorithms (k-means, hierarchical clustering), dimensionality reduction techniques (principal component analysis, t-distributed stochastic neighbor embedding), and association rule learning. Semi-supervised learning combines elements of supervised and unsupervised learning, where the algorithm learns from a mixture of labeled and unlabeled data.

The algorithm leverages the labeled data to guide the learning process and improve performance on tasks such as classification or regression.

3. Reinforcement Learning: Reinforcement learning involves training an agent to make sequential decisions in an environment to maximize cumulative rewards. The agent learns through trial and error, receiving feedback from the environment based on its actions. Reinforcement learning algorithms aim to learn optimal policies or strategies for achieving specific goals in dynamic and uncertain environments.

Machine Learning Models:

1. Linear Regression: Linear regression is a supervised learning algorithm used for predicting continuous output variables based on one or more input features. It models the relationship between the input variables and the output variable using a linear equation.
2. Decision Trees: Decision trees are versatile supervised learning algorithms used for classification and regression tasks. Decision trees partition the input space into regions based on the feature values and make predictions by traversing the tree structure.
3. Neural Networks: Neural networks are a class of deep learning models inspired by the structure and function of the human brain. They consist of interconnected nodes organized into layers, where each node performs a simple computation. Neural networks can learn complex patterns and relationships from data and are used for tasks such as image recognition, natural language processing, and reinforcement learning.
4. Support Vector Machines (SVM): SVM is a supervised learning algorithm used for classification and regression tasks. It works by finding the optimal hyperplane that separates the input data into different classes while maximizing the margin between classes.

Figure 1.5 shows the techniques and algorithms used in machine learning. Most form of unsupervised learning are form of cluster analysis.

Machine Learning Applications:

1. Image Recognition: Machine learning algorithms are widely used for image recognition tasks, such as object detection, classification, and segmentation. Applications include facial recognition systems, autonomous vehicles, medical imaging, and quality control in manufacturing.
2. Natural Language Processing (NLP): Machine learning techniques are used for various NLP tasks, including sentiment analysis, named entity recognition, machine translation, and text generation. NLP applications include virtual assistants, chatbots, language translation services, and document summarization tools.
3. Predictive Analytics: Machine learning algorithms are employed for predictive analytics tasks, such as forecasting future trends, predicting customer behavior, and

Fig. 1.5 Schematic of techniques and algorithms in machine learning

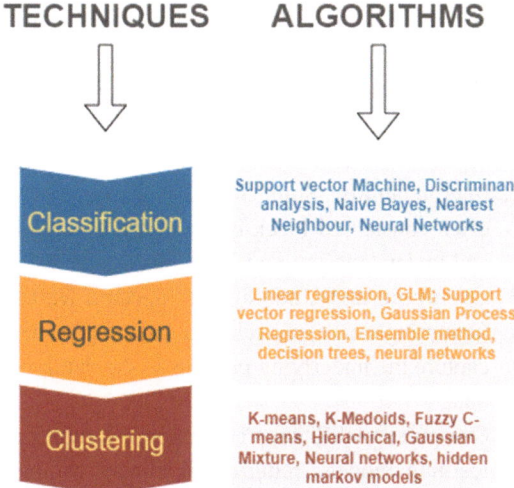

identifying anomalies or outliers in data. Applications include financial forecasting, demand forecasting, predictive maintenance, and fraud detection.

4. Healthcare: Machine learning is revolutionizing healthcare by enabling personalized medicine, disease diagnosis, treatment planning, and drug discovery. Applications include medical imaging analysis, patient risk stratification, genomics, and electronic health record analysis.

5. Financial Services: Machine learning algorithms are used in financial services for tasks such as credit scoring, risk assessment, algorithmic trading, and fraud detection. Applications include loan approval systems, automated trading platforms, fraud detection systems, and customer churn prediction.

Opportunities in Machine Learning:

1. Automation: Machine learning enables automation of repetitive tasks and decision-making processes, leading to increased efficiency, productivity, and scalability across industries.

2. Personalization: Machine learning algorithms can analyze large volumes of data to understand individual preferences, behaviors, and needs, enabling personalized recommendations, services, and experiences.

3. Insights Discovery: Machine learning techniques uncover hidden patterns, trends, and insights from data that can inform strategic decision-making, innovation, and optimization of business processes.

4. Healthcare Innovation: Machine learning contributes to advancements in healthcare by improving diagnostic accuracy, treatment effectiveness, and patient outcomes through personalized medicine, predictive analytics, and precision medicine.

Limitations of Machine Learning:

1. Data Dependency: Machine learning models require large volumes of high-quality data for training, which may not always be available or representative of real-world scenarios. Biased or incomplete data can lead to inaccurate or biased predictions.
2. Interpretability: Some machine learning models, such as deep neural networks, are often considered black-box models, making it challenging to interpret their decisions and understand the underlying factors driving predictions.
3. Overfitting and Underfitting: Machine learning models are susceptible to overfitting (capturing noise or irrelevant patterns in the training data) or underfitting (failing to capture the underlying patterns in the data), which can degrade performance on unseen data.
4. Computational Resources: Training complex machine learning models, especially deep learning models, requires significant computational resources, including processing power, memory, and storage, which can be costly and time-consuming.
5. Ethical and Privacy Concerns: Machine learning algorithms may inadvertently perpetuate or amplify biases present in the training data, leading to unfair or discriminatory outcomes. Additionally, the use of sensitive or personal data in machine learning models raises concerns about privacy, security, and data protection.

Machine learning holds immense potential to transform industries, drive innovation, and improve decision-making processes. However, addressing the challenges and limitations of machine learning requires ongoing research, collaboration, and ethical considerations to ensure responsible and equitable deployment of machine learning technologies.

Expert Systems:

Expert Systems are AI systems designed to emulate the decision-making capabilities of human experts in specific domains. These systems use a knowledge base containing expert knowledge and a set of rules or algorithms to reason and make decisions. Expert Systems are typically used in domains where expertise is valuable but scarce, such as medical diagnosis, financial analysis, and troubleshooting technical problems. Expert Systems have achieved significant milestones and have been applied across various domains, contributing to advancements in decision-making, problem-solving, and knowledge management. While specific statistics regarding the adoption and economic impact of Expert Systems may vary depending on the domain and application, their contributions to efficiency, accuracy, and innovation are evident. In the healthcare sector, Expert Systems have made significant strides in medical diagnosis, treatment planning, and disease management. For instance, studies have shown that Expert Systems for medical diagnosis can achieve accuracy rates ranging from 80 to over 90% in certain conditions. These systems help healthcare professionals make more informed decisions, reduce diagnostic errors, and improve patient outcomes. In terms of economic impact, the global healthcare artificial intelligence market size was valued at approximately $2.5 billion in 2020, with an expected compound annual growth rate (CAGR) of over 41% from 2021 to 2028. In finance, Expert Systems have revolutionized investment advisory services, risk assessment, and fraud detection. By analyzing vast amounts of financial data and market trends, Expert Systems help investors make informed decisions, optimize portfolios, and mitigate risks. The global AI in Fintech market size was valued at around $6.2 billion in 2020, with a projected CAGR of over 23% from 2021 to 2027. In manufacturing and engineering, Expert Systems contribute to process optimization, quality control, and predictive maintenance. By integrating Expert Systems with sensor data and automation technologies, manufacturers can enhance production efficiency, minimize downtime, and improve product quality. The global AI in manufacturing market size was estimated at approximately $1.1 billion in 2020, with an expected CAGR of over 57% from 2021 to 2028. Overall, Expert Systems continue to play a crucial role in various industries, driving innovation, efficiency, and competitiveness. While the economic impact and adoption rates may vary across sectors, the potential for Expert Systems to revolutionize decision-making and problem-solving processes remains significant. As organizations increasingly recognize the value of leveraging expert knowledge and AI technologies, the adoption of Expert Systems is expected to continue to grow, leading to further advancements and opportunities for businesses and society as a whole.

Fuzzy Logic:

Fuzzy Logic is a mathematical framework for representing and reasoning with uncertainty and imprecision. Unlike traditional logic, which operates on binary true or false values, fuzzy logic allows for degrees of truth between 0 and 1. Fuzzy Logic is particularly useful in situations where linguistic variables and vague or ambiguous concepts are involved, such as in natural language processing, control systems, and decision-making processes.

Fuzzy Logic has been widely applied in control systems, particularly in situations where precise mathematical models are difficult to establish or where system dynamics are complex and nonlinear. Fuzzy Logic controllers have been successfully used in various industrial and consumer applications, including automotive systems, household appliances, and robotics. Fuzzy Logic has also been applied in pattern recognition tasks, such as image processing, speech recognition, and handwriting recognition. Fuzzy pattern recognition algorithms can handle uncertainties and variations in input data, making them robust to noise and distortion. Fuzzy Logic-based decision support systems provide a framework for modeling human reasoning and decision-making processes. These systems can handle uncertain or incomplete information and assist decision-makers in evaluating alternatives and selecting optimal courses of action.

Fuzzy Logic-based control systems have improved vehicle stability, traction control, and anti-lock braking systems, leading to safer and more efficient transportation. The global automotive electronics market size was valued at over $200 billion in 2020, with a projected CAGR of over 5% from 2021 to 2028. Fuzzy Logic is used in household appliances such as washing machines, air conditioners, and refrigerators to optimize performance, energy efficiency, and user experience. The global home appliance market size was valued at over $180 billion in 2020, with a projected CAGR of over 5% from 2021 to 2028. Fuzzy Logic-based control systems have improved process control, optimization, and fault detection in industrial automation applications. The global industrial automation market size was valued at over $180 billion in 2020, with a projected CAGR of over 7% from 2021 to 2028.

Overall, Fuzzy Logic continues to play a significant role in shaping modern technologies and systems, driving efficiency, reliability, and innovation across diverse industries. As organizations increasingly recognize the value of flexible and adaptive decision-making approaches, the adoption of Fuzzy Logic is expected to continue to grow, leading to further advancements and opportunities for businesses and society as a whole.

Robotics:

Robotics is a multidisciplinary field that involves the design, construction, operation, and use of robots. Robots are autonomous or semi-autonomous machines that can perform tasks in the physical world. Robotics combines principles from computer science, mechanical engineering, electrical engineering, and artificial intelligence to create robots capable of sensing, perception, decision-making, and action. Applications of robotics include industrial automation, medical robotics, autonomous vehicles, and service robots.

Robotics is a multidisciplinary field that involves the design, construction, operation, and use of robots to perform tasks in the physical world. Robots are autonomous or semi-autonomous machines that can sense, process information, and act upon their environment to achieve specific goals. Robotics combines principles from computer science, mechanical engineering, electrical engineering, and artificial intelligence to create robots capable of performing a wide range of functions across various industries and applications.

Key Components of Robotics:

1. Sensing: Sensing is a fundamental aspect of robotics that enables robots to perceive and interpret information from their environment. Sensors such as cameras, LiDAR (Light Detection and Ranging), ultrasonic sensors, and infrared sensors allow robots to detect objects, obstacles, and changes in their surroundings.
2. Actuation: Actuation refers to the mechanisms that enable robots to move and manipulate objects in their environment. Actuators such as motors, pneumatic systems, and hydraulic systems convert electrical or mechanical energy into motion, allowing robots to perform tasks such as locomotion, manipulation, and navigation.
3. Control: Control systems are responsible for coordinating the actions of robots based on input from sensors and desired objectives. Control algorithms, feedback loops, and motion planning algorithms enable robots to execute tasks accurately and efficiently while adapting to changes in their environment.

4. Intelligence: Intelligence encompasses the cognitive abilities of robots, including perception, reasoning, decision-making, and learning. Artificial intelligence (AI) techniques such as machine learning, computer vision, and natural language processing enable robots to process and interpret complex data, make informed decisions, and learn from experience.

Applications of Robotics:

1. Manufacturing: Robotics plays a crucial role in industrial automation, where robots are used for tasks such as assembly, welding, painting, and packaging in manufacturing plants. Industrial robots improve productivity, quality, and safety in manufacturing processes while reducing costs and cycle times.
2. Healthcare: Robotics is increasingly used in healthcare settings for tasks such as surgery, rehabilitation, and assistance to elderly or disabled individuals. Surgical robots enable minimally invasive procedures with greater precision and accuracy, leading to faster recovery times and reduced complications.
3. Agriculture: Agricultural robots, also known as agribots, automate tasks such as planting, harvesting, spraying pesticides, and monitoring crop health. Agricultural robots improve efficiency, productivity, and sustainability in farming operations while reducing labor costs and environmental impact.
4. Logistics and Transportation: Robotics is revolutionizing logistics and transportation with applications such as warehouse automation, autonomous vehicles, and last-mile delivery robots. Robotics enables faster order fulfillment, efficient inventory management, and safe and reliable transportation of goods.

Overall, robotics continues to advance rapidly, driving innovation, productivity, and safety across various industries and applications. As robotics technologies mature and become more accessible, the potential for robots to transform the way we live and work will continue to grow, leading to new opportunities and challenges in the years to come.

Natural Language Processing:

Natural Language Processing (NLP) is a subfield of artificial intelligence focused on enabling computers to understand, interpret, and generate human language. NLP techniques include text analysis, sentiment analysis, machine translation, and speech recognition. NLP algorithms process and analyze large volumes of text data, extracting meaning, sentiment, and insights from unstructured text. NLP is used in a wide range of applications, including virtual assistants, chatbots, information retrieval, and sentiment analysis.

Key Components of Natural Language Processing:

1. Text Preprocessing: Text preprocessing involves cleaning and transforming raw text data into a format suitable for analysis. This includes tasks such as tokenization (splitting text into words or phrases), stemming (reducing words to their root form), and removing stop words and punctuation.
2. Language Understanding: Language understanding involves extracting meaning from text by analyzing syntax, semantics, and context. Techniques such as part-of-speech tagging, named entity recognition, and dependency parsing help identify linguistic features and relationships within sentences.
3. Language Generation: Language generation involves generating coherent and grammatically correct text based on input data or predefined rules. This includes tasks such as text generation, summarization, and dialogue generation, where NLP models produce human-like responses or summaries of input text.
4. Sentiment Analysis: Sentiment analysis involves determining the sentiment or emotional tone expressed in text, such as positive, negative, or neutral. NLP models analyze textual data to identify sentiment-bearing words and phrases, classify sentiment polarity, and quantify sentiment intensity.
5. Machine Translation: Machine translation involves automatically translating text from one language to another. NLP models use statistical methods or neural machine translation (NMT) techniques to learn mappings between languages and generate accurate translations.

Level of Penetration:

NLP has achieved significant penetration across various industries and applications, driven by advancements in AI and increased availability of data. Some notable areas where NLP is widely used.

1. Search Engines: Search engines employ NLP techniques to understand user queries, analyze web pages, and retrieve relevant search results. NLP helps improve the accuracy and relevance of search results by understanding the user's intent and context.
2. Virtual Assistants: Virtual assistants such as Apple's Siri, Amazon's Alexa, and Google Assistant leverage NLP to understand spoken commands, answer questions, and perform tasks such as setting reminders, sending messages, and providing recommendations.
3. Customer Support: Many companies use NLP-powered chatbots and virtual agents to automate customer support interactions. NLP enables chatbots to understand and respond to customer inquiries, troubleshoot issues, and provide personalized assistance.
4. Social Media Analysis: NLP is used to analyze social media data to extract insights, monitor trends, and identify sentiment towards brands, products, or topics. Social

media platforms employ NLP to filter and categorize content, detect spam or abusive behavior, and deliver personalized recommendations.
5. Healthcare: NLP is increasingly used in healthcare for tasks such as clinical documentation, medical coding, and patient monitoring. NLP models analyze electronic health records (EHRs) and medical literature to extract relevant information, assist in diagnosis, and support clinical decision-making.

Overall, NLP continues to expand its presence across industries and applications, driving innovation and improving efficiency in various domains. As NLP technology advances and becomes more sophisticated, its impact on society and everyday life is expected to continue growing in the coming years.

Components of Artificial Intelligence
Artificial Intelligence (AI) comprises several key components that enable machines to exhibit intelligent behavior and perform tasks traditionally requiring human intelligence. These components are shown in Fig. 1.6.

1. Knowledge Representation: Knowledge representation involves capturing and structuring information in a form that machines can process and manipulate. This includes representing facts, concepts, rules, and relationships within a domain. Various techniques, such as semantic networks, frames, and ontologies, are used to represent knowledge in AI systems.
2. Reasoning: Reasoning refers to the process of drawing conclusions or making inferences based on available knowledge and evidence. AI systems employ various reasoning methods, including deductive reasoning, inductive reasoning, abductive reasoning, and probabilistic reasoning, to derive new information or solve problems.

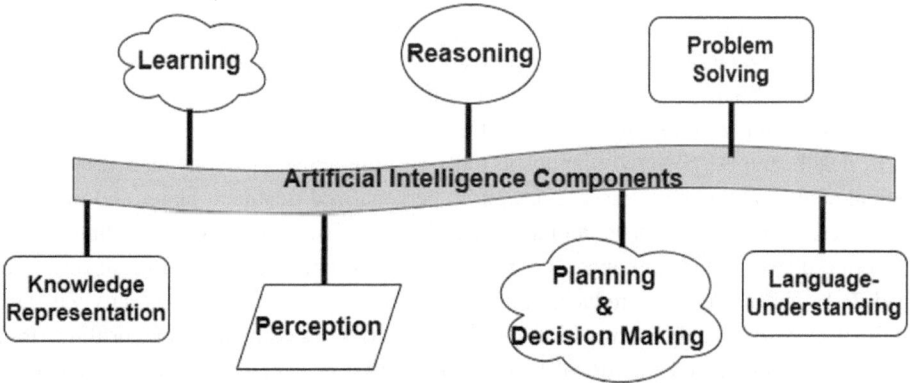

Fig. 1.6 Schematic of key components of artificial intelligence

3. Learning: Learning is a fundamental aspect of AI that enables machines to acquire knowledge and improve performance over time. Machine Learning algorithms enable computers to learn from data and make predictions or decisions without being explicitly programmed. Reinforcement Learning enables agents to learn optimal behaviors through trial and error, while unsupervised learning techniques discover patterns and structures in unlabeled data.

4. Perception: Perception involves the ability of AI systems to sense and interpret information from the environment. This includes tasks such as image recognition, speech recognition, natural language processing, and sensor data processing. Perception enables machines to understand and interact with the world around them.

5. Planning and Decision Making: Planning and decision-making capabilities enable AI systems to formulate goals, generate action sequences, and make decisions to achieve desired outcomes. This includes tasks such as path planning, scheduling, optimization, and decision theory. Planning and decision-making algorithms enable autonomous agents to navigate complex environments and achieve objectives efficiently.

6. Natural Language Processing: Natural Language Processing (NLP) enables machines to understand, interpret, and generate human language. NLP components include tasks such as speech recognition, language understanding, language generation, and machine translation. NLP enables AI systems to interact with users in natural language and process textual information effectively.

7. Problem-Solving: Problem-solving is a core component of AI that involves formulating problems, generating solutions, and evaluating alternatives. AI systems employ algorithms such as search algorithms, constraint satisfaction, and heuristic methods to solve complex problems efficiently. Problem-solving capabilities enable AI systems to tackle a wide range of challenges across various domains.

Artificial Intelligence (AI) stands at the forefront of technological innovation, revolutionizing industries, augmenting human capabilities, and driving unprecedented advancements in automation, decision-making, and problem-solving. With its ability to analyze vast amounts of data, recognize patterns, and learn from experience, AI has achieved remarkable milestones across various domains, unleashing new possibilities and reshaping the way we live and work as summarized in Fig. 1.7.

In healthcare, AI-powered diagnostic tools and predictive analytics have transformed patient care, enabling early disease detection, personalized treatment plans, and improved outcomes. AI algorithms have demonstrated diagnostic accuracy rates exceeding 90% in detecting diseases such as cancer, cardiovascular conditions, and neurological disorders, leading to faster diagnosis, reduced treatment costs, and enhanced patient survival rates. In finance, AI-driven algorithms for fraud detection, risk assessment, and algorithmic trading have revolutionized traditional banking and investment practices. AI-powered fraud detection systems have achieved detection rates of 99% with minimal false positives, saving financial institutions billions of dollars annually by preventing fraudulent transactions

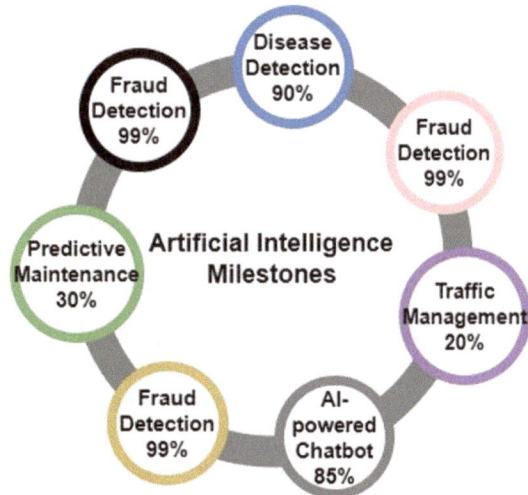

Fig. 1.7 Key milestones of artificial intelligence

and mitigating financial risks. In manufacturing, AI-enabled robotics, automation, and predictive maintenance have optimized production processes, increased operational efficiency, and reduced downtime. AI-driven predictive maintenance solutions have reduced equipment failures by 25–30%, leading to substantial cost savings and productivity gains in industries such as automotive, aerospace, and electronics manufacturing. In transportation, AI algorithms for route optimization, autonomous vehicles, and traffic management have improved safety, efficiency, and sustainability. Autonomous vehicles equipped with AI technologies have demonstrated accident rates up to 10 times lower than human-driven vehicles, while AI-powered traffic management systems have reduced congestion and travel times by 10–20% in urban areas. In customer service and retail, AI-powered chatbots, recommendation engines, and personalized marketing campaigns have enhanced customer experiences and increased sales. AI-driven chatbots have achieved customer satisfaction rates exceeding 85% by providing instant responses to inquiries and resolving issues efficiently, while recommendation engines have boosted sales by 10–30% through personalized product recommendations based on user preferences and behavior. Artificial Intelligence has achieved significant milestones across industries, delivering tangible benefits in terms of efficiency gains, cost savings, and improved outcomes. As AI technologies continue to advance and mature, the potential for further transformative impact on society, economies, and the workforce remains vast, promising a future where intelligent systems augment human capabilities and drive innovation at unprecedented scales.

1.7.3 Big Data Analytics

The proliferation of data in the digital age has underscored the importance of Big Data Analytics in Industry 4.0. By harnessing advanced analytics tools and techniques, organizations can derive meaningful insights from vast datasets generated by IoT devices, production systems, and supply chain operations. These insights inform strategic decision-making, facilitate predictive maintenance, and drive continuous improvement initiatives.

Big Data Analytics has achieved significant milestones across industries, driving transformative changes and delivering tangible benefits in terms of efficiency gains, cost savings, and improved decision-making. Big Data Analytics enables organizations to gain a deeper understanding of their customers' preferences, behaviors, and needs. By analyzing large volumes of customer data, businesses can personalize marketing campaigns, products, and services, leading to increased customer satisfaction and loyalty. Studies have shown that personalized marketing campaigns driven by Big Data Analytics can result in a 10–30% increase in sales and a 20–50% increase in marketing ROI. Big Data Analytics helps organizations optimize operational processes and streamline workflows by identifying inefficiencies, bottlenecks, and areas for improvement. By analyzing production data, supply chain metrics, and equipment performance, companies can reduce costs, minimize downtime, and improve overall efficiency. Studies have found that implementing Big Data Analytics solutions can lead to 15–20% improvement in operational efficiency and 25–30% reduction in maintenance costs. Big Data Analytics plays a crucial role in risk management and fraud detection across various industries, including banking, insurance, and cybersecurity. By analyzing transactional data, network traffic, and user behavior in real-time, organizations can detect anomalies and suspicious activities, leading to faster response times and reduced financial losses. Studies have shown that advanced analytics techniques such as machine learning and anomaly detection can improve fraud detection rates by 20–30% and reduce false positives by 50–70%. Big Data Analytics has the potential to revolutionize healthcare by improving diagnosis accuracy, treatment outcomes, and patient care delivery. By analyzing electronic health records (EHRs), medical images, and genomic data, healthcare providers can identify patterns and trends to personalize treatment plans, predict disease risks, and enhance preventive care strategies. Research has shown that leveraging Big Data Analytics in healthcare can lead to 25–30% reduction in hospital readmissions and 20–40% decrease in healthcare costs. Big Data Analytics enables organizations to implement predictive maintenance strategies by analyzing sensor data, machine performance metrics, and historical maintenance records. By predicting equipment failures before they occur, companies can minimize downtime, extend asset lifecycles, and reduce maintenance costs. Studies have found that implementing predictive maintenance solutions driven by Big Data Analytics can lead to 10–20% reduction in maintenance costs and 25–30% increase in equipment uptime. Big Data Analytics continues to drive innovation, efficiency, and competitive advantage across industries, unlocking new opportunities for organizations to leverage data for strategic decision-making and business growth. As the volume and complexity of data continue to grow, the role of Big Data Analytics will only become more crucial in extracting actionable insights and driving value in the digital age.

1.7.4 Cyber-Physical Systems (CPS)

Cyber-Physical Systems (CPS) represent the integration of computational and physical components, blurring the lines between the digital and physical worlds. In Industry 4.0, CPS orchestrate the seamless interaction between digital systems and physical processes, enabling real-time monitoring, control, and optimization of industrial operations. This convergence fosters greater flexibility, responsiveness, and adaptability in manufacturing environments.

Cyber-Physical Systems (CPS) represent a transformative integration of computational and physical components, blurring the boundaries between the digital and physical worlds. These systems combine advanced computing, communication, and control technologies to monitor, analyze, and interact with physical processes in real-time, enabling unprecedented levels of automation, efficiency, and intelligence across various domains. One of the key achievements of Cyber-Physical Systems is their ability to enhance operational efficiency and productivity in industrial settings. By embedding sensors, actuators, and controllers into physical machinery and infrastructure, CPS enable real-time monitoring, control, and optimization of manufacturing processes, supply chains, and critical infrastructure. Studies have shown that implementing CPS solutions can lead to 15–30% improvement in production efficiency and 20–40% reduction in operational costs. Moreover, Cyber-Physical Systems play a crucial role in enhancing safety and reliability in critical infrastructure and transportation systems. By integrating real-time data monitoring and predictive analytics capabilities, CPS enable early detection of potential failures, anomalies, and safety hazards, allowing for proactive intervention and risk mitigation. Studies have demonstrated that CPS-driven predictive maintenance strategies can reduce equipment failures by 25–40% and improve system reliability by 30–50%. Furthermore, Cyber-Physical Systems are instrumental in enabling the development and deployment of autonomous vehicles and intelligent transportation systems. By combining sensors, communication networks, and decision-making algorithms, CPS enable vehicles to perceive and navigate their environment autonomously, leading to safer, more efficient, and sustainable transportation systems. Research has shown that autonomous vehicles powered by CPS technologies can reduce accident rates by 70–90% and traffic congestion by 10–20%.

In addition to industrial and transportation applications, Cyber-Physical Systems have significant implications for healthcare, smart cities, and environmental monitoring. In healthcare, CPS facilitate remote patient monitoring, personalized treatment plans, and telemedicine services, improving access to healthcare and patient outcomes. In smart cities, CPS enable efficient resource management, energy optimization, and infrastructure resilience, enhancing the quality of life for residents and reducing environmental impact. In environmental monitoring, CPS enable real-time data collection and analysis to monitor air and water quality, detect pollution, and mitigate environmental risks. Overall, Cyber-Physical Systems represent a paradigm shift in how we design, operate, and interact with complex systems, offering unprecedented opportunities to enhance efficiency, safety, and sustainability across various domains. As CPS technologies continue to advance and mature, their potential to drive innovation and address societal challenges will only continue to grow, ushering in a new era of interconnected, intelligent systems that reshape the way we live, work, and interact with the world.

1.7.5   Additive Manufacturing (3D Printing)

Additive Manufacturing, commonly known as 3D printing, is revolutionizing traditional manufacturing processes by enabling the on-demand production of complex geometries with unparalleled precision and efficiency. Industry 4.0 leverages additive manufacturing technologies to streamline prototyping, customize products, and implement decentralized manufacturing models, thereby accelerating time-to-market and reducing production costs. Additive Manufacturing, commonly known as 3D Printing, has emerged as a disruptive technology with the potential to revolutionize traditional manufacturing processes. Unlike conventional subtractive manufacturing methods that involve cutting, shaping, and molding raw materials, additive manufacturing builds objects layer by layer using digital design data, offering unprecedented flexibility, customization, and cost-effectiveness. In the early 1980s, the concept of 3D printing was introduced, laying the foundation for the development of additive manufacturing technologies. In 1986, the first patent for stereolithography (SLA), one of the earliest 3D printing techniques, was filed by Chuck Hull, leading to the commercialization of 3D printing technology. Throughout the 1990s and early 2000s, advancements in materials, software, and printing techniques fueled the growth of 3D printing, expanding its applications across various industries. In 2009, the expiration of key patents paved the way for the democratization of 3D printing technology, leading to increased accessibility and affordability. In recent years, additive manufacturing has witnessed significant advancements in speed, accuracy, and scalability, enabling the production of complex geometries with high precision and efficiency. The additive manufacturing industry has attracted substantial investments from both public and private sectors, fueling research, development, and commercialization efforts. Venture capital funding for 3D printing startups has surged in recent years, with investments totaling billions of dollars globally. Additionally, governments and research institutions have allocated funding for additive manufacturing initiatives to support innovation, workforce development, and economic growth. Additive manufacturing technologies are deployed across various industries, including aerospace, automotive, healthcare, and consumer goods, among others. 3D printing is used to produce lightweight, high-performance components for aircraft and spacecraft, reducing material waste and lead times while improving fuel efficiency and performance. Additive manufacturing enables the production of patient-specific implants, prosthetics, and medical devices tailored to individual anatomies, improving patient outcomes and reducing costs. Automakers utilize 3D printing for rapid prototyping, tooling, and production of customized components, enhancing design flexibility, and accelerating time-to-market. Additive manufacturing is increasingly used in the production of consumer products, including footwear, jewelry, and home goods, enabling customization and on-demand manufacturing. Additive manufacturing has achieved significant milestones and attracted substantial investments, driving innovation and deployment across various industries. As technology continues to advance and adoption expands, the potential for additive manufacturing to disrupt traditional manufacturing paradigms and unlock new opportunities for customization, sustainability, and efficiency is immense.

1.7.6   Cloud Computing

Cloud Computing serves as the backbone of Industry 4.0, providing scalable and on-demand access to computing resources, storage, and services over the internet. By migrating critical workloads and applications to the cloud, organizations can enhance collaboration, scalability, and agility while reducing infrastructure costs and mitigating security risks. Cloud-based platforms also facilitate the integration of disparate systems and enable seamless data exchange across distributed environments. Cloud computing has emerged as a transformative technology paradigm, revolutionizing the way organizations store, manage, and access data, applications, and computing resources. At its core, cloud computing refers to the delivery of computing services over the internet, allowing users to access resources on-demand, scale dynamically, and pay only for what they use. The concept of cloud computing traces back to the 1960s, with the development of utility computing and time-sharing systems. In the late 1990s and early 2000s, companies like Salesforce and Amazon began offering Software as a Service (SaaS) and Infrastructure as a Service (IaaS) solutions, laying the groundwork for modern cloud computing. In 2006, Amazon Web Services (AWS) launched Elastic Compute Cloud (EC2), a key milestone in the development of cloud computing, offering scalable virtual servers on-demand. In subsequent years, other major players such as Microsoft Azure and Google Cloud Platform entered the market, further driving innovation and competition in the cloud computing space. Today, cloud computing has become ubiquitous, with organizations of all sizes leveraging cloud services for a wide range of applications, from data storage and processing to software development and deployment. The cloud computing market has attracted significant investments from both tech giants and startups alike. Companies have invested billions of dollars in data center infrastructure, research, and development to expand their cloud offerings and improve service quality, security, and performance. Additionally, venture capital funding for cloud startups has surged in recent years, reflecting growing investor confidence in the potential of cloud computing to drive innovation and disrupt traditional IT models. Cloud computing is deployed across various industries and use cases, offering organizations greater agility, scalability, and cost-effectiveness. Some common deployment scenarios include. Infrastructure as a Service (IaaS) Organizations leverage IaaS solutions to provision virtual servers, storage, and networking resources on-demand, enabling flexible and scalable IT infrastructure without the need for upfront capital investment. Platform as a Service (PaaS) offerings provide developers with tools and frameworks to build, deploy, and manage applications quickly and efficiently, streamlining the software development lifecycle and accelerating time-to-market. Software as a Service (SaaS) applications are delivered over the internet on a subscription basis, allowing users to access software applications from any device without the need for installation or maintenance. Many organizations adopt hybrid and multi-cloud strategies to leverage the benefits of both public and private cloud environments, optimizing performance, security, and cost-effectiveness based on their specific requirements.

Cloud computing has transformed the IT landscape, offering organizations unprecedented flexibility, scalability, and efficiency in managing their computing resources. As technology continues to evolve and adoption expands, the cloud computing market is poised for continued growth, driving innovation and reshaping the way businesses operate in the digital age. The technological enablers of Industry 4.0 represent a transformative force driving the evolution of manufacturing and industry towards a more connected, intelligent, and agile future. By embracing these cutting-edge technologies, organizations can unlock new opportunities for innovation, growth, and competitive advantage in an increasingly digitized world.

1.8 Economic Implications of Industry 4.0

Industry 4.0, characterized by the integration of digital technologies into manufacturing and industrial processes, brings about profound economic implications that reshape global economies, industries, and labor markets. At its core, Industry 4.0 promises increased productivity, efficiency, and competitiveness through the adoption of advanced technologies such as automation, artificial intelligence (AI), Internet of Things (IoT), and data analytics. Furthermore, it triggers significant changes in supply chain dynamics, driving innovation in logistics, inventory management, and customer relations. These economic implications have far-reaching consequences, impacting businesses, workers, and policymakers worldwide.

One of the primary economic implications of Industry 4.0 is the potential for significant increases in productivity and efficiency across industries. By leveraging automation and digital technologies, businesses can streamline processes, reduce waste, and opti-mize resource allocation. For example, in manufacturing, the adoption of smart factories equipped with IoT sensors and AI-driven analytics enables real-time monitoring of pro-duction lines, predictive maintenance, and adaptive manufacturing processes. This leads to higher output levels, improved quality control, and reduced production costs. Moreover, Industry 4.0 facilitates the implementation of lean manufacturing principles, allowing businesses to achieve greater operational efficiency and responsiveness to customer demands. By integrating data-driven decision-making into their operations, companies can optimize production schedules, minimize downtime, and enhance overall productiv-ity. As a result, Industry 4.0 enables businesses to remain competitive in an increasingly fast-paced and dynamic global marketplace.

Industry 4.0 also drives significant changes in supply chain dynamics, transforming the way goods are sourced, produced, and delivered to customers. The integration of dig-ital technologies into supply chain management enables greater visibility, transparency, and collaboration across the entire value chain. For instance, IoT-enabled sensors pro-vide real-time tracking of inventory levels, shipment status, and transportation routes, allowing businesses to optimize logistics and reduce lead times. Furthermore, Industry 4.0 enables the development of agile and responsive supply chains capable of adapting to changing market conditions and customer preferences. By leveraging data analytics and AI algorithms, companies can forecast demand more accurately, anticipate supply chain disruptions, and optimize inventory levels. This results in improved inventory man-agement, reduced stockouts, and lower carrying costs, ultimately enhancing supply chain resilience and customer satisfaction. Additionally, Industry 4.0 fosters closer collaboration between suppliers, manufacturers, and distributors through digital platforms and ecosys-tems. By connecting disparate stakeholders in the supply chain, businesses can streamline communication, coordinate activities, and drive innovation through co-creation and co-innovation. This collaborative approach to supply chain management enables companies to leverage collective expertise and resources to address complex challenges and seize new opportunities in the global marketplace.

The economic implications of Industry 4.0 have profound implications for businesses of all sizes and sectors. For large corporations, Industry 4.0 presents opportunities to enhance competitiveness, drive innovation, and unlock new revenue streams through dig-ital transformation. By investing in advanced technologies and talent development, large enterprises can position themselves as industry leaders and disruptors in their respec-tive markets. Similarly, small and medium-sized enterprises (SMEs) stand to benefit from Industry 4.0 by leveraging digital technologies to level the playing field and compete more effectively with larger rivals. Cloud computing, Software as a Service (SaaS), and other digital solutions offer SMEs access to scalable and affordable tools for managing operations, marketing products, and reaching customers globally. Additionally, Industry 4.0 enables SMEs to tap into new markets and business models through e-commerce, online marketplaces, and digital platforms. However, the transition to Industry 4.0 also presents challenges for businesses, particularly in terms of workforce adaptation, cyber-security, and data privacy. The automation of routine tasks and the adoption of AI-driven technologies may disrupt traditional job roles and require reskilling or upskilling of the workforce. Furthermore, the increasing reliance on digital systems and interconnected devices exposes businesses to cybersecurity threats and data breaches, necessitating robust security measures and risk management strategies.

The economic implications of Industry 4.0 extend to the workforce, with significant implications for employment, skills, and labor markets. While Industry 4.0 creates new opportunities for high-skilled workers in fields such as data analytics, AI development, and cybersecurity, it also poses challenges for workers in traditional manufacturing and blue-collar industries. The automation of repetitive tasks and the adoption of robotics and AI technologies may lead to job displacement and require workers to adapt to new roles and responsibilities. However, Industry 4.0 also creates opportunities for workers to acquire new skills and pursue careers in emerging fields such as digital marketing, e-commerce, and software development. Lifelong learning and continuous education become essential for workers to remain competitive in the digital economy, requiring investment in train-ing programs and vocational education. Additionally, Industry 4.0 enables remote work and telecommuting opportunities, offering greater flexibility and work-life balance for employees. Furthermore, Industry 4.0 has implications for income inequality and social mobility, with the potential to exacerbate disparities between high-skilled and low-skilled workers. As automation replaces routine tasks and jobs, workers with specialized skills and education may command higher wages and greater job security, while those with lim-ited skills or outdated qualifications face challenges in the labor market. Addressing these disparities requires policies and initiatives to promote inclusive growth, invest in edu-cation and training, and support displaced workers through retraining and social safety nets.

Policymakers play a crucial role in shaping the economic implications of Industry 4.0 and ensuring that the benefits of digital transformation are shared equitably across society. Governments must adopt forward-looking policies and regulations that promote inno-vation, investment, and entrepreneurship while safeguarding workers' rights, consumer protection, and data privacy. This includes initiatives to support research and development, foster digital infrastructure, and incentivize businesses to adopt Industry 4.0 technologies. Moreover, policymakers need to address the social and economic challenges associated with workforce displacement and job transition in the age of automation. This may involve implementing education and training programs to equip workers with the skills needed for the digital economy, supporting lifelong learning initiatives, and promoting work-force mobility and reemployment services. Additionally, governments must ensure that digital infrastructure and broadband connectivity are accessible to all citizens, bridging the digital divide and promoting inclusive economic growth. Furthermore, policymak-ers must address the regulatory and ethical implications of Industry 4.0, particularly in areas such as data governance, cybersecurity, and intellectual property rights. Establishing clear guidelines and standards for data privacy, cybersecurity, and ethical AI ensures trust and confidence in digital technologies while protecting individuals' rights and freedoms. Additionally, international cooperation and collaboration are essential to address global challenges such as cybersecurity threats, data sovereignty, and cross-border data flows in the digital age.

The economic implications of Industry 4.0 are profound and multifaceted, impacting businesses, workers, and policymakers worldwide. While Industry 4.0 presents opportunities for increased productivity, efficiency, and competitiveness, it also poses challenges in terms of job displacement, skills mismatch, and income inequality. Addressing these challenges requires collaboration between stakeholders, investment in education and training, and forward-looking policies that promote inclusive growth and sustainable development in the digital economy. By embracing digital transformation and harnessing the power of Industry 4.0 technologies, societies can unlock new opportunities for innovation, prosperity, and social progress in the twenty-first century.

1.9   Societal Impacts and Challenges

The advent of transformative technologies associated with Industry 4.0, such as automation, artificial intelligence (AI), and robotics, brings about significant societal impacts and challenges that reshape the way we work, interact, and live. These advancements have the potential to drive economic growth, improve quality of life, and address pressing global challenges. However, they also raise concerns about employment shifts, workforce adaptation, and ethical considerations related to automation and AI.

One of the most pressing societal impacts of Industry 4.0 is the shift in employment patterns and the need for workforce adaptation to meet the demands of the digital age. Automation and AI technologies have the potential to disrupt traditional job roles and industries, leading to job displacement and creating the need for workers to acquire new skills and competencies. As industries adopt automation and AI technologies to streamline processes and increase efficiency, jobs that involve routine, repetitive tasks are most at risk of being automated. This includes jobs in manufacturing, logistics, retail, and administrative functions. However, automation also creates new job opportunities in fields such as data analytics, cybersecurity, and software development, which require skills in critical thinking, problem-solving, and digital literacy. To address the challenges of employment shifts and workforce adaptation, stakeholders must invest in education and training programs that equip workers with the skills needed for the digital economy. Lifelong learning initiatives, vocational training, and reskilling programs can help workers transition to new roles and industries and remain competitive in the labor market. Additionally, governments, businesses, and educational institutions must collaborate to develop policies and strategies that promote workforce mobility, upskilling, and employment opportunities in emerging fields.

Alongside the societal benefits of automation and AI technologies come ethical considerations related to their deployment and impact on individuals, communities, and society as a whole. As machines become increasingly autonomous and capable of making decisions, ethical questions arise regarding accountability, transparency, and fairness in AI-driven systems. One of the primary ethical considerations in automation and AI is the potential for bias and discrimination in decision-making algorithms. AI systems learn from historical data, which may contain biases and prejudices that perpetuate inequalities and discrimination. For example, AI algorithms used in hiring processes or loan approvals may inadvertently favor certain demographic groups over others, leading to unfair outcomes and perpetuating existing social disparities. Furthermore, the rise of autonomous systems raises questions about accountability and responsibility in cases of errors or accidents. Who is responsible when an AI-driven vehicle causes an accident, or when an automated decision-making system makes a harmful mistake? Clarifying legal and ethical frameworks for assigning accountability and liability in such cases is essential to ensure that individuals and organizations are held responsible for the consequences of their actions. Additionally, there are concerns about the impact of automation and AI on employment, income inequality, and societal well-being. As automation replaces human labor in certain tasks and industries, there is a risk of widening the gap between skilled and unskilled workers, exacerbating income inequality and social disparities. Moreover, the loss of jobs due to automation may lead to social unrest and economic instability, particularly in communities heavily reliant on traditional industries. Addressing the ethical considerations of automation and AI requires a multi-stakeholder approach involving policymakers, technologists, ethicists, and civil society organizations. Establishing ethical guidelines, standards, and regulations for the development and deployment of AI technologies is crucial to ensure that they align with societal values, human rights, and ethical principles. Furthermore, fostering public dialogue and engagement on ethical issues related to automation and AI promotes transparency, accountability, and trust in the technology.

The societal impacts and challenges associated with Industry 4.0 require proactive and collaborative efforts to address employment shifts, workforce adaptation, and ethical considerations in automation and AI. By investing in education and training, fostering ethical AI development, and promoting inclusive growth, societies can harness the potential of Industry 4.0 technologies to create a more equitable, prosperous, and sustainable future for all.

1.10 Industry 4.0 in Manufacturing

Industry 4.0 has revolutionized the manufacturing sector, ushering in a new era of digital transformation and innovation. At the heart of this revolution are advanced technologies such as automation, robotics, artificial intelligence (AI), and the Internet of Things (IoT), which are reshaping traditional manufacturing processes and driving efficiencies across the entire value chain.

One of the hallmarks of Industry 4.0 in manufacturing is the emergence of smart factories equipped with interconnected sensors, actuators, and intelligent machines. These smart factories leverage IoT technologies to collect real-time data from production equipment, materials, and products, enabling seamless communication and coordination between machines and systems. Through the integration of IoT sensors and AI-driven analytics, manufacturers gain unprecedented visibility and control over their production processes. Sensors embedded in machines and equipment monitor various parameters such as temperature, pressure, and vibration, providing insights into equipment health and performance. AI algorithms analyze this data in real-time to identify patterns, detect anomalies, and optimize production schedules, leading to improved efficiency and productivity. Furthermore, smart factories enable adaptive manufacturing processes that can respond dynamically to changes in demand, supply chain disruptions, and market conditions. By leveraging AI-driven predictive modeling and demand forecasting, manufacturers can adjust production schedules, optimize resource allocation, and minimize inventory levels while meeting customer demands in real-time.

Industry 4.0 facilitates the implementation of predictive maintenance strategies that enable manufacturers to anticipate equipment failures before they occur, thereby reducing downtime and maintenance costs. IoT sensors embedded in machinery and equipment continuously monitor performance metrics and detect early signs of wear, degradation, or malfunctions. Through AI-driven analytics and machine learning algorithms, manufacturers can analyze historical data, sensor readings, and maintenance records to predict when equipment is likely to fail and schedule proactive maintenance activities accordingly. This predictive approach to maintenance ensures that maintenance interventions are performed at the optimal time, minimizing disruptions to production and maximizing equipment uptime. Moreover, Industry 4.0 enables manufacturers to enhance quality control processes through real-time monitoring, inspection, and feedback mechanisms. IoT sensors and cameras deployed throughout the production line capture data on product dimensions, surface defects, and other quality parameters. AI algorithms analyze this data to identify deviations from quality standards and trigger corrective actions, such as adjusting process parameters or rejecting defective products. By leveraging predictive maintenance and quality control capabilities, manufacturers can improve product quality, reduce rework and scrap rates, and enhance customer satisfaction. Additionally, Industry 4.0 enables manufacturers to implement agile and responsive supply chain strategies, allowing them to collaborate closely with suppliers, optimize inventory levels, and meet changing customer demands in real-time. Industry 4.0 is transforming the manufacturing sector by enabling the development of smart factories and production processes that leverage advanced technologies such as IoT, AI, and predictive analytics. By embracing Industry 4.0 principles, manufacturers can enhance operational efficiency, reduce costs, and improve product quality while remaining agile and responsive in an increasingly competitive global marketplace.

1.11   Industry 4.0 in Logistics and Supply Chain Management

Industry 4.0 is revolutionizing logistics and supply chain management, driving efficiencies, reducing costs, and improving overall operational performance. At the core of this transformation are advanced technologies such as the Internet of Things (IoT), artificial intelligence (AI), big data analytics, and automation, which enable real-time visibility, predictive analytics, and optimization across the entire supply chain.

One of the key aspects of Industry 4.0 in logistics is the implementation of smart warehousing and inventory management systems. IoT sensors and devices are deployed throughout warehouses and distribution centers to track the movement, location, and condition of inventory items in real-time. These sensors collect data on factors such as temperature, humidity, and shelf life, providing insights into inventory levels and stock availability. AI-driven analytics processes this data to optimize warehouse layouts, streamline picking and packing processes, and minimize storage space requirements. By analyzing historical demand patterns and seasonal fluctuations, AI algorithms can predict inventory levels and recommend replenishment strategies to ensure optimal stock levels while minimizing carrying costs and stockouts. Furthermore, Industry 4.0 enables the implementation of autonomous robots and drones for inventory management tasks such as picking, sorting, and inventory counting. These robots can navigate autonomously through warehouses, retrieve items, and deliver them to packing stations, reducing manual labor and increasing operational efficiency.

Industry 4.0 facilitates real-time tracking and optimization of goods throughout the supply chain, from manufacturing facilities to distribution centers to end customers. IoT-enabled sensors and devices embedded in products, vehicles, and packaging enable continuous monitoring of shipment status, location, and condition. Through IoT connectivity and data analytics, supply chain stakeholders gain visibility into the movement of goods, allowing them to track shipments in real-time, anticipate delays, and respond quickly to unexpected events such as traffic congestion, adverse weather conditions, or port delays. This real-time visibility enables proactive decision-making, allowing companies to reroute shipments, adjust production schedules, or allocate resources to minimize disruptions and ensure on-time delivery. Moreover, Industry 4.0 enables the optimization of transportation routes and logistics operations through AI-driven predictive analytics and optimization algorithms. By analyzing historical traffic data, demand forecasts, and delivery schedules, AI algorithms can recommend the most efficient routes, modes of transport, and delivery schedules to minimize transportation costs, reduce fuel consumption, and improve delivery performance. Additionally, Industry 4.0 facilitates the implementation of dynamic pricing and demand-responsive logistics strategies, enabling companies to adjust prices and delivery schedules based on real-time demand signals. By leveraging AI-driven pricing algorithms and demand forecasting models, companies can optimize pricing strategies, maximize revenue, and enhance customer satisfaction. Industry 4.0 is transforming logistics and supply chain management by enabling real-time visibility, predictive analytics, and optimization across the entire supply chain. By leveraging advanced technologies such as IoT, AI, and automation, companies can enhance operational efficiency, reduce costs, and improve customer service while remaining agile and responsive in an increasingly complex and dynamic global marketplace.

1.12 Industry 4.0 in Healthcare

Industry 4.0, characterized by the integration of advanced digital technologies into various sectors, including healthcare, is revolutionizing the way medical services are delivered, managed, and accessed. With the adoption of technologies such as telemedicine, remote patient monitoring, personalized medicine, and healthcare analytics, Industry 4.0 is driving improvements in patient care, efficiency, and outcomes.

One of the key applications of Industry 4.0 in healthcare is telemedicine, which involves the delivery of medical services and consultations remotely using telecommunications technology. Telemedicine enables patients to consult with healthcare providers, specialists, or therapists from the comfort of their homes, eliminating the need for in-person visits and reducing travel time and costs. Moreover, Industry 4.0 facilitates remote patient monitoring through the use of connected devices and wearable sensors that track vital signs, medication adherence, and disease progression in real-time. Patients can use wearable devices such as smartwatches, fitness trackers, or home monitoring kits to collect and transmit health data to healthcare providers, allowing for proactive intervention and timely adjustments to treatment plans. Telemedicine and remote patient monitoring not only improve access to healthcare services, particularly for patients in rural or underserved areas but also enable early detection of health issues, proactive management of chronic conditions, and continuity of care outside traditional healthcare settings. By leveraging Industry 4.0 technologies, healthcare providers can deliver personalized, patient-centric care while optimizing resource utilization and reducing healthcare costs.

Industry 4.0 enables the advancement of personalized medicine, which tailors medical treatments and interventions to individual patients' genetic makeup, lifestyle factors, and health profiles. By integrating genomics, proteomics, and other omics data with clinical information, AI-driven analytics can identify biomarkers, predict disease risks, and recommend personalized treatment strategies. Furthermore, healthcare analytics powered by big data and machine learning algorithms enable healthcare providers to analyze large volumes of clinical data, electronic health records (EHRs), and medical imaging data to derive actionable insights and improve clinical decision-making. By identifying patterns, trends, and correlations in healthcare data, AI algorithms can assist in disease diagnosis, treatment planning, and prognosis prediction. Moreover, Industry 4.0 facilitates population health management through the analysis of healthcare data at the population level to identify health trends, risk factors, and opportunities for preventive interventions. By leveraging predictive analytics and risk stratification models, healthcare organizations can target interventions and allocate resources more effectively to improve population health outcomes and reduce healthcare disparities. Industry 4.0 is transforming the healthcare industry by enabling the adoption of telemedicine, remote patient monitoring, personalized medicine, and healthcare analytics. By leveraging advanced digital technologies, healthcare providers can deliver more accessible, efficient, and personalized care while improving patient outcomes and reducing healthcare costs. As Industry 4.0 continues to evolve, the integration of digital technologies into healthcare holds promise for addressing current and future challenges in healthcare delivery, management, and innovation.

1.13   Industry 4.0 in Agriculture

Industry 4.0, with its arsenal of advanced technologies, is making significant strides in modernizing agriculture, thereby enhancing productivity, sustainability, and efficiency. The integration of digital technologies such as precision farming, smart agriculture practices, and IoT-enabled crop monitoring is revolutionizing traditional agricultural methods, paving the way for a more sustainable and productive future.

Precision Farming and Smart Agriculture Practices:

Industry 4.0 has ushered in the era of precision farming, which involves the use of advanced technologies to optimize crop production while minimizing inputs such as water, fertilizers, and pesticides. Precision farming techniques leverage data-driven insights to tailor agricultural practices to specific field conditions, crop requirements, and environmental factors, thereby maximizing yields and resource efficiency. One of the key components of precision farming is the use of satellite imagery, drones, and GPS technology to collect data on soil moisture, nutrient levels, and crop health. This data is then analyzed using AI algorithms to generate detailed maps of field variability, allowing farmers to make informed decisions about planting, irrigation, and fertilization. By precisely targeting inputs based on localized conditions, farmers can optimize resource use, reduce waste, and improve crop yields. Moreover, smart agriculture practices leverage IoT sensors, actuators, and automated systems to monitor and control various aspects of crop production in real-time. IoT-enabled devices installed in fields, greenhouses, and irrigation systems collect data on environmental conditions, plant health, and water usage, allowing farmers to remotely monitor and manage operations from anywhere using mobile devices or computers.

IoT-enabled Crop Monitoring and Management:

Industry 4.0 facilitates IoT-enabled crop monitoring and management, which enables farmers to track and optimize crop growth, health, and yield throughout the growing season. IoT sensors embedded in the soil, plants, and agricultural equipment continuously collect data on factors such as temperature, humidity, pH levels, and nutrient concentrations. This real-time data is transmitted wirelessly to a centralized platform where it is analyzed using AI-driven analytics to provide insights into crop health, growth patterns, and resource requirements. Farmers can use this information to make data-driven decisions about irrigation scheduling, pest management, and nutrient application, optimizing crop yields while minimizing environmental impact. Furthermore, Industry 4.0 enables the development of autonomous agricultural machinery and robotic systems that can perform tasks such as planting, harvesting, and weeding with precision and efficiency. These autonomous systems leverage AI algorithms, computer vision, and robotic actuators to navigate fields, identify crops and weeds, and perform targeted interventions, reducing labor costs and increasing operational efficiency.

Industry 4.0 is transforming agriculture by enabling the adoption of precision farming, smart agriculture practices, and IoT-enabled crop monitoring and management. By leveraging advanced technologies such as AI, IoT, and robotics, farmers can optimize crop production, conserve resources, and improve sustainability while meeting the growing demand for food in a rapidly changing world. As Industry 4.0 continues to evolve, the integration of digital technologies into agriculture holds promise for addressing current and future challenges in food security, environmental sustainability, and rural development.

1.14 Industry 4.0 in Energy Sector

Industry 4.0 is reshaping the energy sector, driving innovation, efficiency, and sustainability through the adoption of advanced technologies such as smart grids, energy management systems, and renewable energy integration. These technologies enable utilities, grid operators, and energy companies to optimize energy generation, distribution, and consumption, thereby enhancing reliability, resilience, and environmental sustainability.

Smart Grids and Energy Management Systems:

One of the key components of Industry 4.0 in the energy sector is the development of smart grids and energy management systems. Smart grids leverage digital communication and automation technologies to monitor, control, and optimize the flow of electricity across the grid in real-time. By integrating sensors, meters, and intelligent devices, smart grids enable utilities to detect and respond to changes in demand, supply, and grid conditions instantaneously. Energy management systems (EMS) complement smart grids by providing utilities and grid operators with advanced analytics and decision support tools to optimize energy generation, transmission, and distribution. EMS leverage AI algorithms and predictive analytics to forecast energy demand, identify grid constraints, and optimize the dispatch of generation resources, including conventional power plants, renewable energy sources, and energy storage systems. Moreover, Industry 4.0 enables the implementation of demand response programs that incentivize consumers to adjust their electricity usage in response to grid conditions and pricing signals. Smart meters and IoT-enabled devices enable consumers to monitor their energy consumption in real-time and participate in demand-side management initiatives, such as time-of-use pricing, peak shaving, and load shifting, thereby reducing costs and improving grid stability.

Renewable Energy Integration and Optimization:

Industry 4.0 is driving the integration and optimization of renewable energy sources such as solar, wind, and hydroelectric power into the energy mix. Renewable energy technologies have experienced significant advancements in recent years, becoming more efficient, cost-effective, and scalable, thanks to innovations in materials, manufacturing, and digital control systems. Smart grids play a crucial role in enabling the integration of renewable energy into the grid by providing the flexibility and resilience needed to accommodate variable generation from sources such as solar and wind. Advanced forecasting models and predictive analytics algorithms help utilities anticipate renewable energy output and balance supply and demand in real-time, ensuring grid stability and reliability. Furthermore, Industry 4.0 facilitates the deployment of distributed energy resources (DERs) such as rooftop solar panels, wind turbines, and energy storage systems at the local level. IoT-enabled devices and grid-edge technologies enable consumers to generate, store, and manage their electricity locally, reducing reliance on centralized generation and improving energy resilience. Additionally, AI-driven optimization algorithms enable utilities to maximize the value of renewable energy assets by optimizing their operation and maintenance schedules, minimizing downtime, and maximizing energy yield. Predictive maintenance techniques leverage real-time data and machine learning algorithms to detect equipment failures and performance degradation before they occur, reducing maintenance costs and improving asset reliability.

In conclusion, Industry 4.0 is transforming the energy sector by enabling the development of smart grids, energy management systems, and renewable energy integration solutions. By leveraging advanced technologies such as AI, IoT, and predictive analytics, utilities and energy companies can optimize energy generation, distribution, and consumption, thereby enhancing reliability, resilience, and environmental sustainability. As Industry 4.0 continues to evolve, the integration of digital technologies into the energy sector holds promise for accelerating the transition to a cleaner, more sustainable energy future.

1.15   Industry 4.0 in Transportation and Mobility

Industry 4.0 is revolutionizing the transportation and mobility sector, driving innovation, efficiency, and safety through the integration of advanced technologies such as autonomous vehicles, connected infrastructure, and intelligent transportation systems. These technologies are reshaping the way people and goods move, optimizing transportation networks, and enhancing the overall mobility experience.

Autonomous Vehicles and Connected Infrastructure:

One of the most transformative aspects of Industry 4.0 in transportation is the development and deployment of autonomous vehicles (AVs) and connected infrastructure. AVs, equipped with sensors, cameras, and advanced computing systems, have the ability to perceive their surroundings, navigate autonomously, and make real-time decisions without human intervention. Industry 4.0 enables the integration of AVs with connected infrastructure, including smart traffic lights, road sensors, and communication systems, to create a seamless and efficient transportation ecosystem. Connected infrastructure provides real-time data on road conditions, traffic flow, and weather conditions, enabling AVs to anticipate and respond to potential hazards and optimize their routes accordingly. Moreover, Industry 4.0 facilitates vehicle-to-everything (V2X) communication, allowing AVs to exchange data with other vehicles, pedestrians, and infrastructure elements in their vicinity. V2X communication enables collaborative maneuvers, such as platooning and cooperative merging, which improve traffic flow, reduce congestion, and enhance safety on the roads.

Autonomous vehicles (AVs) and connected infrastructure represent revolutionary technologies that are poised to transform the future of transportation. Autonomous vehicles, also known as self-driving cars, rely on a combination of sensors, cameras, and artificial intelligence (AI) algorithms to navigate roads and make driving decisions without human intervention. Connected infrastructure, on the other hand, involves equipping roads, traffic signals, and other infrastructure elements with sensors and communication technologies to facilitate real-time data exchange with vehicles.

The integration of autonomous vehicles and connected infrastructure holds the potential to revolutionize transportation systems by improving safety, efficiency, and sustainability while reducing congestion and emissions. However, the widespread adoption of these technologies also raises a myriad of challenges and considerations, ranging from technical and regulatory hurdles to societal and ethical implications. One of the primary challenges facing autonomous vehicles is ensuring their safety and reliability. While AVs have the potential to reduce human error, which is a leading cause of accidents on roads, they must navigate complex and unpredictable environments with a high degree of precision and accuracy. Achieving this level of performance requires robust sensor technologies, sophisticated AI algorithms, and rigorous testing and validation processes to ensure that AVs can operate safely under various conditions and scenarios. Moreover, the development and deployment of autonomous vehicles raise legal and regulatory challenges related to liability, insurance, and accountability. In the event of an accident involving an AV, questions may arise regarding the responsibility of the vehicle manufacturer, the software developer, the vehicle operator (if any), and other parties involved. Establishing clear guidelines and regulations for AVs is essential to address these concerns and ensure accountability while fostering innovation and adoption. Connected infrastructure also presents challenges related to data privacy, cybersecurity, and interoperability. As vehicles and infrastructure elements exchange data in real-time, there is a need to safeguard sensitive information and prevent unauthorized access or tampering. Additionally, ensuring seamless communication and interoperability between different systems and technologies is essential for maximizing the benefits of connected infrastructure while minimizing disruptions and compatibility issues.

Despite these challenges, autonomous vehicles and connected infrastructure offer numerous opportunities for improving transportation systems and enhancing quality of life. One of the most significant benefits of AVs is their potential to reduce traffic congestion and improve mobility for people of all ages and abilities. By optimizing routes, reducing travel times, and minimizing delays, AVs can make transportation more efficient and accessible, particularly in urban areas. Furthermore, autonomous vehicles have the potential to revolutionize logistics and freight transportation by enabling more efficient and cost-effective delivery services. With self-driving trucks and drones, companies can streamline supply chain operations, reduce delivery times, and lower transportation costs, leading to increased productivity and competitiveness. Connected infrastructure also offers opportunities for improving road safety and efficiency through real-time traffic management and optimization. By equipping roads with sensors and communication technologies, traffic signals can adjust their timing and sequencing in response to changing traffic conditions, reducing congestion and minimizing delays. Additionally, connected infrastructure enables vehicles to receive real-time updates on road conditions, hazards, and detours, allowing for safer and more informed driving decisions. Moreover, autonomous vehicles and connected infrastructure have the potential to enhance sustainability and reduce environmental impact by optimizing vehicle routing, reducing fuel consumption, and promoting the use of electric and alternative fuel vehicles. By minimizing idling, optimizing acceleration and deceleration patterns, and reducing stop-and-go traffic, AVs and connected infrastructure can contribute to lower emissions and improved air quality in urban areas. Autonomous vehicles and connected infrastructure represent transformative technologies that have the potential to revolutionize transportation systems and improve quality of life. While they present numerous challenges and considerations, including safety, regulation, and interoperability, the benefits of these technologies are immense. By addressing technical, regulatory, and societal challenges, stakeholders can unlock the full potential of autonomous vehicles and connected infrastructure and create a safer, more efficient, and sustainable transportation future.

Intelligent Transportation Systems:

Industry 4.0 enables the development of intelligent transportation systems (ITS) that leverage data-driven insights and advanced analytics to optimize transportation networks, improve safety, and enhance the overall mobility experience. ITS encompasses a wide range of technologies and applications, including traffic management, congestion pricing, and predictive maintenance. One key application of ITS is traffic management, which uses real-time data from sensors, cameras, and GPS devices to monitor traffic flow, detect congestion, and optimize signal timing. AI algorithms analyze this data to predict traffic patterns, identify bottlenecks, and recommend adaptive traffic control strategies, such as dynamic signal timing and lane management, to improve throughput and reduce delays. Furthermore, Industry 4.0 enables the implementation of predictive maintenance programs for transportation infrastructure, such as roads, bridges, and railways. IoT sensors embedded in infrastructure assets collect data on structural health, vibration levels, and environmental conditions, allowing transportation agencies to monitor asset condition in real-time and detect potential maintenance issues before they escalate. Additionally, Industry 4.0 facilitates the development of multimodal transportation solutions that integrate different modes of transport, such as public transit, ridesharing, and micro-mobility services, to provide seamless and efficient mobility options for passengers. Mobility-as-a-Service (MaaS) platforms leverage digital technologies to integrate and coordinate various transportation options, enabling travelers to plan, book, and pay for their journeys seamlessly using a single app or platform. Industry 4.0 is transforming the transportation and mobility sector by enabling the development of autonomous vehicles, connected infrastructure, and intelligent transportation systems. By leveraging advanced technologies such as AI, IoT, and data analytics, transportation agencies and mobility providers can optimize transportation networks, improve safety, and enhance the overall mobility experience for passengers and goods. As Industry 4.0 continues to evolve, the integration of digital technologies into transportation holds promise for creating more efficient, sustainable, and resilient mobility systems for the future.

Intelligent Transportation Systems (ITS) encompass a range of technologies and applications aimed at improving the safety, efficiency, and sustainability of transportation systems. From advanced traffic management systems to connected vehicles and smart infrastructure, ITS leverages data, communication, and automation to enhance mobility and address the challenges facing modern transportation networks. One of the key components of ITS is advanced traffic management systems (ATMS), which utilize sensors, cameras, and data analytics to monitor and control traffic flow on roadways. ATMS can detect congestion, incidents, and adverse weather conditions in real-time, allowing transportation agencies to implement dynamic traffic control measures such as signal timing adjustments, lane closures, and variable speed limits to alleviate congestion and improve safety. Furthermore, ITS enables the implementation of adaptive traffic signal control systems, which adjust signal timing based on real-time traffic conditions and demand. Adaptive signal control algorithms optimize signal timing patterns to minimize delays, reduce travel times, and improve traffic flow efficiency at intersections, particularly during peak periods. By reducing the frequency and severity of traffic congestion, adaptive signal control systems help enhance overall mobility and reduce fuel consumption and emissions.

Another key component of ITS is connected vehicle technology, which enables vehicles to communicate with each other and with infrastructure elements such as traffic signals, signs, and roadside sensors. Connected vehicles exchange real-time data on vehicle position, speed, and trajectory, allowing for cooperative collision avoidance, intersection safety, and platooning, where vehicles travel closely together in a coordinated manner to reduce aerodynamic drag and fuel consumption. Moreover, ITS facilitates the deployment of autonomous or self-driving vehicles, which utilize sensors, cameras, and artificial intelligence algorithms to navigate roads and make driving decisions without human intervention. Autonomous vehicles have the potential to revolutionize transportation by improving safety, reducing congestion, and enhancing mobility for people of all ages and abilities. However, their widespread adoption and integration into existing transportation systems require addressing technical, regulatory, and societal challenges, including safety, liability, privacy, and public acceptance. Furthermore, ITS encompasses traveler information systems, which provide real-time information on traffic conditions, transit schedules, and alternative routes to help travelers make informed decisions and plan their trips more efficiently. Traveler information systems include variable message signs, mobile applications, and websites that provide up-to-date information on road and weather conditions, construction zones, and traffic incidents, allowing travelers to avoid congestion and delays.

Additionally, ITS enables the implementation of electronic toll collection systems, which use radio-frequency identification (RFID) or other technologies to automatically collect tolls from vehicles as they pass through tolling points, eliminating the need for manual toll booths and reducing congestion and emissions. Electronic toll collection systems improve traffic flow, enhance tolling efficiency, and provide a seamless travel experience for motorists. Moreover, ITS supports the development and deployment of smart infrastructure, including intelligent transportation management centers (TMCs), connected traffic signals, and integrated transportation networks. Smart infrastructure integrates data from various sources, including sensors, cameras, and connected vehicles, to monitor and manage traffic flow, optimize transportation operations, and enhance safety and security on roadways. Intelligent Transportation Systems (ITS) play a critical role in improving the safety, efficiency, and sustainability of transportation systems. By leveraging advanced technologies such as traffic management systems, connected vehicles, and smart infrastructure, ITS enables transportation agencies and stakeholders to address the challenges facing modern mobility and create a safer, more efficient, and sustainable transportation future. However, realizing the full potential of ITS requires collaboration, innovation, and investment from government agencies, private sector partners, and other stakeholders to deploy and integrate these technologies effectively and ensure their benefits are realized by all users of the transportation system.

1.16   Regulatory Frameworks and Policy Considerations

Regulatory frameworks and policy considerations play a crucial role in shaping the adoption and deployment of Industry 4.0 technologies, ensuring that they are deployed responsibly, ethically, and in compliance with legal and regulatory requirements. As Industry 4.0 technologies continue to advance and permeate various sectors, policymakers face the challenge of balancing innovation and economic growth with concerns related to data privacy, security, and ethical implications. Two key areas of focus in regulatory frameworks and policy considerations for Industry 4.0 are data privacy and security regulations and standards and interoperability.

Regulatory frameworks and policy considerations play a crucial role in shaping the adoption and deployment of Industry 4.0 technologies, ensuring that they are deployed responsibly, ethically, and in compliance with legal and regulatory requirements. Industry 4.0, often referred to as the fourth industrial revolution, encompasses a range of advanced technologies such as artificial intelligence (AI), internet of things (IoT), big data analytics, robotics, and additive manufacturing, which are transforming manufacturing and other sectors by enabling automation, connectivity, and data-driven decision-making. One of the primary challenges facing regulators and policymakers in the context of Industry 4.0 is keeping pace with the rapid technological advancements and their implications for society, economy, and governance. As Industry 4.0 technologies continue to evolve and disrupt traditional business models and practices, regulators must proactively anticipate and address emerging risks and opportunities to ensure that these technologies are harnessed for the benefit of society while minimizing potential harms. One key area of focus for regulatory frameworks and policy considerations in Industry 4.0 is data privacy and security. With the proliferation of IoT devices and interconnected systems, vast amounts of data are generated, collected, and analyzed to drive decision-making and innovation. However, this data often contains sensitive information about individuals, businesses, and organizations, raising concerns about privacy, confidentiality, and cybersecurity. To address these concerns, regulators have implemented data protection laws and regulations, such as the European Union's General Data Protection Regulation (GDPR) and similar laws in other jurisdictions, which establish principles and requirements for the collection, processing, and storage of personal data. These regulations require organizations to obtain informed consent from individuals for data processing activities, implement security measures to protect data from unauthorized access and breaches, and provide mechanisms for individuals to exercise their data rights, such as the right to access, rectify, or delete their personal data. Moreover, regulators are increasingly focused on ensuring transparency and accountability in AI and algorithmic decision-making systems. AI technologies have the potential to make decisions that impact individuals' lives in various domains, including finance, healthcare, criminal justice, and employment. However, there are concerns about bias, discrimination, and lack of explainability in AI algorithms, which may result in unfair or unethical outcomes. To address these concerns, regulators are exploring ways to promote fairness, accountability, and transparency in AI systems through regulatory oversight, industry standards, and best practices. For example, some jurisdictions have proposed regulations requiring AI developers and users to document and disclose information about the data, algorithms, and decision-making processes used in AI systems to enable independent auditing, validation, and accountability.

Furthermore, regulatory frameworks and policy considerations are essential for ensuring safety and reliability in autonomous systems and robotics, which are key components of Industry 4.0. Autonomous vehicles, drones, and industrial robots have the potential to improve efficiency, productivity, and safety in various sectors. Still, there are concerns about accidents, malfunctions, and unintended consequences arising from the deployment of these technologies. Regulators are developing standards, guidelines, and certification processes to ensure that autonomous systems and robotics meet safety and performance requirements and adhere to ethical principles. For example, in the automotive industry, regulators are developing safety standards and testing protocols for autonomous vehicles to ensure that they can operate safely under various conditions and scenarios. Similarly, in the healthcare sector, regulators are establishing guidelines for the use of medical robots and AI-driven diagnostic tools to ensure patient safety and regulatory compliance.

Moreover, regulatory frameworks and policy considerations play a crucial role in shaping the workforce of the future in the context of Industry 4.0. As automation and AI technologies transform the nature of work and skills requirements, policymakers must develop strategies to promote lifelong learning, reskilling, and upskilling to ensure that workers can adapt to changing job roles and technological advancements. Furthermore, regulators are exploring ways to address the potential impact of automation on employment, income inequality, and social inclusion. For example, some jurisdictions have proposed policies such as universal basic income (UBI), job guarantee programs, and wage subsidies to support workers affected by automation and ensure that the benefits of technological progress are shared equitably across society. Regulatory frameworks and policy considerations play a critical role in shaping the adoption and deployment of Industry 4.0 technologies, ensuring that they are deployed responsibly, ethically, and in compliance with legal and regulatory requirements. By addressing issues such as data privacy and security, transparency and accountability in AI, safety and reliability in autonomous systems, and workforce development, policymakers can help maximize the benefits of Industry 4.0 while mitigating potential risks and challenges. However, achieving this balance requires collaboration, innovation, and adaptive governance approaches to navigate the complexities of the digital age effectively.

Data Privacy and Security Regulations:

One of the primary concerns surrounding Industry 4.0 is the collection, storage, and use of vast amounts of data generated by interconnected devices, sensors, and systems. Data privacy regulations aim to protect individuals' rights and freedoms concerning the processing of their personal data and ensure that data is handled in a transparent, fair, and lawful manner. Regulatory frameworks such as the European Union's General Data Protection Regulation (GDPR) and the California Consumer Privacy Act (CCPA) impose strict requirements on organizations regarding the collection, processing, and storage of personal data. These regulations require organizations to obtain explicit consent from individuals before collecting their data, implement security measures to protect data from unauthorized access or disclosure, and provide individuals with rights to access, correct, and delete their personal information. Furthermore, data security regulations require organizations to implement robust cybersecurity measures to safeguard data against cyber threats, such as data breaches, ransomware attacks, and malware infections. Compliance with data security standards, such as the ISO/IEC 27001, NIST Cybersecurity Framework, and Payment Card Industry Data Security Standard (PCI DSS), helps organizations mitigate cybersecurity risks and demonstrate their commitment to protecting sensitive information.

One of the primary concerns surrounding Industry 4.0 is the collection, storage, and use of vast amounts of data generated by interconnected devices, sensors, and systems. Industry 4.0, also known as the fourth industrial revolution, is characterized by the integration of digital technologies such as artificial intelligence (AI), internet of things (IoT), big data analytics, and cloud computing into manufacturing and other sectors. While these technologies offer numerous benefits, including increased efficiency, productivity, and innovation, they also raise significant challenges related to data privacy, security, and governance.

In the context of Industry 4.0, data is often referred to as the "new oil" or "lifeblood" of the digital economy, as it fuels decision-making, optimization, and innovation across various domains. However, the widespread adoption of IoT devices, connected sensors, and smart systems has led to an explosion of data volume, velocity, and variety, creating challenges for organizations in terms of data management, storage, and analysis. One of the primary challenges associated with data in Industry 4.0 is ensuring its privacy and security. With the proliferation of interconnected devices and systems, vast amounts of sensitive data are generated, transmitted, and stored, raising concerns about unauthorized access, breaches, and misuse. Organizations must implement robust security measures, such as encryption, access controls, and intrusion detection systems, to protect data from cyber threats and ensure compliance with data protection regulations.

Moreover, data privacy regulations, such as the European Union's General Data Protection Regulation (GDPR) and similar laws in other jurisdictions, impose requirements on organizations regarding the collection, processing, and storage of personal data. These regulations establish principles such as data minimization, purpose limitation, and transparency, requiring organizations to obtain informed consent from individuals for data processing activities, implement security measures to protect data, and provide mechanisms for individuals to exercise their data rights. Furthermore, data governance is a critical aspect of managing data in Industry 4.0, ensuring that data is accurate, reliable, and accessible for decision-making and analysis. Data governance involves establishing policies, procedures, and controls for data management, including data quality, metadata management, and data lifecycle management. By implementing effective data governance practices, organizations can ensure that data is used effectively and ethically to drive business outcomes and create value.

Additionally, data analytics plays a crucial role in extracting insights and intelligence from data in Industry 4.0, enabling organizations to make informed decisions and optimize operations. Advanced analytics techniques such as machine learning, predictive modeling, and prescriptive analytics are used to analyze large volumes of data and uncover patterns, trends, and correlations. However, the success of data analytics initiatives depends on the availability of high-quality data, robust analytics tools, and skilled data scientists and analysts. Moreover, data integration and interoperability are key challenges in Industry 4.0, as organizations must integrate data from disparate sources and systems to enable seamless communication and collaboration. With the proliferation of IoT devices and connected systems, data silos and fragmentation can occur, hindering data sharing and analysis. Organizations must invest in data integration technologies such as application programming interfaces (APIs), data warehouses, and middleware to enable interoperability and facilitate data exchange across systems and platforms. Furthermore, data ethics and responsible AI are emerging considerations in the context of Industry 4.0, as organizations grapple with ethical dilemmas and societal impacts associated with data-driven decision-making and AI technologies. Issues such as bias, discrimination, and fairness in algorithms, as well as concerns about surveillance, privacy, and autonomy, raise questions about the ethical use of data and AI in society. Organizations must adopt ethical frameworks, guidelines, and practices to ensure that data and AI technologies are deployed responsibly and ethically, with due consideration for human rights, societal values, and the public interest. The collection, storage, and use of data in Industry 4.0 present significant challenges and considerations related to privacy, security, governance, analytics, integration, and ethics. By addressing these challenges and adopting best practices for data management, organizations can harness the power of data to drive innovation, enhance competitiveness, and create value in the digital economy. However, achieving this requires a holistic approach that considers technical, regulatory, ethical, and societal dimensions of data in Industry 4.0.

Standards and Interoperability:

Another critical aspect of regulatory frameworks for Industry 4.0 is the development and adoption of standards and interoperability protocols that ensure compatibility, consistency, and interoperability between different technologies, systems, and devices. Standards provide guidelines, specifications, and best practices for the design, implementation, and operation of Industry 4.0 solutions, promoting consistency and interoperability across diverse ecosystems. Industry consortia, standards organizations, and regulatory bodies play a vital role in developing and promoting standards for Industry 4.0 technologies, such as IoT, AI, cloud computing, and cybersecurity. Standards such as the OPC Unified

Architecture (OPC UA), MQTT, and ISO 9001 provide protocols and frameworks for data exchange, communication, and interoperability between industrial devices and systems. Moreover, interoperability standards enable seamless integration and connectivity between different components of Industry 4.0 systems, such as sensors, actuators, controllers, and enterprise systems. By adhering to interoperability standards, organizations can avoid vendor lock-in, reduce integration costs, and facilitate collaboration and information exchange across the entire value chain. Regulatory frameworks and policy considerations are essential for ensuring the responsible and ethical deployment of Industry 4.0 technologies while addressing concerns related to data privacy, security, and interoperability. By establishing clear guidelines, standards, and regulations, policymakers can foster trust, confidence, and innovation in Industry 4.0, promoting sustainable growth and development in the digital economy. As Industry 4.0 continues to evolve, policymakers must remain vigilant and proactive in addressing emerging challenges and opportunities to ensure that Industry 4.0 technologies benefit society as a whole.

Another critical aspect of regulatory frameworks for Industry 4.0 is the development and adoption of standards and interoperability protocols that ensure compatibility, consistency, and interoperability between different technologies, systems, and devices. Industry 4.0, also known as the fourth industrial revolution, is characterized by the integration of digital technologies such as artificial intelligence (AI), internet of things (IoT), big data analytics, and cloud computing into manufacturing and other sectors. These technologies enable automation, connectivity, and data-driven decision-making, transforming business models, operations, and supply chains. Standards play a crucial role in facilitating the interoperability and integration of diverse technologies and systems in Industry 4.0. Standards define common technical specifications, protocols, and interfaces that enable different devices, platforms, and applications to communicate and work together seamlessly. By adhering to standards, organizations can ensure that their products and services are compatible with those of other vendors and suppliers, enabling interoperability and reducing complexity and fragmentation in the digital ecosystem.

One of the key challenges in the development of standards for Industry 4.0 is the dynamic and evolving nature of technology. As new technologies emerge and existing technologies evolve, standards must be updated and revised to reflect changes in capabilities, requirements, and best practices. This requires collaboration and coordination among industry stakeholders, standards organizations, regulatory bodies, and research institutions to develop consensus-based standards that address the needs of diverse stakeholders and promote innovation and interoperability. Moreover, interoperability protocols are essential for enabling communication and data exchange between different devices, systems, and platforms in Industry 4.0. Interoperability protocols define rules and procedures for encoding, transmitting, and interpreting data in a standardized format, ensuring that information can be exchanged and understood across heterogeneous environments. Common interoperability protocols used in Industry 4.0 include communication protocols such as MQTT (Message Queuing Telemetry Transport), CoAP (Constrained Application

Protocol), OPC UA (Open Platform Communications Unified Architecture), and standards such as ISO 27001 (Information Security Management System), ISO 9001 (Quality Management System), and ISO 14001 (Environmental Management System).

Furthermore, regulatory frameworks and policy considerations play a critical role in promoting the adoption and implementation of standards and interoperability protocols in Industry 4.0. Regulators and policymakers can incentivize compliance with standards through regulatory mandates, incentives, and certification programs, encouraging organizations to adopt best practices and adhere to industry standards. Moreover, regulators can establish guidelines and requirements for interoperability testing and certification to ensure that products and services meet interoperability requirements and can seamlessly integrate with other systems and devices. Additionally, international collaboration and harmonization are essential for promoting global interoperability and standards adoption in Industry 4.0. As Industry 4.0 technologies transcend national borders and markets, harmonized standards and interoperability protocols are critical for enabling cross-border trade, collaboration, and innovation. Organizations such as the International Organization for Standardization (ISO), the International Electrotechnical Commission (IEC), and the Institute of Electrical and Electronics Engineers (IEEE) play a crucial role in developing and promoting international standards and interoperability protocols for Industry 4.0.

Moreover, industry consortia and alliances are driving efforts to develop industry-specific standards and interoperability frameworks tailored to the needs of specific sectors and applications. For example, the Industrial Internet Consortium (IIC), the OPC Foundation, and the Automotive Industry Action Group (AIAG) are developing standards and interoperability protocols for industrial IoT, industrial automation, and automotive manufacturing, respectively. These industry-led initiatives bring together stakeholders from across the value chain to develop consensus-based standards and interoperability solutions that address industry-specific challenges and requirements. Standards and interoperability protocols are essential for enabling compatibility, consistency, and interoperability between different technologies, systems, and devices in Industry 4.0. Regulatory frameworks and policy considerations play a critical role in promoting the adoption and implementation of standards by incentivizing compliance, establishing guidelines and requirements, and fostering international collaboration and harmonization. By adhering to standards and interoperability protocols, organizations can ensure that their products and services are compatible, interoperable, and future-proof, enabling seamless integration and collaboration in the digital ecosystem.

1.17   Future Trends and Outlook for Industry 4.0

The future of Industry 4.0 holds great promise, characterized by the continued evolution and integration of emerging technologies and innovations that drive advancements in productivity, efficiency, and sustainability across various sectors. While Industry 4.0 has already transformed industries, there are still numerous opportunities and challenges for adoption and implementation as we look towards the future.

Emerging Technologies and Innovations

The future of Industry 4.0 will be shaped by the continued development and integration of emerging technologies that push the boundaries of what is possible. Technologies such as artificial intelligence (AI), machine learning, blockchain, augmented reality (AR), and quantum computing are expected to play a significant role in driving innovation and enabling new applications in Industry 4.0. AI and machine learning algorithms will become increasingly sophisticated, enabling predictive analytics, autonomous decision-making, and adaptive learning in manufacturing, logistics, healthcare, and other industries. Blockchain technology will revolutionize supply chain management, enabling transparent, secure, and decentralized transactions and record-keeping. Augmented reality (AR) and virtual reality (VR) technologies will transform workforce training, remote collaboration, and maintenance and repair activities by providing immersive, interactive experiences. Additionally, advancements in quantum computing will enable organizations to solve complex optimization and simulation problems at unprecedented speeds, revolutionizing fields such as materials science, drug discovery, and financial modeling.

The future of Industry 4.0 will be shaped by the continued development and integration of emerging technologies that push the boundaries of what is possible. Industry 4.0, also known as the fourth industrial revolution, represents a paradigm shift in manufacturing and other sectors, driven by the convergence of digital technologies such as artificial intelligence (AI), internet of things (IoT), big data analytics, robotics, and additive manufacturing. These technologies enable automation, connectivity, and data-driven decision-making, transforming business models, operations, and supply chains. One of the key trends shaping the future of Industry 4.0 is the advancement of artificial intelligence and machine learning. AI technologies such as machine learning, natural language processing, and computer vision are enabling machines to learn from data, adapt to changing environments, and perform tasks that were once the exclusive domain of humans. In the context of Industry 4.0, AI is being used to optimize manufacturing processes, predict equipment failures, and improve product quality through predictive maintenance and quality control. Moreover, AI-powered robotics and autonomous systems are transforming industrial operations by enabling flexible and adaptive manufacturing processes, collaborative robots (cobots), and autonomous vehicles in warehouses and logistics operations. Furthermore, the internet of things (IoT) is driving the proliferation of connected devices and sensors that collect and transmit data in real-time, enabling remote monitoring, control, and optimization of industrial processes and assets. In the future, the IoT will continue to expand its reach into new domains and applications, including smart cities, agriculture, healthcare, and energy management. With the growth of edge computing and 5G connectivity, IoT devices will become increasingly powerful and autonomous, enabling real-time data processing and decision-making at the edge of the network. More-over, big data analytics and cloud computing are transforming how organizations collect, store, and analyze vast amounts of data generated by interconnected devices, sensors, and systems. In the future, advancements in cloud computing, edge computing, and distributed computing will enable organizations to leverage big data analytics to derive actionable insights and intelligence from data in real-time. These insights can be used to opti-mize operations, improve decision-making, and drive innovation across various domains, including manufacturing, supply chain management, and customer service.

Additionally, additive manufacturing, also known as 3D printing, is revolutionizing how products are designed, prototyped, and manufactured. In the future, advancements in additive manufacturing technologies and materials will enable the production of complex, customizable, and lightweight parts and components for a wide range of applications, including aerospace, automotive, healthcare, and consumer goods. With the growth of digital manufacturing platforms and distributed manufacturing networks, additive manu-facturing will become more accessible and affordable, enabling small and medium-sized enterprises (SMEs) and individual makers to innovate and create value. Furthermore, cybersecurity and data privacy will continue to be critical considerations in the future of Industry 4.0. With the proliferation of connected devices and digital systems, the risk of cyber threats and data breaches is increasing, posing significant challenges for organi-zations in terms of protecting sensitive information, ensuring regulatory compliance, and maintaining trust and confidence in digital technologies. In the future, organizations will need to invest in robust cybersecurity measures, such as encryption, authentication, and access controls, to safeguard data and systems from cyber attacks and breaches. More-over, regulatory frameworks and standards for cybersecurity and data privacy will play a crucial role in shaping the future of Industry 4.0 by establishing guidelines and require-ments for organizations to adhere to in order to mitigate risks and protect data and privacy rights. The future of Industry 4.0 will be shaped by the continued development and inte-gration of emerging technologies that push the boundaries of what is possible. From artificial intelligence and internet of things to big data analytics and additive manufac-turing, these technologies hold the potential to revolutionize how products are designed, manufactured, and delivered, creating new opportunities for innovation, growth, and com-petitiveness. However, realizing the full potential of Industry 4.0 requires addressing challenges such as cybersecurity, data privacy, and regulatory compliance, and fostering collaboration and partnerships among stakeholders to drive adoption and implementation of digital technologies.

Opportunities and Challenges for Adoption and Implementation:

While Industry 4.0 offers immense opportunities for innovation, growth, and competitive-ness, there are also challenges that must be addressed to realize its full potential. One of the key challenges is the need for skilled talent capable of designing, implementing, and managing Industry 4.0 technologies and systems. As Industry 4.0 technologies evolve, there will be an increasing demand for data scientists, AI engineers, cybersecurity experts, and other specialized roles. Furthermore, organizations face challenges related to data governance, interoperability, and cybersecurity when integrating diverse systems, devices, and platforms in the Industry 4.0 ecosystem. Ensuring data privacy, security, and compliance with regulatory requirements will be paramount to building trust and confidence in Industry 4.0 technologies. Moreover, there are challenges related to the scalability and sustainability of Industry 4.0 solutions, particularly in resource-constrained environments and developing economies. Addressing issues such as digital divide, infrastructure limitations, and environmental impact will be essential to ensuring that the benefits of Industry 4.0 are accessible and equitable for all. Despite these challenges, Industry 4.0 presents significant opportunities for organizations to drive innovation, improve efficiency, and create value across the entire value chain. By embracing emerging technologies, fostering collaboration and partnerships, and investing in talent development and digital infrastructure, organizations can position themselves for success in the future of Industry 4.0. The future of Industry 4.0 is characterized by continuous innovation, driven by emerging technologies and a growing ecosystem of digital capabilities. While there are challenges to overcome, the opportunities for organizations to transform their operations, drive growth, and create positive societal impact are immense. By embracing the opportunities and addressing the challenges, organizations can navigate the evolving landscape of Industry 4.0 and unlock new possibilities for the future.

1.18 Summary

Industry 4.0 represents a transformative era in manufacturing, driven by digitalization, connectivity, automation, and advanced technologies. By embracing the key drivers of Industry 4.0, organizations can unlock new opportunities for efficiency, agility, and competitiveness in the global marketplace. As the Fourth Industrial Revolution continues to unfold, the synergy between technology, innovation, and human ingenuity will shape the future of industry in unprecedented ways.

Industry 4.0, also known as the Fourth Industrial Revolution, is revolutionizing manufacturing and production processes worldwide by integrating advanced digital technologies, automation, and data-driven insights. This paradigm shift is driven by several key factors. Industry 4.0 harnesses a range of cutting-edge technologies including artificial intelligence, Internet of Things, robotics, additive manufacturing, augmented reality, and cloud computing. These innovations digitize and automate manufacturing processes, enhancing efficiency, flexibility, and scalability. The proliferation of IoT devices and sensors enables real-time data collection from manufacturing equipment, products, and supply chains. Data analytics and machine learning algorithms derive actionable insights from this data, optimizing production processes, predicting maintenance needs, and improving product quality. Industry 4.0 blurs the lines between physical and digital realms through cyber-physical systems (CPS). These systems integrate sensors, actuators, and control systems to monitor and control physical processes, facilitating real-time adjustments and autonomous decision-making. Industry 4.0 enables mass customization and personalized production by leveraging digital technologies to tailor products and services to individual customer preferences. Flexible manufacturing systems and agile processes support rapid changeovers and batch-size-one production runs. Digitization and integration of supply chains enhance end-to-end visibility, traceability, and collaboration among stakeholders. Digital supply networks facilitate seamless communication, inventory optimization, and predictive logistics, bolstering supply chain resilience and agility. Industry 4.0 requires a skilled workforce capable of leveraging digital technologies and collaborating with automated systems. Upskilling initiatives and human–machine interfaces empower workers to interact with intelligent machines, enhancing productivity and innovation. Industry 4.0 promotes sustainable manufacturing practices by optimizing resource utilization, reducing waste, and minimizing environmental impact. Digital twins, energy management systems, and predictive maintenance solutions enable proactive resource management and energy efficiency improvements.

In summary, Industry 4.0 represents a convergence of technological innovation, data-driven insights, and collaborative ecosystems. By embracing digital transformation and adopting Industry 4.0 principles, businesses can unlock new opportunities for innovation, competitiveness, and sustainable growth in the global marketplace.

References

Armbrust, M., Fox, A., Griffith, R., Joseph, A., Katz, R., Konwinski, A., Lee, G., Patterson, D., Rabkin, A., Stoica, I., & Zaharia, M. (2010). A view of cloud computing. *Communications of the Acm, 53*(4), 50–58. https://doi.org/10.1145/1721654.1721672

Arrieta, A., Díaz-Rodríguez, N., Ser, J., Bennetot, A., Tabik, S., Barbado, A., Garcia, S., Gil-Lopez, S., Molina, D., Benjamins, R., Chatila, R., & Herrera, F. (2020). Explainable artificial intelligence (XAI): Concepts, taxonomies, opportunities and challenges toward responsible AI. *Information Fusion, 58*, 82–115. https://doi.org/10.1016/j.inffus.2019.12.012

Autor, D. (2015). Why are there still so many jobs? the history and future of workplace automation. *Journal of Economic Perspectives, 29*(3), 3–30. https://doi.org/10.1257/jep.29.3.3

Fouad, N. (2022). The security economics of edtech: Vendors' responsibility and the cybersecurity challenge in the education sector. *Digital Policy Regulation and Governance, 24*(3), 259–273. https://doi.org/10.1108/dprg-07-2021-0090

Ho, P., Albajez, J., Santolaria, J., & Yagüe-Fabra, J. (2022). Study of augmented reality based manufacturing for further integration of quality control 4.0: a systematic literature review. *Applied Sciences, 12*(4), 1961. https://doi.org/10.3390/app12041961

Mertanen, K., Vainio, S., & Brunila, K. (2021). Educating for the future? mapping the emerging lines of precision education governance. *Policy Futures in Education, 20*(6), 731–744. https://doi.org/10.1177/14782103211049914

Nawaz, A., Merces, L., Ferro, L., Sonar, P., & Bufon, C. (2023). Impact of planar and vertical organic field-effect transistors on flexible electronics. *Advanced Materials, 35*(11). https://doi.org/10.1002/adma.202204804

Paschen, J., Kietzmann, J., & Kietzmann, T. (2019). Artificial intelligence (AI) and its implications for market knowledge in B2B marketing. *Journal of Business and Industrial Marketing, 34*(7), 1410–1419. https://doi.org/10.1108/jbim-10-2018-0295

Perminova, S. (2021). Creation of edtech startupises as a factor of online education development. *Efektyvna Ekonomika,* (3). https://doi.org/10.32702/2307-2105-2021.3.82

Sethi, S., Saxena, S., & Singh, M. (2022). A nexus of market web traffic and investor's behavior in the edtech market: Evidence of performance from us and india. *Benchmarking an International Journal, 30*(9), 3150–3167. https://doi.org/10.1108/bij-05-2022-0317

Wei, Q., Chen, Y., Wang, Z., Yu, D., Wang, W., Qu, Y., Li, J. -Q., Chen, L. -H., Li, Y., & Su, B. (2022). Light-assisted semi-hydrogenation of 1,3-butadiene with water. *Angewandte Chemie, 134*(38). https://doi.org/10.1002/ange.202210573

Williamson, B., Macgilchrist, F., & Potter, J. (2021). Covid-19 controversies and critical research in digital education. *Learning Media and Technology, 46*(2), 117–127. https://doi.org/10.1080/17439884.2021.1922437

Thin Films: History, Properties and Emerging Trends

2

2.1 Introduction

Thin films refer to thin layers of material deposited onto a substrate, typically with thicknesses ranging from a few nanometers to several micrometers. These films can be composed of a variety of materials, including metals, semiconductors, oxides, polymers, and ceramics. Thin films find applications across various fields, including electronics, optics, energy, biomedical engineering, and coatings.

Thin films have a rich history spanning various disciplines, from physics to materials science and engineering. The history of thin films can be traced back to the early twentieth century when scientists began depositing thin layers of materials onto substrates for various applications. Thin films gained prominence in the 1950s with the rise of the semiconductor industry. Techniques like chemical vapor deposition (CVD) and physical vapor deposition (PVD) were developed to deposit thin layers of semiconductors for electronic devices. Thin films found applications in optics and photonics, particularly in the development of anti-reflective coatings, mirrors, and filters. Thin films played a crucial role in the advancement of magnetic recording technology, enabling the development of devices like magnetic tapes and hard disk drives. In recent decades, thin films have become integral to various emerging technologies, including flexible electronics, solar cells, and microelectromechanical systems (MEMS).

Thin films exhibit unique properties that can be tailored for specific applications. These properties include electrical conductivity, optical transparency or reflectivity, mechanical flexibility, chemical stability, and magnetic behavior. By controlling the deposition process parameters and material composition, researchers and engineers can manipulate these properties to design thin films with desired characteristics. Thin films are characterized by their thickness, typically ranging from a few nanometers to several micrometers. Uniformity of thickness and composition across the film is essential for many applications,

K. Ukoba and T.-C. Jen, *Shaping Tomorrow: Thin Films and 3D Printing in the Fourth Industrial Revolution 1*, Synthesis Lectures on Mechanical Engineering, https://doi.org/10.1007/978-3-031-84124-8_2

ensuring consistent performance. Thin films can exhibit different crystalline structures, such as polycrystalline or amorphous, depending on the deposition process and conditions. Thin films can be engineered to exhibit specific optical properties, including transparency, reflectivity, and absorption, making them suitable for optical coatings and devices. The electrical properties of thin films, such as conductivity and resistivity, are crucial for electronic and semiconductor applications. Thin films can have tailored mechanical properties, including flexibility and hardness, making them suitable for flexible electronics and protective coatings.

Thin films are often deposited using techniques such as Physical Vapor Deposition (PVD) in which material is evaporated or sputtered from a solid source and deposited onto a substrate, forming a thin film. Common PVD methods include evaporation and sputtering. Chemical Vapor Deposition (CVD) involves the chemical reaction of precursor gases on a substrate surface to deposit a thin film. Variations of CVD include plasma-enhanced CVD (PECVD) and atomic layer deposition (ALD). Sol–Gel Deposition method involves the conversion of precursor solutions into solid thin films through a chemical process, typically at relatively low temperatures. Spin Coating in which a liquid precursor is deposited onto a substrate, which is then spun at high speeds to spread the material evenly and form a thin film. Langmuir–Blodgett Technique involves the transfer of molecules from the surface of a liquid onto a solid substrate, resulting in a thin film with precise molecular arrangement.

Applications of thin films are diverse, ranging from electronic devices (such as transistors, sensors, and solar cells) to optical components (such as lenses, mirrors, and waveguides), from protective coatings (for corrosion resistance or scratch resistance) to biomedical implants and drug delivery systems. Thin films continue to be an area of active research and technological advancement, driving innovations across multiple industries.

The integration of nanomaterials into thin films is a burgeoning trend, enabling the development of nanoscale devices and structures with unique properties. Thin-film technologies are enabling the development of flexible and stretchable electronics for applications in wearable devices, healthcare, and robotics. Thin films are being used in the development of next-generation solar cells, batteries, and capacitors for efficient energy harvesting and storage. The emergence of two-dimensional (2D) materials like graphene and transition metal dichalcogenides (TMDs) is opening up new possibilities for thin-film electronics and optoelectronics. Additive manufacturing techniques, such as inkjet printing and aerosol deposition, are being employed to deposit thin films with enhanced precision and scalability. Novel deposition techniques, such as atomic layer deposition (ALD) and molecular beam epitaxy (MBE), are enabling precise control over thin-film properties at the atomic scale.

Thin films have a storied history and continue to be at the forefront of technological innovation, with emerging trends focusing on nanotechnology, flexible electronics, energy applications, and advanced deposition techniques. These advancements hold the promise of enabling new generations of devices with enhanced performance and functionality.

2.2   History of Thin Films

Thin films have a rich history deeply intertwined with the progress of science and technology. These ultra-thin layers of material, ranging from nanometers to micrometers in thickness, have found applications across a wide array of fields, from electronics to optics, and from energy to biomedical engineering. Understanding the historical journey of thin films illuminates their significance and sheds light on their diverse applications in modern society.

Thin films began to capture scientific interest in the early twentieth century, marked by the pioneering work of researchers exploring the deposition of materials onto substrates. The initial experiments focused on simple techniques to create thin layers, laying the groundwork for future advancements. These early developments were primarily driven by fundamental scientific curiosity, with little anticipation of the vast technological impact that thin films would later have. The pivotal moment for thin films came with the rise of the semiconductor industry in the mid-twentieth century. As electronic devices evolved from bulky vacuum tubes to compact solid-state components, the demand for miniaturization and improved performance grew exponentially. Thin films emerged as a crucial enabling technology, facilitating the fabrication of semiconductor devices such as transistors, diodes, and integrated circuits. Techniques like physical vapor deposition (PVD) and chemical vapor deposition (CVD) were developed to deposit thin layers of semiconductors onto substrates with precision, revolutionizing the electronics industry and paving the way for modern computing and telecommunications.

In parallel with their role in electronics, thin films made significant contributions to the field of optics and photonics. By manipulating the deposition process and material composition, researchers could engineer thin films with specific optical properties, such as transparency, reflectivity, and absorption. This capability led to the development of advanced optical coatings used in lenses, mirrors, and filters, enhancing the performance of optical devices and systems across various applications, from cameras and microscopes to lasers and telecommunications networks. Another area where thin films left an indelible mark is magnetic recording technology. With the advent of magnetic tapes and hard disk drives, thin film magnetic materials became indispensable for storing and retrieving vast amounts of data efficiently. Thin film heads, fabricated using advanced deposition techniques, played a crucial role in reading and writing data with high precision, enabling the digital revolution and the proliferation of information technology in the modern world. In recent decades, thin films have continued to evolve, driven by emerging technologies and the quest for novel applications. Flexible electronics, for example, leverage thin film transistors and interconnects to create bendable and stretchable devices for wearable electronics, healthcare monitoring, and smart textiles. Thin film solar cells have emerged as a promising alternative to traditional silicon-based photovoltaics, offering lightweight and flexible solutions for renewable energy generation. Furthermore, advancements in deposition techniques, such as atomic layer deposition (ALD) and molecular beam epitaxy

(MBE), have enabled precise control over thin film properties at the atomic scale, opening up new frontiers in nanotechnology and quantum devices.

Looking ahead, the future of thin films appears promising, with ongoing research focused on pushing the boundaries of materials science and engineering. From the development of novel materials with unique properties to the exploration of innovative deposition techniques, scientists and engineers continue to innovate, driven by the desire to unlock new possibilities and address pressing societal challenges. The history of thin films is a testament to human ingenuity and the relentless pursuit of scientific knowledge. From their humble beginnings as experimental curiosities to their pervasive presence in modern technology, thin films have come a long way, leaving an indelible mark on society and shaping the course of technological progress. As we stand on the cusp of a new era of innovation, the journey of thin films serves as a reminder of the transformative power of interdisciplinary collaboration and the boundless potential of human creativity.

Role of Thin Films in Semiconductor Industry

Thin films have played a pivotal role in the semiconductor industry, particularly in the fabrication of electronic devices. With the advent of solid-state electronics in the mid-twentieth century, there was a growing demand for miniaturization and improved performance in electronic components. Thin films emerged as a critical technology to meet these demands, offering a means to deposit precise layers of semiconductor materials onto substrates.

In the semiconductor manufacturing process, thin films are used for various purposes, including the formation of thin film transistors (TFTs), diodes, and integrated circuits (ICs). Techniques such as physical vapor deposition (PVD) and chemical vapor deposition (CVD) allow for the controlled deposition of semiconductor materials like silicon, gallium arsenide, and indium phosphide onto silicon or glass substrates. These thin film devices serve as the building blocks for modern electronic systems, including microprocessors, memory chips, and sensors. Thin film technology has enabled the continuous miniaturization of electronic components, leading to the development of smaller, faster, and more energy-efficient devices. The ability to fabricate thin film structures with nanoscale dimensions has driven advancements in semiconductor manufacturing, pushing the limits of Moore's Law and enabling the integration of billions of transistors on a single chip. Moreover, thin films have facilitated the development of novel semiconductor materials and device architectures, such as heterostructures, quantum dots, and nanowires, which exhibit unique electronic properties and functionalities. These advancements have expanded the capabilities of electronic devices, enabling new applications in areas such as quantum computing, optoelectronics, and wearable electronics.

Contributions of Thin Films to Optical Coatings

Thin films have made significant contributions to the field of optics and photonics through the development of advanced optical coatings. By precisely depositing thin layers of materials onto optical components, researchers can manipulate the transmission, reflection, and absorption of light, enabling the creation of optical coatings with tailored properties.

One of the most notable contributions of thin films to optics is in the area of anti-reflective coatings (ARCs). ARCs are thin film coatings designed to reduce unwanted reflections from the surface of optical elements, such as lenses and windows, thereby improving the transmission of light and enhancing the performance of optical systems. Thin film ARCs are widely used in cameras, microscopes, eyeglasses, and solar panels to increase optical efficiency and minimize glare. Thin films are also employed in the fabrication of interference coatings, which exploit the wave nature of light to selectively enhance or suppress specific wavelengths. Interference coatings are used in optical filters, mirrors, and beamsplitters to manipulate the spectral properties of light for various applications, including spectroscopy, telecommunications, and laser systems. Additionally, thin film coatings are utilized in the development of high-reflectivity mirrors, dichroic filters, and polarizing films, enabling precise control over light polarization and spectral properties. These coatings find applications in lasers, display technologies, and optical communication systems, where precise optical performance is essential. Overall, thin film optical coatings play a critical role in enhancing the performance of optical devices and systems across a wide range of applications, from consumer electronics to scientific instruments, enabling advances in imaging, communications, and photonics research.

Advancements in Magnetic Recording

Thin films have played a crucial role in the advancement of magnetic recording technology, which has revolutionized data storage and retrieval in the digital age. Magnetic recording relies on the ability to store and retrieve information by manipulating the magnetization of thin film magnetic materials deposited onto recording media. One of the key contributions of thin films to magnetic recording is the development of high-density recording media. Thin film magnetic materials, such as iron oxide and cobalt alloys, exhibit superior magnetic properties compared to bulk materials, allowing for higher data densities and improved signal-to-noise ratios. Thin film deposition techniques, such as sputtering and evaporation, enable the fabrication of thin magnetic layers with precise control over composition, grain size, and magnetic anisotropy, optimizing the performance of magnetic recording media. Thin film heads are another critical component of magnetic recording systems, responsible for reading and writing data on magnetic storage media. Thin film heads consist of multiple layers of materials deposited onto substrates, including magnetic, insulating, and conducting layers. These complex structures require advanced thin film deposition techniques, such as ion beam deposition and chemical vapor deposition, to achieve the necessary magnetic and electrical properties for efficient data recording and playback. Advancements in thin film magnetic materials and head design have led to significant improvements in the capacity, speed, and reliability of magnetic storage devices, such as hard disk drives (HDDs) and magnetic tapes. Thin film technology has enabled the continuous scaling of storage densities, allowing for the development of higher-capacity storage solutions for applications ranging from personal computers and servers to data centers and cloud storage. In addition to traditional magnetic recording, thin films are also being explored for emerging magnetic storage technologies, such as spintronics and magnetic random-access memory (MRAM), which promise even higher performance and energy efficiency. These innovative thin film-based storage solutions hold the potential to address the growing demand for data storage in the era of big data and cloud computing, driving further advancements in magnetic recording technology.

Recent Innovations and Emerging Technologies of Thin Films

In recent years, thin films have continued to evolve, driven by emerging technologies and the pursuit of novel applications. The most significant recent innovations and emerging trends in thin films include flexible electronics, thin films solar cells, and the likes.

Thin film technology has enabled the development of flexible and stretchable electronic devices for applications in wearable electronics, healthcare monitoring, and soft robotics. Flexible thin film transistors, sensors, and interconnects can be fabricated on flexible substrates such as plastics and elastomers, enabling conformal integration with curved or irregular surfaces. Thin film photovoltaic devices offer lightweight, flexible, and cost-effective alternatives to traditional silicon-based solar cells. Thin film solar cells can be fabricated using a variety of materials, including cadmium telluride (CdTe), copper indium gallium selenide (CIGS), and organic photovoltaic (OPV) materials, allowing for versatile deployment in various applications, such as building-integrated photovoltaics, portable electronics, and off-grid power systems. The integration of two-dimensional (2D) materials, such as graphene, transition metal dichalcogenides (TMDs), and black phosphorus, into thin film structures has opened up new opportunities for nanotechnology and quantum devices. Thin film heterostructures composed of 2D materials exhibit unique electronic, optical, and mechanical properties, offering unprecedented control over material properties at the atomic scale. Additive manufacturing techniques, such as inkjet printing, aerosol deposition, and spray coating, are being explored for the deposition of thin films with enhanced precision and scalability. Additive manufacturing enables the direct deposition of functional materials onto substrates, eliminating the need for expensive vacuum deposition equipment and enabling rapid prototyping and customization of thin film devices. Novel thin film deposition techniques, such as atomic layer deposition (ALD), molecular beam epitaxy (MBE), and pulsed laser deposition (PLD), offer precise control over thin film growth and composition at the atomic scale. These advanced deposition techniques enable the fabrication of thin film structures with tailored properties and complex geometries, facilitating the development of next-generation electronic, optical, and energy devices.

In summary, recent innovations and emerging technologies in thin films hold the promise of revolutionizing various industries and enabling new applications in electronics, photonics, energy, and nanotechnology. From flexible electronics and thin film solar cells to 2D materials and additive manufacturing, the future of thin films is bright with possibilities, driving continued advancements in materials science and engineering.

2.3 Properties of Thin Films

Thin films possess a myriad of properties that make them incredibly versatile and suitable for a wide range of applications. Understanding these properties in detail is essential for tailoring thin films to meet specific requirements across various industries.

2.3.1 Thickness and Uniformity

The thickness of a thin film is a critical parameter that directly influences its optical, electrical, and mechanical properties. Thin films can range from a few nanometers to several micrometers in thickness, and precise control over this parameter is essential for achieving desired functionality. Uniformity in thickness across the entire surface of the thin film is equally crucial. Non-uniform thickness distribution can lead to variations in optical transmission, electrical conductivity, and mechanical properties, impacting device performance and reliability. Techniques such as chemical vapor deposition (CVD), physical vapor deposition (PVD), and atomic layer deposition (ALD) are employed to deposit thin films with high uniformity and precise thickness control.

The thickness and uniformity of thin films are critical parameters that significantly influence their performance and functionality across various applications. The thickness of a thin film refers to the distance between its top and bottom surfaces, typically measured in nanometers (nm) or micrometers (μm). Precise control over thin film thickness is essential for achieving desired optical, electrical, and mechanical properties.

The essential aspect of thin film thickness are here examined.

i. Controlled Deposition Techniques: Thin films are deposited onto substrates using various techniques such as physical vapor deposition (PVD), chemical vapor deposition (CVD), atomic layer deposition (ALD), and sputtering. These techniques allow for precise control over the thickness of the deposited film.
ii. Film Growth Rate: The rate at which thin films grow during deposition plays a crucial role in determining their final thickness. By adjusting deposition parameters such as temperature, pressure, and deposition rate, researchers can control the growth rate and achieve the desired film thickness.

iii. Thickness Measurement: Several techniques are available for measuring thin film thickness, including ellipsometry, profilometry, spectroscopic techniques, and atomic force microscopy (AFM). These methods provide accurate and non-destructive means of determining the thickness of thin films.

iv. Uniformity: While achieving the desired thickness is essential, ensuring uniformity across the entire surface of the thin film is equally crucial. Non-uniform thickness distribution can lead to variations in optical, electrical, and mechanical properties, affecting device performance and reliability.

Uniformity refers to the consistency of thin film thickness across the entire surface of the substrate. Achieving uniformity is essential for ensuring consistent performance and reliability of thin film-based devices. The essential factors affecting the uniformity of thin films are discussed.

i. Deposition Techniques: The choice of deposition technique significantly impacts the uniformity of thin films. Techniques such as ALD and spin coating offer high levels of uniformity due to their ability to deposit thin films evenly over large areas.

ii. Substrate Preparation: The condition of the substrate surface plays a crucial role in achieving uniform thin film deposition. Proper substrate cleaning and surface treatment help to minimize surface irregularities and promote uniform film growth.

iii. Deposition Parameters: Deposition parameters such as temperature, pressure, deposition rate, and precursor flow rates can affect the uniformity of thin films. Optimizing these parameters ensures consistent deposition across the substrate surface.

iv. Monitoring and Control: Real-time monitoring and control of deposition parameters during thin film deposition help to maintain uniformity. Feedback control systems adjust deposition parameters in response to deviations from the desired film thickness, ensuring uniform film growth.

v. Post-Deposition Processing: Post-deposition processing steps such as annealing, etching, and surface treatments can also influence the uniformity of thin films. These processes help to improve film quality and homogeneity, further enhancing uniformity.

In summary, achieving precise control over the thickness and uniformity of thin films is essential for optimizing their performance and functionality in various applications. By utilizing advanced deposition techniques, monitoring systems, and post-deposition processing methods, researchers can tailor thin films to meet specific requirements and drive innovation across diverse industries.

2.3.2 Structural Properties

The structural properties of thin films refer to the arrangement and organization of atoms or molecules within the film. Thin films can exhibit different structural characteristics, including crystalline, polycrystalline, or amorphous structures, depending on deposition parameters and material composition.

Crystalline thin films consist of well-defined atomic arrangements with long-range order, exhibiting properties such as high electrical conductivity and optical transparency. Polycrystalline thin films contain multiple crystalline domains separated by grain boundaries, which can influence mechanical strength and electrical resistance. Amorphous thin films lack long-range order and exhibit isotropic properties, making them suitable for applications requiring optical transparency and mechanical flexibility.

The structural properties of thin films refer to the arrangement and organization of atoms or molecules within the film. Understanding these properties is essential for tailoring thin films to meet specific application requirements.

1. **Crystalline Structure**:
   - **Definition**: Crystalline thin films exhibit a well-defined atomic arrangement with long-range order, similar to that of bulk crystals.
   - **Characteristics**: Crystalline thin films have sharp diffraction peaks in X-ray diffraction (XRD) patterns, indicating the presence of specific crystallographic planes.
   - **Properties**: Crystalline thin films often exhibit anisotropic properties, with mechanical, electrical, and optical properties varying with crystallographic orientation.
   - **Applications**: Crystalline thin films are used in semiconductor devices, optical coatings, and magnetic storage media, where precise control over material properties is essential.

The crystalline structure is a fundamental concept in materials science that describes the arrangement of atoms or molecules in a solid material. It plays a crucial role in determining the physical, chemical, mechanical, and electronic properties of materials, influencing their behavior and performance in various applications. Understanding the crystalline structure of materials is essential for designing and engineering materials with desired properties for specific applications. At the heart of crystalline structure lies the concept of periodic arrangement of atoms or molecules in three-dimensional space. In a crystalline solid, the constituent particles are arranged in a regular, repeating pattern known as a crystal lattice. This orderly arrangement results in the formation of well-defined crystallographic planes, directions, and symmetry elements, which govern the overall structure and properties of the material. One of the defining characteristics of crystalline materials is the presence of long-range order, meaning that the arrangement of atoms or molecules extends over large distances, typically on the order of nanometers to micrometers. This long-range order gives rise to distinct crystallographic features, such as crystal faces, edges, and corners, which can be observed through techniques like X-ray diffraction, electron microscopy, and atomic force microscopy. Crystalline materials can be classified into different crystal systems based on the symmetry of their crystal lattice. The seven basic crystal systems include cubic, tetragonal, orthorhombic, monoclinic, triclinic, hexagonal, and rhombohedral, each characterized by specific geometric parameters and symmetry elements. Within each crystal system, different crystallographic structures, such as cubic, hexagonal, and tetragonal, can exist, depending on the arrangement of atoms or molecules in the crystal lattice. The crystalline structure of a material can have a profound impact on its properties and behavior. For example, materials with a cubic crystal structure, such as diamond and silicon, exhibit isotropic properties, meaning that their properties are independent of direction. In contrast, materials with anisotropic crystal structures, such as graphite and mica, exhibit directional dependencies in their properties, with different properties along different crystallographic directions. Furthermore, defects and imperfections in the crystalline structure, such as vacancies, dislocations, and grain boundaries, can influence the mechanical strength, electrical conductivity, and optical properties of materials. Understanding and controlling these defects is essential for tailoring the properties of materials for specific applications, such as strengthening metallic alloys, enhancing the efficiency of semiconductor devices, and improving the performance of optical coatings. In summary, the crystalline structure is a fundamental aspect of materials science that governs the properties and behavior of crystalline materials. By understanding the arrangement of atoms or molecules in the crystal lattice and its influence on material properties, researchers and engineers can design and engineer materials with tailored properties for a wide range of applications, from electronics and photonics to aerospace and biomedical devices.

2. **Polycrystalline Structure**:
   - **Definition**: Polycrystalline thin films consist of multiple crystalline domains separated by grain boundaries.
   - **Characteristics**: Polycrystalline thin films exhibit diffraction peaks in XRD patterns, but the peaks are broadened due to the presence of grain boundaries.
   - **Properties**: The properties of polycrystalline thin films are influenced by grain size, orientation, and distribution. Grain boundaries can affect mechanical strength, electrical conductivity, and optical transparency.
   - **Applications**: Polycrystalline thin films are used in thin film transistors, solar cells, and magnetic recording media, where a balance between performance and manufacturability is desired.

Polycrystalline structure is a common arrangement found in many materials, characterized by the presence of multiple crystalline grains or domains within the material. Unlike single-crystal materials, which possess a uniform and continuous crystal lattice throughout, polycrystalline materials consist of numerous crystalline regions with different orientations and grain boundaries separating them. This structure has significant implications for the mechanical, electrical, thermal, and optical properties of the material. The formation of polycrystalline structure typically occurs during the solidification process of a material from its molten state. As the material cools and solidifies, nucleation and growth of individual crystal grains occur simultaneously in different regions of the material. Each crystal grain grows independently, adopting its own orientation and lattice structure dictated by the local thermodynamic conditions and crystallographic preferences. One of the defining features of polycrystalline materials is the presence of grain boundaries, which are interfaces separating adjacent crystal grains. Grain boundaries can vary in orientation, thickness, and chemical composition, depending on factors such as processing conditions, impurities, and strain-induced deformation. These grain boundaries act as barriers to the propagation of dislocations and defects, influencing the mechanical properties, deformation behavior, and fracture resistance of the material. The presence of grain boundaries also affects the electrical and thermal properties of polycrystalline materials. In metals and semiconductors, grain boundaries can impede the movement of charge carriers, leading to increased electrical resistivity and reduced conductivity compared to single-crystal counterparts. Similarly, grain boundaries can impede the flow of heat through the material, resulting in reduced thermal conductivity and increased thermal resistance. Furthermore, the optical properties of polycrystalline materials can be influenced by the presence of grain boundaries and grain size variations. Light scattering and absorption at grain boundaries can affect the transparency, reflectivity, and color of the material, particularly in transparent ceramics, thin films, and optical coatings. Grain size control and grain boundary engineering are essential strategies for optimizing the optical properties of polycrystalline materials for specific applications. Despite the presence of grain boundaries, polycrystalline materials offer several advantages over single-crystal counterparts. For example, polycrystalline materials are typically easier and less expensive to produce on a large scale, making them attractive for industrial applications. Additionally, the presence of grain boundaries can enhance certain properties, such as mechanical strength, fracture toughness, and corrosion resistance, by acting as barriers to crack propagation and enhancing grain boundary sliding mechanisms. In summary, polycrystalline structure is a common arrangement found in many materials, characterized by the presence of multiple crystalline grains separated by grain boundaries. This structure has significant implications for the mechanical, electrical, thermal, and optical properties of the material, influencing its behavior and performance in various applications. Understanding and controlling the grain structure and grain boundaries are essential for optimizing the properties of polycrystalline materials for specific engineering and technological applications.

3. **Amorphous Structure**:
   - **Definition**: Amorphous thin films lack long-range order and exhibit a disordered atomic arrangement.
   - **Characteristics**: Amorphous thin films do not produce sharp diffraction peaks in XRD patterns, instead showing a broad hump indicative of short-range disorder.

- **Properties**: Amorphous thin films have isotropic properties, making them suitable for applications requiring optical transparency, mechanical flexibility, and low electrical conductivity.
- **Applications**: Amorphous thin films are used in thin film transistors, display panels, and protective coatings, where flexibility, transparency, and uniformity are important.

Amorphous structure refers to the arrangement of atoms or molecules in a solid material that lacks long-range order characteristic of crystalline materials. In contrast to crystalline materials, which exhibit a regular and repeating three-dimensional lattice structure, amorphous materials possess a disordered and random arrangement of atoms or molecules. This lack of long-range order results in unique physical, chemical, mechanical, and electrical properties that distinguish amorphous materials from their crystalline counterparts. One of the defining characteristics of amorphous materials is their short-range order, wherein atoms or molecules are arranged in a local structure that may exhibit some degree of spatial correlation but lacks the periodicity and symmetry observed in crystalline materials. This short-range order extends over distances typically on the order of a few atomic or molecular diameters, resulting in a structure that appears isotropic and homogeneous on a microscopic scale. The lack of long-range order in amorphous materials arises from the absence of a well-defined crystal lattice, allowing atoms or molecules to occupy random positions within the material without forming regular patterns or coordination arrangements. As a result, amorphous materials often exhibit isotropic properties, meaning that their properties are independent of direction, unlike the anisotropic properties observed in crystalline materials. Amorphous materials can be formed through various processing techniques, such as rapid cooling, solid-state amorphization, and chemical vapor deposition. Rapid cooling of a molten material prevents the formation of a crystalline structure by freezing the atoms or molecules in a disordered state before they can arrange into a regular lattice. Solid-state amorphization involves subjecting a crystalline material to mechanical deformation, irradiation, or chemical reactions to disrupt its crystal lattice and induce an amorphous transformation. Chemical vapor deposition allows for the deposition of thin films of amorphous materials onto substrates through controlled chemical reactions in the vapor phase. The lack of long-range order in amorphous materials gives rise to several unique properties and behaviors. For example, amorphous materials often exhibit high optical transparency, low mechanical hardness, and low electrical conductivity compared to crystalline materials. These properties make amorphous materials suitable for various applications, including optical coatings, protective coatings, thin film electronics, and photovoltaic devices. Amorphous materials also exhibit unique mechanical properties, such as low fracture toughness, high ductility, and ease of deformation. This makes them suitable for applications requiring flexibility, such as in flexible electronics, packaging materials, and biomedical devices. Additionally, the disordered nature of amorphous materials allows for the incorporation of dopant atoms or molecules, enabling the tailoring of their properties for specific applications, such as semiconductor devices, sensors, and catalytic materials. Despite their unique properties and advantages, amorphous materials also have limitations, such as low thermal stability, susceptibility to chemical degradation, and limited long-term reliability. These limitations can be addressed through careful material selection, processing optimization, and surface modification techniques.

In summary, amorphous structure is a characteristic arrangement of atoms or molecules in a solid material that lacks long-range order observed in crystalline materials. This structure gives rise to unique physical, chemical, mechanical, and electrical properties that distinguish amorphous materials from their crystalline counterparts. Understanding and exploiting the properties of amorphous materials are essential for designing and engineering advanced materials for a wide range of applications in various industries.

4. Epitaxial Growth:
   - **Definition**: Epitaxial thin films are deposited with a specific crystallographic orientation on a substrate that serves as a template.
   - **Characteristics**: Epitaxial thin films exhibit a well-defined relationship between crystallographic planes of the film and substrate, resulting in high-quality interfaces.
   - **Properties**: Epitaxial thin films often have superior electronic and optical properties compared to polycrystalline or amorphous films due to the absence of grain boundaries and defects.
   - **Applications**: Epitaxial thin films are used in semiconductor devices, heterostructures, and quantum wells, where precise control over material properties and interfaces is crucial for device performance.

Epitaxial growth is a key process in materials science and semiconductor technology, involving the deposition of thin crystalline layers on a substrate with a well-defined crystal structure and orientation. This technique allows for the precise control of crystallographic orientation, lattice matching, and film thickness, enabling the fabrication of high-quality thin films with tailored properties and functionalities. Epitaxial growth plays a crucial role in various technological applications, including microelectronics, optoelectronics, photonics, and spintronics. The epitaxial growth process typically occurs through one of two mechanisms: homoepitaxy or heteroepitaxy. In homoepitaxy, the thin film material is deposited onto a substrate of the same material, resulting in the growth of a single-crystal film with the same crystal structure and orientation as the substrate. This process is commonly used in semiconductor manufacturing, where epitaxial layers of silicon are grown on silicon substrates to create integrated circuits and electronic devices. In heteroepitaxy, the thin film material is deposited onto a substrate of a different material, leading to the growth of a crystalline film with a different crystal structure and/or lattice constant from the substrate. Despite the mismatch in crystal structure and lattice parameters, heteroepitaxy can still occur if the lattice mismatch is sufficiently small and can be accommodated by the formation of defects, such as dislocations and misfit dislocations, at the interface between the film and the substrate. Heteroepitaxy is widely used in the fabrication of compound semiconductor devices, such as light-emitting diodes (LEDs), laser diodes, and photovoltaic cells, where different semiconductor materials with specific bandgaps and electronic properties are combined to achieve desired device functionalities. The epitaxial growth process can be carried out using various deposition techniques, including molecular beam epitaxy (MBE), metalorganic chemical vapor deposition (MOCVD), chemical vapor deposition (CVD), and pulsed laser deposition (PLD). These techniques offer precise control over deposition parameters, such as temperature, pressure, and precursor flux, allowing for the growth of epitaxial layers with atomic-level precision and uniformity.

One of the key challenges in epitaxial growth is achieving high-quality interfaces between the thin film and the substrate, as defects and dislocations at the interface can degrade the performance and reliability of the resulting devices. Strategies for improving interface quality include substrate surface preparation, such as cleaning, etching, and annealing, to remove contaminants and defects that may nucleate during epitaxial growth. Additionally, buffer layers or interlayers with lattice-matching properties can be used to reduce the strain and minimize defects at the interface between the film and the substrate. Advances in epitaxial growth technology have led to the development of novel materials and devices with enhanced performance and functionality. For example, strained-layer epitaxy, where thin films are grown under controlled strain conditions, can modify the electronic band structure and enhance carrier mobility in semiconductor devices. Moreover, selective epitaxy techniques, such as selective area growth and epitaxial lateral overgrowth (ELO), enable the fabrication of complex device architectures and integrated circuits with improved device performance and reduced parasitic effects. In summary, epitaxial growth is a versatile and powerful technique for fabricating high-quality thin films with tailored properties and functionalities for a wide range of technological applications. By controlling the deposition parameters and interface properties, researchers and engineers can exploit epitaxial growth to design and engineer advanced materials and devices with unprecedented performance and reliability.

5. **Morphology and Surface Roughness**:
   - **Definition**: Morphology refers to the surface structure and topography of thin films, while surface roughness quantifies deviations from a perfectly flat surface.
   - **Characteristics**: Morphology and surface roughness are influenced by deposition parameters such as substrate temperature, deposition rate, and film thickness.
   - **Properties**: Morphology and surface roughness can affect thin film adhesion, optical reflectivity, and electrical conductivity, particularly in thin film coatings and thin film-based devices.
   - **Applications**: Control over morphology and surface roughness is important in applications such as anti-reflective coatings, microelectronics, and MEMS devices, where surface properties impact device performance and reliability.

Morphology and surface roughness are fundamental characteristics of materials and surfaces that play crucial roles in determining their physical, chemical, mechanical, and optical properties. Understanding and controlling these properties are essential for various technological applications, including thin film deposition, surface engineering, and device fabrication. Morphology refers to the overall shape, size, and structure of a material or surface, including its surface features, crystallographic orientation, and grain boundaries. The morphology of a material can vary widely depending on factors such as processing conditions, composition, and environmental factors. In crystalline materials, morphology is influenced by the arrangement of atoms or molecules in the crystal lattice and the presence of defects, dislocations, and grain boundaries. Different crystallographic orientations and growth conditions can result in various morphologies, including facets, grains, dendrites, and twinned structures. In amorphous materials, morphology is characterized by a lack of long-range order and the presence of short-range order, resulting in a random and irregular arrangement of atoms or molecules. Amorphous materials can exhibit a wide range of morphologies, including smooth surfaces, rough surfaces, and porous structures, depending on factors such as deposition method, substrate temperature, and deposition rate. The morphology of a material can have significant implications for its properties and behavior. For example, in thin film deposition, the morphology of the substrate surface can influence nucleation and growth processes, affecting the microstructure, adhesion, and film quality of the deposited thin films. Similarly, in materials science and engineering, the morphology of surfaces and interfaces can influence adhesion, friction, wear resistance, and corrosion resistance properties.

Surface roughness refers to deviations in surface topography from an ideal flat surface at various length scales, ranging from nanometers to millimeters. It is commonly quantified using parameters such as Ra (average roughness), Rq (root mean square roughness), Rmax (maximum height), and Rz (average maximum peak-to-valley height). Surface roughness can arise from various factors, including surface defects, machining processes, material properties, and environmental factors. In machining processes such as grinding, milling, and polishing, surface roughness is directly influenced by factors such as cutting speed, feed rate, tool geometry, and cutting conditions. In materials processing techniques such as thin film deposition, surface roughness can be controlled by adjusting parameters such as deposition rate, substrate temperature, and deposition method. Surface roughness plays a critical role in determining the performance and functionality of materials and surfaces in various applications. For example, in tribology, surface roughness affects friction, wear, and lubrication properties by influencing the contact area, adhesion, and interfacial interactions between mating surfaces. In optics and photonics, surface roughness can affect light scattering, reflection, and transmission properties by altering surface reflectivity, scattering efficiency, and diffraction patterns.

Morphology and surface roughness are essential characteristics of materials and surfaces that influence their properties, behavior, and performance in various applications.

Understanding and controlling these properties are crucial for designing and engineering materials and surfaces with tailored properties for specific technological applications, ranging from thin film deposition and surface engineering to device fabrication and functional coatings.

In summary, the structural properties of thin films, including crystallinity, grain structure, amorphousness, epitaxial growth, morphology, and surface roughness, play a crucial role in determining their properties and behavior in various applications. By understanding and controlling these structural properties, researchers can tailor thin films to meet specific performance requirements and drive innovation across diverse industries.

2.3.3 Optical Properties

Thin films can be engineered to exhibit specific optical properties, including transparency, reflectivity, absorption, and refractive index. These properties are crucial for applications in optics, photonics, and display technologies. Transparent thin films are used in applications such as optical windows, displays, and solar cells, where high transmission of light is essential. Reflective thin films are employed in mirrors, optical coatings, and laser resonators, where precise control over reflection properties is desired. Absorptive thin films are utilized in optical filters, photodetectors, and thermal imaging devices, where selective absorption of light is required. The refractive index of thin films determines their optical density and dispersion, affecting their behavior in optical devices and systems. Optical properties are among the most crucial characteristics of thin films, determining their behavior and applications in various fields such as optics, photonics, and optoelectronics. Understanding and controlling these properties are essential for designing thin films for specific optical applications.

Transmission and Absorption: Transmission and absorption are fundamental optical properties that play crucial roles in determining the behavior of thin films when interacting with light. These properties are of significant importance in various applications, ranging from optical filters to coatings and windows. Understanding and controlling transmission and absorption spectra are essential for designing thin films with desired optical characteristics. Transmission refers to the ability of a thin film to allow light to pass through it without significant attenuation. When light encounters a thin film, some of it is transmitted through the film, while the rest may be reflected or absorbed. The amount of light transmitted through the film depends on several factors, including the film thickness, material composition, and the optical constants of the materials involved. On the other hand, absorption occurs when the thin film absorbs some or all of the incident light energy. Absorption can occur due to various mechanisms, including electronic transitions, vibrational excitations, and scattering processes within the material. The absorbed energy is converted into other forms, such as heat or fluorescence, depending on the specific properties of the material. The transmission and absorption spectra of thin films are characterized by their dependence on wavelength. Different materials exhibit unique absorption and transmission behaviors across the electromagnetic spectrum, leading to variations in their optical properties. For example, some materials may have high transmission in the visible range but absorb strongly in the ultraviolet or infrared regions. By engineering the thickness, material composition, and optical constants of thin films, researchers can tailor their optical transmission and absorption characteristics to meet specific application requirements. For instance, in optical filters, thin films can be designed to selectively transmit or block certain wavelengths of light, allowing for the creation of filters with precise spectral filtering capabilities. Similarly, in coatings and windows, thin films can be optimized to provide specific levels of transparency and opacity to light. For example, anti-reflective coatings are designed to minimize reflection and enhance transmission through glass or other transparent substrates, improving the optical performance of optical systems and devices. In summary, transmission and absorption are fundamental optical properties of thin films that depend on factors such as film thickness, material composition, and optical constants. By engineering these parameters, researchers can tailor thin films to exhibit desired optical transmission and absorption characteristics, making them suitable for a wide range of applications in optics, photonics, and materials science. Whether it's designing optical filters with precise spectral filtering capabilities or optimizing coatings for enhanced transparency, the ability to control transmission and absorption in thin films opens up new possibilities for innovation in optical technology.

Reflection and Refraction: Reflection and refraction are phenomena that occur when light encounters the interface between a thin film and another medium, such as air or a substrate. Reflection refers to the bouncing back of light waves from the film interface, while refraction involves the bending of light as it passes through the interface. The reflection and refraction properties of thin films are governed by the film's refractive index, which determines the speed of light propagation and the angle of incidence. Thin films with high refractive indices are often used in optical coatings and anti-reflection coatings to control light reflection and enhance optical performance.

Reflection occurs when light waves encounter the interface between a thin film and another medium and bounce back into the original medium. This bouncing back of light waves results in the formation of reflected rays. The amount of light reflected depends on the properties of the thin film and the angle of incidence. When light strikes the interface between the thin film and the surrounding medium, a portion of the incident light is reflected back into the medium from which it originated. This reflection can be influenced by factors such as the angle of incidence, the refractive indices of the thin film and the surrounding medium, and the polarization state of the incident light. In optical systems and devices, controlling reflection is essential for maximizing the efficiency and performance of light transmission. Thin films with specific optical properties, such as high reflectivity or low reflectivity, can be designed and engineered to meet the requirements of various applications.

Refraction occurs when light waves pass through the interface between a thin film and another medium and change direction. This change in direction is due to the difference in the speed of light between the two media, which causes the light waves to bend as they enter or exit the thin film. The degree to which light is refracted depends on the angle of incidence and the refractive indices of the thin film and the surrounding medium. Snell's law describes the relationship between the angles of incidence and refraction and the refractive indices of the two media. In optical systems and devices, controlling refraction is essential for manipulating the path of light and achieving desired optical effects. Thin films with specific refractive indices can be used to bend or focus light, correct aberrations, and enhance optical performance.

The reflection and refraction properties of thin films are crucial for a wide range of applications in optics, photonics, and materials science. In optical coatings, thin films with high reflectivity are used to create mirrors and reflective surfaces for lasers, telescopes, and optical instruments. Conversely, thin films with low reflectivity are employed in anti-reflection coatings to minimize unwanted reflections and improve optical transmission. In photovoltaic devices, thin films with tailored reflection and refraction properties are used to enhance light absorption and maximize energy conversion efficiency. By optimizing the optical properties of thin films, researchers and engineers can improve the performance of solar cells and increase their power output. In display technologies, thin films with specific refractive indices are used to control the direction and intensity of light emitted by display panels, such as liquid crystal displays (LCDs) and organic light-emitting diodes (OLEDs). By manipulating reflection and refraction, manufacturers can enhance the brightness, contrast, and color accuracy of display devices. Reflection and refraction are fundamental optical phenomena that govern the behavior of light when it encounters the interface between a thin film and another medium. By engineering the reflection and refraction properties of thin films, researchers and engineers can develop materials and coatings with tailored optical characteristics for a wide range of applications in optics, photonics, and materials science. Whether it's enhancing light transmission, minimizing unwanted reflections, or optimizing energy conversion efficiency, the ability to control reflection and refraction in thin films opens up new possibilities for innovation in optical technology.

Optical Constants: Optical constants, encompassing the refractive index (n) and extinction coefficient (k), serve as pivotal parameters defining the optical behavior of thin films. These constants dictate how light interacts with and propagates through a medium, profoundly impacting the optical transparency, color, and reflectance of thin films. The refractive index governs the speed of light within the film, while the extinction coefficient quantifies the absorption of light within the material. Their manipulation allows researchers to engineer thin films with precisely tailored optical properties, opening avenues for applications such as lenses, mirrors, and waveguides.

Refractive Index and Extinction Coefficient:

The refractive index (n) of a material characterizes the speed of light propagation within it relative to the speed in a vacuum. It governs how light bends or refracts when transitioning between media with different refractive indices, such as when entering or exiting a thin film. The refractive index is typically wavelength-dependent, resulting in dispersive effects where the speed of light varies with wavelength. Complementing the refractive index, the extinction coefficient (k) quantifies the attenuation or absorption of light within the material. It accounts for the reduction in light intensity as it traverses through the thin film due to absorption processes. Higher values of the extinction coefficient indicate greater light absorption within the material, resulting in reduced optical transparency.

Variations with Wavelength and Material Composition:

Both the refractive index and extinction coefficient are wavelength-dependent, exhibiting variations across the electromagnetic spectrum. This wavelength dependence arises from the interaction of light with the electronic and vibrational states of atoms and molecules within the material. As a result, thin films may exhibit distinct optical behaviors at different wavelengths, influencing their color, transparency, and reflective properties. Additionally, the optical constants of thin films are influenced by their material composition and microstructure. Variations in elemental composition, crystal structure, and defect density can alter the refractive index and extinction coefficient, thereby modulating the optical properties of the thin film. For instance, changes in the chemical composition of a thin film can lead to shifts in its absorption spectra and refractive index dispersion.

Engineering Tailored Optical Properties:

Precise control over the optical constants of thin films enables the design and engineering of materials with tailored optical properties for specific applications. By selecting appropriate materials and deposition techniques, researchers can manipulate the refractive index and extinction coefficient to achieve desired optical effects. For example, in lens and optical systems, thin films with high refractive indices are employed to control the refraction of light, enabling the focusing and dispersion of light rays. Conversely, low-refractive-index films are utilized in anti-reflective coatings to minimize light reflection and enhance optical transmission. In mirrors and reflective surfaces, thin films with high extinction coefficients are preferred to maximize light absorption and minimize reflective losses. By optimizing the optical constants, researchers can design mirrors with high reflectivity across specific wavelength ranges, making them suitable for applications such as lasers and optical sensors.

Applications in Lenses, Mirrors, and Waveguides:

Thin films with tailored optical properties find widespread applications in various optical devices and systems. In lenses, gradient-index thin films are utilized to achieve precise control over light refraction, enabling the correction of aberrations and the formation of high-quality images. Mirrors coated with dielectric thin films exhibit high reflectivity and low optical losses, making them indispensable components in lasers, optical cavities, and astronomical telescopes. Waveguides constructed from thin films with controlled refractive indices guide and confine light within optical circuits, facilitating data transmission and signal processing in telecommunications and integrated photonics. Optical constants such as the refractive index and extinction coefficient are vital parameters that govern the optical behavior of thin films. By engineering these parameters, researchers can design materials with tailored optical properties for a myriad of applications, including lenses, mirrors, and waveguides. The ability to control the optical constants of thin films offers unprecedented opportunities for innovation in optical technology, driving advancements in fields ranging from telecommunications to photonics.

Dispersion and Birefringence: Dispersion and birefringence are fundamental optical phenomena that profoundly influence the behavior of light as it interacts with thin films. These phenomena play pivotal roles in various optical devices, enabling precise control over optical polarization, color dispersion, and chromatic aberration. Understanding and engineering dispersion and birefringence in thin films are essential for the design and optimization of optical devices such as polarizers, retarders, and waveplates. Dispersion refers to the variation of the refractive index of a material with wavelength. In other words, different wavelengths of light propagate at different speeds through the material, causing them to bend or refract by varying amounts. This wavelength-dependent refractive index leads to phenomena such as chromatic dispersion, where different colors of light are refracted at different angles, resulting in color separation. The dispersion characteristics of a material are quantified by its dispersion curve, which describes how the refractive index changes with wavelength across the electromagnetic spectrum. Materials with strong dispersion exhibit significant variations in refractive index over a wide range of wavelengths, while materials with weak dispersion display minimal changes. In thin films, dispersion can manifest as chromatic aberration, where different wavelengths of light focus at different distances, causing blurring and color fringing in images. By carefully controlling the dispersion properties of thin films, researchers can minimize chromatic aberration and enhance the optical performance of lenses, optical systems, and imaging devices.

Birefringence occurs when the refractive index of a material varies with the polarization state of light. In other words, the material exhibits different refractive indices for light polarized in different directions. This polarization-dependent optical behavior arises from the anisotropic structure or symmetry of the material, leading to distinct optical properties along different axes. When light passes through a birefringent material, it splits into two orthogonal polarization components, each traveling at a different speed and experiencing a different phase shift. This differential phase shift results in changes to the polarization state of the light, leading to polarization-dependent effects such as optical rotation and elliptical polarization. In thin films, birefringence can be induced by various factors, including stress, strain, and anisotropic microstructure. By controlling these factors during film deposition or post-processing, researchers can engineer thin films with tailored birefringent properties for specific applications.

Thin films exhibiting dispersion and birefringence find widespread applications in optical devices where precise control over optical polarization and dispersion is required. For example, polarizers are optical devices that selectively transmit light of a specific polarization state while blocking light of orthogonal polarization. By exploiting the birefringent properties of thin films, polarizers can be constructed to selectively absorb or transmit light based on its polarization state, enabling applications such as glare reduction, LCD displays, and photography filters. Retarders, also known as wave plates, are optical devices that introduce a controlled phase delay between orthogonal polarization components of light. Thin films with birefringent properties are utilized to fabricate retarders with precise phase delay values, allowing for polarization control and manipulation in optical systems, telecommunications, and spectroscopy. Waveplates, or retarders, are another class of optical devices that exploit the birefringent properties of thin films. Waveplates introduce a controlled phase delay between orthogonal polarization components of light, enabling applications such as optical modulation, polarization control, and polarization analysis in optical systems, telecommunications, and spectroscopy.

In summary, dispersion and birefringence are fundamental optical phenomena that influence the behavior of light in thin films. By engineering the dispersion and birefringent properties of thin films, researchers can design and optimize optical devices such as polarizers, retarders, and waveplates for a wide range of applications requiring precise control over optical polarization and dispersion. Whether it's reducing chromatic aberration in lenses or manipulating polarization in optical systems, the ability to control dispersion and birefringence in thin films opens up new opportunities for innovation in optical technology.

Optical Coatings and Thin Film Stacks: Thin films are commonly used in optical coatings and thin film stacks to manipulate the transmission, reflection, and absorption of light. Optical coatings consist of multiple thin film layers with tailored optical properties, such as anti-reflection coatings, high-reflectivity mirrors, and bandpass filters. By designing thin film stacks with specific layer thicknesses and refractive indices, researchers can create coatings that enhance light transmission, minimize reflection, or selectively filter wavelengths, enabling applications in imaging, spectroscopy, and telecommunications. Thin films play a pivotal role in optical coatings and thin film stacks, offering precise control over the transmission, reflection, and absorption of light. These coatings typically comprise multiple layers of thin films, each engineered to possess specific optical properties. By strategically designing the thicknesses and refractive indices of these layers, researchers can fabricate coatings tailored to various applications, including anti-reflection coatings, high-reflectivity mirrors, and bandpass filters. These versatile coatings find extensive use in fields such as imaging, spectroscopy, and telecommunications, where precise control over light manipulation is paramount.

Optical Coatings and Thin Film Stacks:

Optical coatings are composite structures consisting of multiple thin film layers deposited onto a substrate. Each layer is engineered to interact with light in a specific manner, thereby achieving desired optical effects. By stacking these layers together, optical coatings can be tailored to exhibit various optical properties, such as transmission enhancement, reflection suppression, or wavelength selectivity. The design of optical coatings relies on the principles of interference and thin film optics. Interference occurs when light waves interact constructively or destructively, leading to the amplification or cancellation of specific wavelengths. In thin film stacks, interference phenomena are exploited to manipulate the transmission, reflection, and absorption of light.

Applications of Optical Coatings:

One of the most common applications of optical coatings is in anti-reflection coatings, which are designed to minimize unwanted reflections from optical surfaces. Anti-reflection coatings typically consist of multiple thin film layers with alternating high and low refractive indices. By exploiting interference effects, these coatings can suppress reflection over a broad range of wavelengths, improving the optical performance of lenses, displays, and camera lenses. High-reflectivity mirrors are another essential application of optical coatings, used in lasers, optical cavities, and astronomical telescopes. These mirrors comprise thin film stacks optimized to maximize reflection at specific wavelengths while minimizing absorption losses. By precisely controlling the layer thicknesses and refractive indices, high-reflectivity mirrors can achieve near-total reflectivity, making them indispensable components in optical systems requiring high optical power. Bandpass filters are optical coatings designed to selectively transmit light within a specific wavelength range while blocking all other wavelengths. These filters are widely used in spectroscopy, optical sensing, and telecommunications to isolate spectral bands of interest. By engineering thin film stacks with precisely tuned optical properties, bandpass filters can be tailored to pass specific wavelengths with high transmission efficiency while attenuating unwanted spectral components.

Design Considerations of optical coatings and thin film stacks
The design of optical coatings and thin film stacks requires careful consideration of various factors, including the refractive indices of the constituent materials, the thicknesses of individual layers, and the desired optical performance. Optimization techniques, such as numerical simulation and thin film design software, are employed to identify optimal layer configurations that meet specific application requirements. Layer thicknesses and refractive indices are critical parameters that determine the optical properties of thin film stacks. By adjusting these parameters, researchers can control the interference effects within the stack, thereby tailoring its transmission, reflection, and absorption characteristics. Advanced deposition techniques, such as physical vapor deposition (PVD) and chemical vapor deposition (CVD), are employed to deposit thin film layers with precise

thicknesses and compositions. Thin films are indispensable components in optical coatings and thin film stacks, offering unparalleled control over the transmission, reflection, and absorption of light. By designing coatings with tailored optical properties, researchers can achieve a wide range of optical effects, from anti-reflection coatings that minimize surface reflections to bandpass filters that isolate specific spectral bands. The versatility of optical coatings makes them invaluable in various fields, including imaging, spectroscopy, and telecommunications, where precise control over light manipulation is essential for device performance and functionality.

In summary, the optical properties of thin films play a critical role in determining their performance and functionality in various optical applications. By understanding and engineering these properties, researchers can design thin films with tailored optical characteristics, enabling advancements in optics, photonics, and optoelectronics.

2.3.4   Electrical Properties

The electrical properties of thin films are vital for their application in electronic devices and circuits. Thin films can exhibit a wide range of electrical behaviors, including conductivity, resistivity, mobility, and carrier concentration, depending on material composition and structural characteristics. Conductive thin films are used in applications such as electrodes, interconnects, and sensors, where low electrical resistance and high charge carrier mobility are essential. Insulating thin films are employed as dielectric layers in capacitors, transistors, and integrated circuits, where high electrical resistance and low leakage current are required. Semiconducting thin films are utilized in electronic devices such as solar cells, photodetectors, and transistors, where controlled conductivity and bandgap properties are critical.

Electrical properties constitute a fundamental aspect of thin films, influencing their performance across a wide spectrum of electronic and optoelectronic applications. The nature of electrical behavior in thin films varies widely based on factors such as material composition, structure, and fabrication techniques.

Conductivity: Conductivity is a pivotal electrical property characterizing the ease with which electric current flows through a material. Thin films exhibit diverse conductivity types viz Metallic Thin Films, Semiconducting Thin Films, and Insulating Thin Films.

Metals possess high electrical conductivity due to the presence of delocalized electrons that can freely move through the material. Metallic thin films find utility in applications demanding low resistance and high conductivity, including electrical interconnects, electrodes, and current collectors. Semiconductors lie between metallic conductors and insulators concerning electrical conductivity. Their conductivity can be modulated by factors such as temperature, doping, and applied electric fields. Semiconducting thin films are extensively employed in electronic devices such as transistors, diodes, and solar cells, where controlled electrical properties are pivotal. Thin films designed to exhibit insulating behavior possess exceedingly low electrical conductivity, effectively impeding the flow of electric current. These films serve as dielectric layers in capacitors, insulation layers in integrated circuits, and barrier coatings in electronic devices necessitating electrical isolation and insulation.

Carrier Mobility
Carrier mobility denotes the capability of charge carriers (electrons or holes) to navigate through a material under the influence of an applied electric field. High carrier mobility in thin films corresponds to efficient charge transport, thereby enhancing the performance of electronic devices.

High Mobility Semiconductors: Thin films crafted from high mobility semiconductor materials, such as silicon (Si), gallium arsenide (GaAs), and organic semiconductors, exhibit substantial carrier mobility. These materials are instrumental in fabricating high-speed transistors, photodetectors, and light-emitting diodes (LEDs), thereby facilitating advancements in electronic and optoelectronic technologies.

Bandgap: The bandgap, an intrinsic property of semiconducting thin films, delineates the energy disparity between the highest occupied energy level (valence band) and the lowest unoccupied energy level (conduction band). Wide bandgap materials serve as insulators, while narrow bandgap materials exhibit semiconductor behavior.

Bandgap Engineering: Manipulating the bandgap of semiconducting thin films via materials design and synthesis enables tailoring of their electrical properties to suit specific applications. Wide bandgap semiconductors find utility in high-power electronic devices and UV photodetectors, whereas narrow bandgap semiconductors are preferred for IR photodetectors and solar cells.

Resistivity and Sheet Resistance: Resistivity and sheet resistance are metrics quantifying a material's resistance to electric current flow. Low resistivity and sheet resistance are desirable in applications necessitating efficient current conduction. Thin films composed of materials such as indium tin oxide (ITO), transparent conductive oxides (TCOs), and graphene exhibit low resistivity. They are commonly employed in transparent electrodes, touchscreens, and OLEDs, where simultaneous electrical conductivity and optical transparency are imperative.

In essence, understanding and fine-tuning the electrical properties of thin films are paramount for harnessing their potential in electronic and optoelectronic devices. By leveraging these properties through meticulous design and fabrication, researchers can pave the way for transformative advancements in various technological domains.

2.3.5 Mechanical Properties

Thin films exhibit unique mechanical properties that can be tailored for specific applications, such as flexibility, hardness, adhesion, and elasticity. These properties are influenced by factors such as film thickness, substrate material, deposition technique, and post-processing treatments.

Flexible thin films are used in applications such as flexible electronics, wearable devices, and stretchable sensors, where mechanical flexibility and conformability are essential. Hard thin films are employed as protective coatings in applications such as cutting tools, automotive components, and aerospace materials, where resistance to wear and abrasion is critical. Adhesive thin films are utilized in bonding applications, such as packaging materials, adhesives, and surface treatments, where strong adhesion to substrates is required. Elastic thin films exhibit reversible deformation under applied stress, making them suitable for applications such as MEMS devices, microactuators, and flexible displays.

Mechanical properties are crucial characteristics of thin films, dictating their structural integrity, durability, and performance in a myriad of applications. These properties encompass a spectrum of behaviors, including strength, flexibility, adhesion, and wear resistance, all of which are influenced by factors such as film composition, thickness, and deposition technique.

Strength and Toughness: Strength denotes a material's ability to withstand applied stress without deformation or failure, while toughness refers to its resistance to fracture when subjected to external forces. Thin films often exhibit enhanced strength compared to bulk materials due to size effects and microstructural features. The strength and toughness of thin films are critical in applications requiring mechanical robustness, such as protective coatings, microelectromechanical systems (MEMS), and wear-resistant surfaces.

Flexibility and Elasticity: Flexibility and elasticity describe a material's ability to bend, stretch, or deform reversibly under applied stress. Thin films can exhibit remarkable flexibility and elasticity, particularly when deposited onto flexible substrates or engineered to possess compliant microstructures. Flexible thin films find applications in flexible electronics, wearable devices, and biomedical implants, where conformal and bendable materials are essential to accommodate complex geometries and dynamic mechanical environments.

Adhesion and Cohesion: Adhesion refers to the strength of bonding between a thin film and its substrate or neighboring layers, while cohesion denotes the internal strength of the film itself. Adequate adhesion and cohesion are vital for ensuring the stability and reliability of thin film coatings, interfaces, and multilayer structures. Various factors, including surface chemistry, interfacial energy, and deposition parameters, influence the adhesion and cohesion properties of thin films, which can be tailored to achieve strong and durable bonding in specific applications.

Wear Resistance and Friction: Wear resistance characterizes a material's ability to withstand surface damage and degradation when subjected to friction, abrasion, or mechanical wear. Thin films with high wear resistance are employed in applications such as protective coatings, cutting tools, and magnetic recording media, where durability and longevity are paramount. Additionally, the frictional behavior of thin films influences their performance in applications such as micro- and nanoelectromechanical systems (MEMS/NEMS) and lubricant coatings, where minimizing friction and wear is essential for device functionality and efficiency.

Creep and Stress Relaxation: Creep refers to the gradual deformation of a material under constant stress over time, while stress relaxation entails the reduction of stress in a material subjected to a constant strain. Understanding and mitigating creep and stress relaxation phenomena are crucial in applications involving thin films subjected to prolonged mechanical loading or thermal cycling, such as structural coatings, electronic packaging, and MEMS devices. By optimizing film composition, microstructure, and processing conditions, researchers can minimize creep and stress relaxation effects and enhance the mechanical stability and longevity of thin film-based systems.

In summary, the mechanical properties of thin films play a vital role in determining their suitability for various applications, ranging from structural coatings and electronic devices to biomedical implants and wear-resistant surfaces. By comprehensively characterizing and engineering these properties, researchers can harness the full potential of thin films and drive innovation across diverse technological domains.

2.3.6 Chemical Properties

Thin films can exhibit diverse chemical properties, including corrosion resistance, chemical stability, surface reactivity, and biocompatibility. These properties are influenced by factors such as material composition, surface chemistry, and environmental conditions.

Corrosion-resistant thin films are used in applications such as protective coatings, barrier layers, and anti-corrosion treatments, where resistance to chemical degradation and environmental exposure is essential. Chemically stable thin films are employed in applications such as sensors, membranes, and catalytic converters, where stability in harsh chemical environments is required. Surface-reactive thin films are utilized in applications such as chemical sensors, biosensors, and catalytic coatings, where interaction with specific molecules or ions is desired. Biocompatible thin films are employed in medical implants, drug delivery systems, and tissue engineering scaffolds, where compatibility with biological systems is critical.

In summary, the properties of thin films are diverse and multifaceted, encompassing aspects such as thickness, uniformity, structure, optics, electrical conductivity, mechanical flexibility, and chemical stability. Understanding and controlling these properties are essential for the design and fabrication of thin film-based devices and systems across various industries, from electronics and optics to energy and biomedical engineering. Chemical properties encompass a broad range of characteristics that govern the chemical reactivity, stability, and interactions of thin films with their environment. These properties are crucial for understanding the behavior of thin films in various applications, including corrosion resistance, chemical sensing, catalysis, and surface modification.

i. **Chemical Reactivity**: The chemical reactivity of thin films is a crucial aspect that determines their propensity to undergo chemical reactions with other substances, including gases, liquids, or solid surfaces. This reactivity can vary significantly depending on factors such as the composition of the thin film, its surface morphology, and the prevailing environmental conditions. Reactive thin films play a pivotal role in numerous applications, including catalysis, gas sensing, and chemical vapor deposition (CVD), where their ability to interact with specific molecules or species is exploited for functional purposes.

Chemical Reactivity of Thin Films:

The chemical reactivity of thin films arises from their surface properties, which can significantly differ from those of bulk materials due to the presence of surface defects, dangling bonds, or adsorbed species. These surface features can act as active sites for chemical reactions, facilitating interactions with surrounding molecules or species. The reactivity of thin films is highly influenced by their composition, with different materials exhibiting varying degrees of reactivity. For example, metals and metal oxides are often highly reactive due to their ability to readily donate or accept electrons, making them suitable for catalytic applications. In contrast, materials such as silicon dioxide ($SiO2$) may exhibit lower reactivity, especially when they form dense, passivated surfaces. Surface morphology also plays a critical role in determining the chemical reactivity of thin films. Nanostructured or rough surfaces can provide increased surface area and more active sites for chemical reactions, enhancing reactivity compared to smooth or flat surfaces. Additionally, surface defects, such as vacancies or step edges, can serve as preferential sites for chemical adsorption and reaction. Environmental conditions, including temperature, pressure, and the presence of reactive species, can significantly influence the chemical reactivity of thin films. For instance, elevated temperatures can enhance reaction kinetics by providing sufficient thermal energy for bond breaking and formation, while the presence of specific gases or vapors can induce surface reactions through adsorption and desorption processes.

Applications of Reactive Thin Films:

Reactive thin films find widespread use in various applications where their chemical reactivity is exploited for functional purposes. Thin films are utilized as catalysts to facilitate chemical reactions by providing active sites for substrate adsorption and reaction. Catalyst thin films can exhibit enhanced reactivity compared to bulk materials, leading to improved catalytic activity, selectivity, and stability. Examples include metal and metal oxide thin films used in heterogeneous catalysis for applications such as pollutant degradation, hydrogen production, and chemical synthesis. Thin films are employed in gas sensing devices to detect and quantify the presence of specific gases or vapors in the surrounding environment. Gas-sensitive thin films undergo chemical reactions with target molecules, resulting in changes in electrical, optical, or mechanical properties that can be measured and correlated with gas concentration. Metal oxide thin films, such as tin oxide (SnO_2) and zinc oxide (ZnO), are commonly used in gas sensors for detecting gases such as hydrogen, carbon monoxide, and volatile organic compounds.

Chemical Vapor Deposition (CVD): Thin films are deposited onto substrates using CVD techniques, where chemical reactions occur between precursor gases and the substrate surface. Reactive thin films can be selectively grown on substrates by controlling the composition and flow rate of precursor gases, enabling the deposition of complex materials with tailored properties. CVD is widely used in semiconductor manufacturing, thin film coating, and materials synthesis for applications such as microelectronics, photovoltaics, and protective coatings. The chemical reactivity of thin films plays a pivotal role in various applications, including catalysis, gas sensing, and chemical vapor deposition. Understanding and controlling the chemical reactivity of thin films are essential for optimizing their performance and functionality in diverse technological applications. By exploiting their reactivity, researchers can develop innovative materials and devices with enhanced capabilities for addressing pressing challenges in areas such as energy, environmental remediation, and sensing technology.

ii. **Corrosion Resistance**: Corrosion resistance is a critical chemical property of thin films, particularly in applications exposed to harsh or corrosive environments. Thin films can be engineered to resist corrosion by selecting materials with inherent corrosion resistance or by applying protective coatings or surface treatments. Corrosion-resistant thin films find applications in industries such as aerospace, automotive, and marine, where protection against oxidation, moisture, and chemical attack is essential for prolonging component lifespan and performance. Corrosion resistance stands as a paramount chemical property of thin films, especially in applications facing harsh or corrosive environments. Through strategic material selection or the application of protective coatings and surface treatments, thin films can be engineered to withstand corrosion effectively. The deployment of corrosion-resistant thin films finds extensive utility across industries such as aerospace, automotive, and marine, where safeguarding against oxidation, moisture, and chemical attack is vital for extending component lifespan and ensuring optimal performance.

The Significance of Corrosion Resistance:

Corrosion poses a significant threat to the integrity and functionality of materials exposed to corrosive environments, leading to degradation, structural weakness, and ultimately, failure. In industries such as aerospace, automotive, and marine, where components are subjected to aggressive conditions such as moisture, saltwater exposure, and chemical contaminants, corrosion resistance emerges as a critical requirement to ensure prolonged service life and operational reliability.

Engineering Corrosion-Resistant Thin Films:

Thin films can be tailored to exhibit robust corrosion resistance through various engineering approaches. Choosing materials with inherent corrosion resistance forms the foundation of engineering corrosion-resistant thin films. Certain metals, metal alloys, and ceramics possess natural resistance to corrosion, making them suitable candidates for thin film deposition in corrosive environments. For instance, stainless steel, titanium, and aluminum alloys are renowned for their resistance to corrosion and are commonly utilized as thin film coatings in aerospace and automotive applications.: Applying protective coatings onto substrate surfaces is an effective strategy to enhance corrosion resistance. These coatings act as barriers, shielding the underlying substrate from exposure to corrosive agents such as moisture, chemicals, and saltwater. Thin film coatings such as metallic coatings (e.g., zinc, chromium), ceramic coatings (e.g., alumina, zirconia), and polymer coatings (e.g., epoxy, polyurethane) are widely employed to impart corrosion resistance to components in diverse industries. Surface treatments, including chemical treatments, thermal treatments, and surface modification techniques, can be employed to alter the surface chemistry and structure of thin films, thereby enhancing their corrosion resistance. Processes such as anodizing, passivation, and ion implantation can introduce protective oxide layers, alter surface energy, or induce compressive stresses within thin films, all of which contribute to improved corrosion resistance.

Applications in Aerospace, Automotive, and Marine Industries:

Corrosion resistant thin films find extensive utilization in critical components and structures across various industries. In the aerospace industry, where aircraft components are subjected to extreme environmental conditions, corrosion-resistant thin films are indispensable for protecting against oxidation, moisture ingress, and saltwater exposure. Thin film coatings applied to aircraft fuselages, wings, and engine components safeguard against corrosion, ensuring the structural integrity and performance of aerospace systems. Automotive components are exposed to diverse corrosive environments, including road salt, moisture, and acidic pollutants. Corrosion-resistant thin films are employed in automotive coatings to protect against rust formation, corrosion-induced degradation, and aesthetic deterioration. Thin film coatings applied to vehicle chassis, body panels, and underbody components enhance durability and longevity, thereby extending vehicle lifespan and maintaining aesthetic appeal. Marine environments, characterized by saltwater exposure and atmospheric corrosion, pose significant challenges to marine structures and equipment. Corrosion-resistant thin films applied to marine vessels, offshore platforms, and underwater equipment provide robust protection against corrosion, preventing structural deterioration and ensuring operational reliability in marine applications.

Corrosion resistance is a crucial chemical property of thin films, particularly in applications exposed to harsh or corrosive environments. Through material selection, protective coatings, and surface treatments, thin films can be engineered to withstand corrosion effectively, thereby prolonging component lifespan and ensuring optimal performance. The deployment of corrosion-resistant thin films finds widespread use in industries such as aerospace, automotive, and marine, where protection against oxidation, moisture, and chemical attack is essential for maintaining structural integrity and operational reliability. By leveraging the inherent corrosion resistance of materials and employing advanced thin film technologies, engineers and researchers continue to develop innovative solutions to address corrosion-related challenges and enhance the durability and sustainability of critical infrastructure and equipment across diverse industrial sectors.

iii. **Surface Modification**: Thin films can be chemically modified to alter their surface properties, such as wettability, adhesion, and bioactivity. Surface modification techniques, including functionalization, grafting, and self-assembly, enable the introduction of specific chemical functionalities onto the film surface, thereby tailoring its interactions with biological, chemical, or physical stimuli. Chemically modified thin films are employed in diverse applications such as biomaterials, biosensors, and microfluidic devices, where precise control over surface properties is critical for controlling cellular adhesion, biomolecule immobilization, and fluid manipulation.

Thin films possess remarkable versatility, as they can undergo chemical modification to alter their surface properties, including wettability, adhesion, and bioactivity. Various surface modification techniques, such as functionalization, grafting, and self-assembly, facilitate the introduction of specific chemical functionalities onto the film surface, enabling precise control over interactions with biological, chemical, or physical stimuli. Chemically modified thin films find wide-ranging applications in fields such as biomaterials, biosensors, and microfluidic devices, where the ability to tailor surface properties is crucial for controlling cellular adhesion, biomolecule immobilization, and fluid manipulation.

Surface Modification of Thin Films:

Surface modification of thin films involves altering their chemical composition or structure to impart desired surface properties. This can be achieved through a range of techniques. Functionalization involves attaching functional groups or molecules to the surface of thin films, imparting specific chemical properties. Functional groups such as hydroxyl (-OH), amino (-NH2), and carboxyl (-COOH) groups can be introduced onto the film surface through chemical reactions or surface treatments, enabling tailored interactions with biomolecules, ions, or other surfaces. Grafting involves covalently attaching polymer chains or molecules to the film surface, forming a graft layer that modifies surface properties. Grafting techniques, such as graft polymerization and surface-initiated polymerization, allow for precise control over the density, length, and composition of the grafted chains, enabling tunable surface properties such as wettability, lubricity, and biocompatibility. Self-assembly techniques utilize spontaneous molecular organization to create ordered structures on the film surface. Molecular self-assembly processes, such as Langmuir–Blodgett deposition and layer-by-layer assembly, enable the controlled deposition of molecules or nanoparticles onto the film surface, resulting in functional surface coatings with tailored properties such as surface charge, roughness, and porosity.

Applications of Chemically Modified Thin Films:

Chemically modified thin films find diverse applications across various fields. In biomaterials applications, chemically modified thin films are used to control cellular adhesion, proliferation, and differentiation. Surface modification techniques such as functionalization with cell-adhesive peptides or growth factors enable the creation of bioactive surfaces that promote tissue regeneration and integration in medical implants, scaffolds, and tissue engineering constructs. Chemically modified thin films serve as sensing platforms in biosensor devices for detecting biomolecules, ions, or pathogens. Surface modification techniques such as immobilization of capture molecules or receptors enable selective recognition and binding of target analytes, resulting in sensitive and specific detection of biological or chemical species in clinical diagnostics, environmental monitoring, and food safety. In microfluidic devices, chemically modified thin films are used to control fluid flow, surface wetting, and biomolecule transport. Surface modification techniques such as patterning with hydrophilic or hydrophobic regions enable precise manipulation of fluid behavior, facilitating applications such as sample preconcentration, cell sorting, and drug delivery in lab-on-a-chip systems and microfluidic-based assays. As research advances in surface modification techniques and materials science, the potential applications of chemically modified thin films continue to expand. Emerging areas such as nanomedicine, wearable electronics, and environmental sensing present new opportunities for utilizing chemically modified thin films to address complex challenges in healthcare, technology, and sustainability. By harnessing the versatility and tunability of surface modification techniques, researchers can develop innovative solutions with tailored surface properties for a wide range of applications, driving advancements in diverse fields and improving quality of life.

iv. **Chemical Sensing**: Thin films are widely utilized in chemical sensing applications due to their ability to selectively interact with target analytes and produce measurable responses. Chemical sensors based on thin films exploit various transduction mechanisms, including electrical, optical, and mass-sensitive, to detect changes in chemical composition or concentration. Thin film sensors find applications in environmental monitoring, industrial process control, and medical diagnostics, where real-time detection and quantification of specific gases, vapors, or analytes are essential for safety, quality control, and healthcare applications.

Thin films play a pivotal role in chemical sensing applications owing to their capability to selectively interact with target analytes and generate measurable responses. Chemical sensors based on thin films harness diverse transduction mechanisms, including electrical, optical, and mass-sensitive methods, to detect alterations in chemical composition or concentration. These thin film sensors find widespread utilization in environmental monitoring, industrial process control, and medical diagnostics, where real-time detection and quantification of specific gases, vapors, or analytes are crucial for ensuring safety, quality control, and healthcare advancements.

Thin films serve as the sensing element in chemical sensors, where they interact with target analytes through surface adsorption, diffusion, or reaction processes. The selective binding of analyte molecules to the thin film surface induces changes in the film's properties, such as electrical conductivity, optical absorption, or mass loading, which can be transduced into measurable signals indicative of the analyte concentration or presence.

Thin film sensors exploit various transduction mechanisms to convert chemical signals into measurable responses:

1. **Electrical**: Electrical transduction involves monitoring changes in the electrical properties of thin films in response to chemical interactions. For instance, chemiresistive sensors measure changes in film resistance due to adsorption or desorption of analyte molecules, while field-effect transistors detect alterations in the conductance or capacitance of thin film channels upon analyte binding.
2. **Optical**: Optical transduction relies on changes in the optical properties of thin films, such as absorbance, fluorescence, or refractive index, induced by chemical interactions. Optical sensors utilize techniques such as surface plasmon resonance (SPR), surface-enhanced Raman spectroscopy (SERS), and fluorescence quenching to detect shifts in optical signals corresponding to analyte binding events.
3. **Mass-Sensitive**: Mass-sensitive transduction involves monitoring changes in the mass or density of thin films resulting from analyte adsorption or surface reactions. Quartz crystal microbalance (QCM) sensors and surface acoustic wave (SAW) sensors measure shifts in resonance frequency or propagation velocity caused by mass loading on thin film surfaces, enabling real-time detection of analyte concentration changes.

Applications of Thin Film Sensors:

Thin film sensors find diverse applications across various sectors:

1. **Environmental Monitoring**: Thin film sensors are employed for detecting and monitoring pollutants, gases, and volatile organic compounds (VOCs) in environmental settings. These sensors enable real-time detection of air and water quality parameters, facilitating pollution control, environmental remediation, and public health protection.
2. **Industrial Process Control**: In industrial settings, thin film sensors are used for monitoring and controlling chemical processes, ensuring product quality, and optimizing production efficiency. These sensors enable continuous monitoring of process parameters, such as gas concentration, temperature, and humidity, enabling proactive maintenance, process optimization, and quality assurance.
3. **Medical Diagnostics**: Thin film sensors play a crucial role in medical diagnostics, enabling rapid and sensitive detection of biomarkers, pathogens, and disease-related molecules. These sensors find applications in point-of-care diagnostics, clinical laboratory testing, and wearable health monitoring devices, facilitating early disease detection, personalized treatment, and remote patient monitoring.

As advancements in materials science, nanotechnology, and sensor technology continue, the capabilities and applications of thin film sensors are expected to expand further. Emerging trends such as wearable sensors, internet of things (IoT) integration, and artificial intelligence (AI) analytics present new opportunities for deploying thin film sensors in diverse applications, from personalized healthcare to smart environmental monitoring systems. By harnessing the versatility and sensitivity of thin film sensors, researchers and engineers can develop innovative solutions to address complex challenges in safety, sustainability, and healthcare, driving advancements in sensor technology and improving quality of life.

v. **Catalytic Activity**: Thin films can exhibit catalytic activity, facilitating chemical reactions by lowering the activation energy barrier and increasing reaction rates. Catalytic thin films are employed in heterogeneous catalysis, where they provide active sites for chemical transformations on their surfaces. Catalyst thin films find applications in industries such as petrochemicals, pharmaceuticals, and environmental remediation, where they enable efficient and selective conversion of reactants into desired products while minimizing energy consumption and waste generation.

Thin films possess the remarkable ability to exhibit catalytic activity, thereby facilitating chemical reactions by reducing the activation energy barrier and enhancing reaction rates. Catalytic thin films play a pivotal role in heterogeneous catalysis, where they serve as active sites for chemical transformations on their surfaces. These catalyst thin films find extensive applications across industries such as petrochemicals, pharmaceuticals, and environmental remediation, where they enable efficient and selective conversion of reactants into desired products while minimizing energy consumption and waste generation.

Catalytic Activity of Thin Films:

Thin films can display catalytic activity due to their unique surface properties, which provide active sites for chemical reactions to occur. Catalyst thin films facilitate reactions by adsorbing reactant molecules onto their surfaces, bringing them into close proximity and facilitating bond-breaking and bond-forming processes. By lowering the activation energy barrier, catalytic thin films accelerate reaction rates, enabling more efficient conversion of reactants into products.

Catalytic thin films find diverse applications across various industries:

1. **Petrochemicals**: In the petrochemical industry, catalytic thin films are used in processes such as hydrocracking, catalytic reforming, and Fischer–Tropsch synthesis to produce fuels, chemicals, and intermediates. Catalyst thin films enable the conversion of crude oil, natural gas, and biomass feedstocks into high-value products such as gasoline, diesel, and olefins, contributing to the production of clean and sustainable energy sources.
2. **Pharmaceuticals**: Catalytic thin films play a crucial role in pharmaceutical synthesis, where they are employed in reactions such as hydrogenation, oxidation, and coupling to produce active pharmaceutical ingredients (APIs) and intermediates. Catalyst thin films enable the synthesis of complex molecules with high selectivity and efficiency, facilitating the development of novel drugs and pharmaceutical formulations for treating various medical conditions.
3. **Environmental Remediation**: Catalytic thin films are utilized in environmental remediation applications to degrade pollutants, detoxify wastewater, and mitigate air pollution. Catalyst thin films enable the conversion of harmful contaminants such as volatile organic compounds (VOCs), nitrogen oxides (NOx), and heavy metals into non-toxic or less harmful species, contributing to environmental sustainability and pollution control efforts.

Catalytic thin films offer several advantages over traditional catalyst forms:

1. **High Surface Area**: Thin films possess high surface area-to-volume ratios, providing a large number of active sites for catalytic reactions to occur. This high surface area enhances catalyst efficiency and enables higher catalytic activity per unit mass compared to bulk catalyst materials.
2. **Tunability**: The properties of catalytic thin films, such as composition, morphology, and surface chemistry, can be precisely controlled during deposition, allowing for

tailored catalytic performance. By adjusting deposition parameters and substrate conditions, researchers can optimize catalyst thin films for specific reaction pathways, selectivity, and stability.

3. **Reduced Mass Transfer Limitations**: Thin films exhibit short diffusion lengths for reactant molecules, minimizing mass transfer limitations and facilitating rapid transport of reactants to active sites. This characteristic enhances reaction kinetics and enables efficient utilization of catalyst materials, leading to improved process efficiency and productivity.

As research in catalytic thin films continues to advance, new opportunities emerge for developing innovative catalyst materials and processes for diverse applications. Emerging trends such as nanotechnology, surface science, and computational modeling offer insights into catalyst design principles and enable the development of next-generation catalytic thin films with enhanced performance, selectivity, and sustainability. By harnessing the catalytic activity of thin films, researchers and engineers can address pressing challenges in energy production, environmental protection, and chemical synthesis, driving advancements in catalysis and contributing to a more sustainable and resilient future.

In summary, the chemical properties of thin films play a pivotal role in determining their behavior and functionality in various applications. By understanding and manipulating these properties, researchers can design thin films with tailored chemical reactivity, corrosion resistance, surface modification capabilities, and sensing functionalities, enabling advancements in fields such as materials science, nanotechnology, and chemical engineering.

2.4   Deposition Techniques for Thin Films

Deposition techniques play a crucial role in the fabrication of thin films, influencing their properties, uniformity, and performance. Some common deposition techniques used for thin film fabrication are here discussed. The classification of thin films deposition techniques based on techniques is shown in Fig. 2.1.

Physical Vapor Deposition (PVD): Physical vapor deposition (PVD) involves the deposition of thin films through the physical processes of evaporation or sputtering. In evaporation, a material is heated in a vacuum chamber until it reaches its vaporization temperature, and the vaporized atoms condense onto a substrate, forming a thin film. In sputtering, energetic ions are used to dislodge atoms from a target material, which then deposit onto a substrate. PVD techniques offer excellent control over film thickness, composition, and uniformity and are widely used in semiconductor manufacturing, optical coatings, and magnetic recording media.

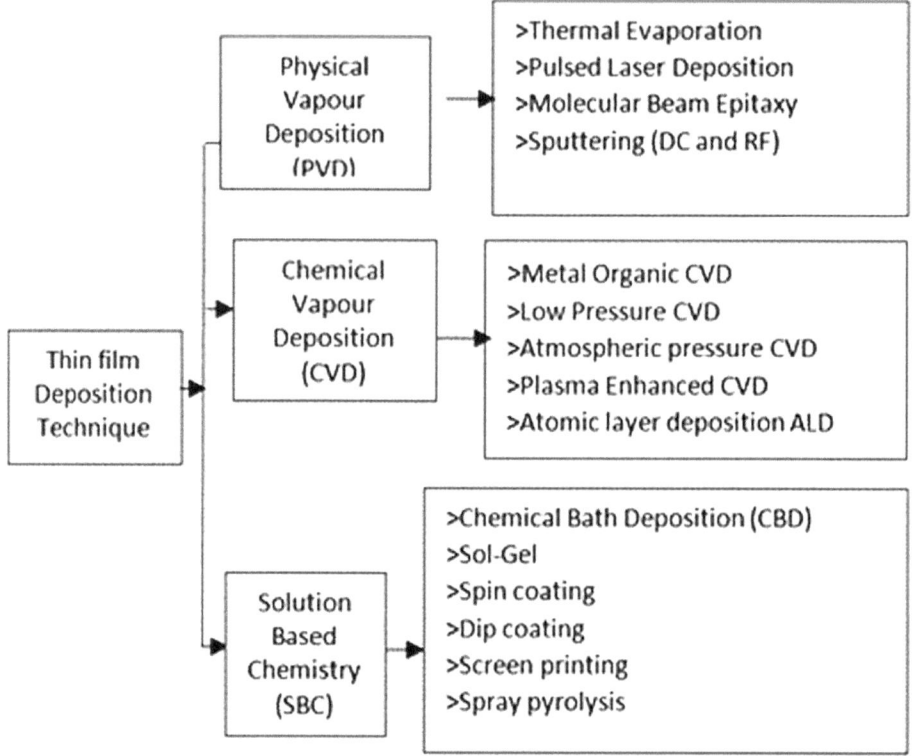

Fig. 2.1 Thin films classification

Principles of Physical Vapor Deposition:

Physical Vapor Deposition (PVD) is based on the principles of condensation and nucleation, where atoms or molecules of a material are evaporated from a solid source and deposited onto a substrate to form a thin film. The process typically takes place in a vacuum environment to minimize contamination and unwanted reactions with air. There are several techniques used in PVD, including evaporation, sputtering, and cathodic arc deposition as shown in Fig. 2.2, each with its unique advantages and applications.

i. Evaporation: In the evaporation process, the material to be deposited (known as the evaporant) is heated in a vacuum chamber until it reaches its vaporization temperature, causing it to transition from a solid to a vapor phase. The vaporized atoms or molecules then travel through the vacuum chamber and condense onto the substrate, where they form a thin film. Evaporation can be achieved using various methods, including resistive heating, electron beam heating, or thermal evaporation.

Fig. 2.2 Schematic of types of
physical vapour deposition

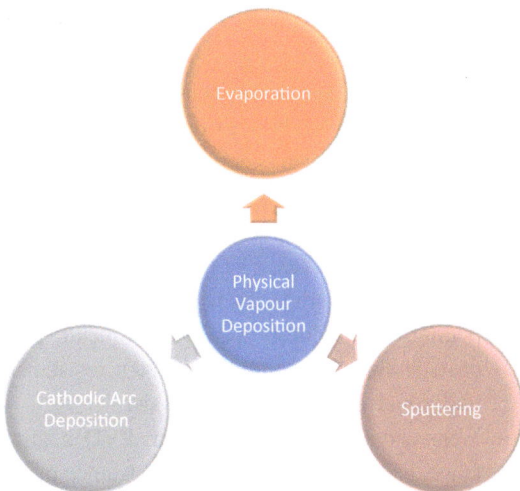

ii. Sputtering: Sputtering is another common technique used in PVD, where material is
    ejected from a solid target through bombardment with energetic ions. In sputtering, a
    high-voltage electrical field is applied between the target (cathode) and the substrate
    (anode) in a vacuum chamber filled with a low-pressure inert gas, such as argon.
    The energetic ions collide with the target surface, dislodging atoms or molecules,
    which then deposit onto the substrate to form a thin film. Sputtering can be achieved
    using various methods, including dc sputtering, rf sputtering, magnetron sputtering,
    and reactive sputtering.

iii. Cathodic Arc Deposition: Cathodic arc deposition is a PVD technique that involves
     the vaporization of material from a solid target through the application of a high-
     voltage electrical arc discharge. In this process, a high-current, low-voltage arc is
     generated between the target material and an anode in a vacuum chamber filled with
     an inert gas. The intense heat and energy of the arc vaporize atoms or ions from the
     target surface, which then condense onto the substrate to form a thin film. Cathodic
     arc deposition is known for its high deposition rates, uniform film thickness, and
     excellent adhesion to substrates.

Applications of Physical Vapor Deposition:

Physical Vapor Deposition (PVD) finds numerous applications across various industries,
where thin films are used to enhance surface properties, modify material properties, and
create functional coatings with specific characteristics. Some common applications of
PVD include:

i. Semiconductor Manufacturing: In the semiconductor industry, PVD is used to deposit
   thin films of materials such as aluminum, titanium, and tungsten onto silicon wafers

to create metal interconnects, diffusion barriers, and dielectric layers in integrated circuits and microelectronics.

ii. Optical Coatings: PVD is used to deposit thin films of materials such as oxides, nitrides, and metals onto optical components such as lenses, mirrors, and filters to control light transmission, reflection, and absorption properties. Optical coatings are used in applications such as anti-reflection coatings, mirror coatings, and optical filters in cameras, telescopes, and spectrophotometers.

iii. Decorative Finishes: PVD coatings are used in the automotive, architectural, and consumer goods industries to create decorative finishes such as chrome plating, gold plating, and colored coatings on metal, plastic, and glass surfaces. PVD coatings provide superior durability, scratch resistance, and corrosion resistance compared to traditional electroplating methods.

iv. Wear and Corrosion Protection: PVD coatings are used to improve the wear resistance, hardness, and corrosion resistance of components in automotive, aerospace, and tooling applications. Thin films of materials such as titanium nitride (TiN), chromium nitride (CrN), and diamond-like carbon (DLC) are deposited onto cutting tools, engine components, and bearings to prolong service life and enhance performance in harsh operating environments.

v. Biomedical Applications: PVD coatings are used in biomedical devices and implants to improve biocompatibility, wear resistance, and corrosion resistance. Thin films of materials such as titanium, titanium nitride, and hydroxyapatite are deposited onto orthopedic implants, dental implants, and surgical instruments to promote tissue integration, reduce friction, and prevent bacterial adhesion.

Advancements in Physical Vapor Deposition:

Over the years, advancements in Physical Vapor Deposition (PVD) technology have led to improvements in deposition rates, film quality, and process control, expanding the range of applications and capabilities of PVD coatings. Some notable advancements in PVD include:

i. High-Speed Deposition: Advances in PVD equipment design, plasma generation, and target materials have enabled higher deposition rates and increased productivity in PVD processes. High-speed deposition techniques such as magnetron sputtering and cathodic arc deposition allow for rapid coating of large-area substrates with uniform film thickness and excellent adhesion.

ii. Multilayer Coatings: PVD technology enables the deposition of multilayer coatings with tailored properties and functionalities by stacking multiple thin films of different materials or compositions. Multilayer coatings offer enhanced performance characteristics such as improved hardness, wear resistance, and thermal stability, making them ideal for demanding applications in aerospace, automotive, and cutting tool industries.

iii. Nanostructured Coatings: Recent developments in PVD techniques such as atomic layer deposition (ALD) and vapor-phase epitaxy (VPE) have enabled the deposition of nanostructured coatings with precise control over film thickness and microstructure. Nanostructured coatings exhibit unique properties such as superhydrophobicity, self-cleaning, and antibacterial activity, opening up new opportunities in energy, environmental, and biomedical applications.

iv. Functional Coatings: PVD coatings can be engineered to provide specific functionalities such as electrical conductivity, optical transparency, or thermal insulation, depending on the requirements of the application. Functional coatings are used in diverse applications such as touchscreens, photovoltaic cells, and heat sinks to enhance performance, reliability, and efficiency.

v. Environmentally Friendly Processes: Advances in PVD technology have led to the development of environmentally friendly processes with reduced energy consumption, waste generation, and emissions. Techniques such as reactive sputtering, pulsed laser deposition (PLD), and electron-beam evaporation (EBE) offer improved process efficiency, material utilization, and environmental sustainability compared to traditional PVD methods.

Physical Vapor Deposition (PVD) is a versatile and indispensable thin-film deposition technique that enables the creation of functional coatings with tailored properties and functionalities for a wide range of applications. From semiconductor manufacturing and optical coatings to decorative finishes and biomedical implants, PVD plays a critical role in enhancing the performance, durability, and reliability of components and products across industries. With ongoing advancements in PVD technology, materials, and processes, the future holds exciting possibilities for innovative coatings with enhanced functionality, sustainability, and performance.

Chemical Vapor Deposition (CVD): Chemical vapor deposition (CVD) involves the deposition of thin films through the chemical reaction of precursor gases on a substrate surface. In CVD, precursor gases containing the desired film constituents are introduced into a reaction chamber, where they react and deposit onto the substrate surface. CVD techniques offer precise control over film composition, thickness, and morphology and are commonly used in semiconductor device fabrication, thin film coatings, and surface modification processes.

Chemical Vapor Deposition (CVD) is a versatile and widely used thin-film deposition technique that involves the chemical reaction of precursor gases to deposit solid materials onto substrates. CVD plays a crucial role in various industries, including semiconductor manufacturing, electronics, optics, aerospace, and surface engineering. This comprehensive discourse explores the principles, techniques, applications, and advancements in Chemical Vapor Deposition, highlighting its significance in modern manufacturing and technology. Chemical Vapor Deposition (CVD) is a highly versatile and widely used thin-film deposition technique that plays a crucial role in various industries, including

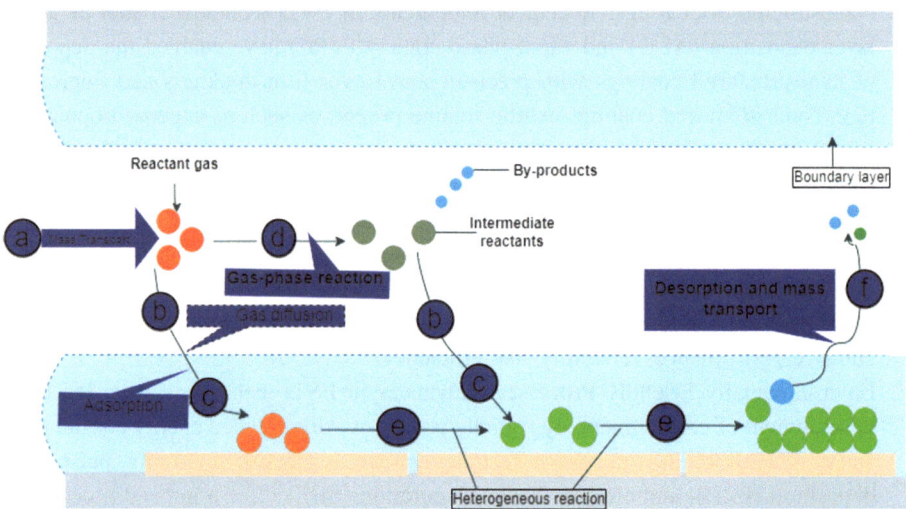

Fig. 2.3 Schematic of chemical vapour deposition

semiconductor manufacturing, electronics, optics, aerospace, and surface engineering. CVD involves the chemical reaction of precursor gases to deposit solid materials onto substrates, resulting in the formation of thin films with precise control over composition, thickness, and microstructure. This comprehensive discourse explores the principles, techniques, applications, and advancements in Chemical Vapor Deposition, highlighting its significance in modern manufacturing and technology. The representation of a chemical vapour deposition is shown in Fig. 2.3.

Principles of Chemical Vapor Deposition:

Chemical Vapor Deposition (CVD) is based on the principles of chemical reactions between precursor gases and a substrate surface to deposit thin films of solid materials. The process typically takes place in a vacuum or controlled atmosphere environment to minimize impurities and ensure uniform film growth. There are several techniques used in CVD, including thermal CVD, plasma-enhanced CVD (PECVD), and atomic layer deposition (ALD) as shown in Fig. 2.4, each with its unique advantages and applications.

i. Thermal Chemical Vapor Deposition: Thermal CVD is one of the most common techniques used in chemical vapor deposition, where precursor gases are thermally decomposed on a heated substrate surface to form solid thin films. In thermal CVD, precursor gases are introduced into a reaction chamber, where they are heated to temperatures typically ranging from several hundred to over a thousand degrees Celsius. The high temperature causes the precursor molecules to break apart and react with the substrate surface, resulting in the deposition of a thin film of the desired material.

Fig. 2.4 Schematics of chemical vapor deposition technique

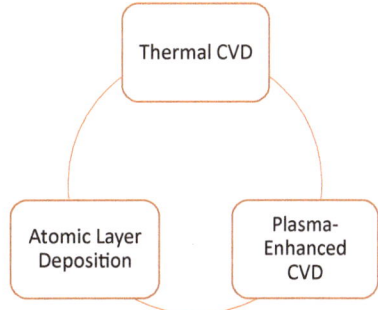

Thermal Chemical Vapor Deposition (CVD) is a prominent technique utilized in the deposition of thin films onto substrates. It operates on the principle of thermal decomposition of precursor gases to form solid thin films through chemical reactions. In this process, precursor gases are introduced into a vacuum chamber, where they undergo thermal decomposition upon exposure to elevated temperatures, typically ranging from several hundred to over a thousand degrees Celsius. As the precursor molecules decompose, they react with the substrate surface, leading to the deposition of a thin film of the desired material. Thermal CVD offers several advantages, including precise control over film thickness, composition, and microstructure, as well as compatibility with a wide range of materials and substrates. The technique allows for the deposition of uniform, high-quality films with excellent adhesion to substrates, making it suitable for various applications in semiconductor manufacturing, optics, and surface engineering.

In thermal CVD, the choice of precursor gases and process parameters such as temperature, pressure, and gas flow rates plays a crucial role in determining the properties and characteristics of the deposited thin films. By carefully controlling these parameters, researchers and engineers can tailor the composition, structure, and properties of the thin films to meet specific application requirements. Thermal CVD finds numerous applications across various industries. In semiconductor manufacturing, it is used to deposit thin films of materials such as silicon dioxide, silicon nitride, and polysilicon onto silicon wafers to create insulating layers, diffusion barriers, and conductive layers in integrated circuits and microelectronics. In the optics industry, thermal CVD is employed to deposit thin films of materials such as oxides, nitrides, and metals onto optical components such as lenses, mirrors, and filters to control light transmission, reflection, and absorption properties. Overall, Thermal Chemical Vapor Deposition (CVD) is a versatile and widely used technique for the deposition of thin films onto substrates. It offers precise control over film properties and characteristics, making it suitable for a wide range of applications in various industries. With ongoing advancements in CVD technology and process optimization, thermal CVD continues to play a critical role in advancing materials science and technology.

ii. Plasma-Enhanced Chemical Vapor Deposition (PECVD):

Plasma-enhanced chemical vapor deposition (PECVD) is a variation of CVD that involves the use of plasma to enhance the chemical reaction between precursor gases and the substrate surface. In PECVD, precursor gases are introduced into a vacuum chamber containing a plasma source, such as a radiofrequency (RF) generator or microwave generator. The plasma generates highly reactive species, such as ions, radicals, and electrons, which facilitate the dissociation and ionization of precursor molecules, leading to more efficient and controlled film growth.

Plasma-Enhanced Chemical Vapor Deposition (PECVD) stands at the forefront of thin-film deposition techniques, offering unique advantages in precision, uniformity, and versatility. Rooted in the principles of chemical reactions in plasma environments, PECVD leverages plasma to enhance the deposition process, enabling the creation of thin films with tailored properties and characteristics. This comprehensive technique has found wide-ranging applications across industries, from semiconductor manufacturing to optics, surface engineering, and beyond. At the core of PECVD lies the utilization of plasma, a state of matter characterized by ionized gases containing free electrons, ions, and neutral particles. Plasma serves as a highly reactive medium that facilitates the dissociation and ionization of precursor gases, leading to enhanced chemical reactions and more controlled deposition processes compared to conventional thermal CVD techniques. In a typical PECVD setup, precursor gases are introduced into a vacuum chamber containing a plasma source, such as a radiofrequency (RF) generator or microwave generator. The precursor gases are then subjected to an electric field, causing them to ionize and dissociate, resulting in the formation of highly reactive species such as ions, radicals, and electrons. These reactive species interact with the substrate surface, initiating chemical reactions that lead to the deposition of thin films.

One of the key advantages of PECVD is its ability to operate at lower temperatures compared to thermal CVD techniques. The energetic species generated in the plasma environment facilitate the deposition process even at relatively low temperatures, minimizing thermal stress on sensitive substrates and enabling the deposition of thin films on temperature-sensitive materials such as plastics and polymers.

Moreover, PECVD offers superior control over film properties and characteristics, including thickness, composition, and microstructure. By adjusting process parameters such as plasma power, gas flow rates, and substrate temperature, practitioners can tailor the properties of deposited thin films to meet specific application requirements. This level of control allows for the deposition of high-quality films with uniform thickness and excellent adhesion to substrates. PECVD finds widespread applications across various industries, owing to its versatility and efficacy in depositing thin films with tailored properties. In semiconductor manufacturing, PECVD is commonly used to deposit insulating and passivating layers, such as silicon dioxide (SiO_2) and silicon nitride (Si_3N_4), onto silicon wafers. These thin films serve critical functions in the fabrication of integrated circuits, providing electrical insulation, protection against contaminants, and surface passivation to improve device performance and reliability. In the optics industry, PECVD is employed to deposit optical coatings onto glass and other transparent substrates. Thin films of materials such as silicon dioxide (SiO_2), titanium dioxide (TiO_2), and tantalum oxide (Ta_2O_5) are deposited using PECVD to control light transmission, reflection, and absorption properties. These optical coatings find applications in lenses, mirrors, filters, and other optical components used in imaging systems, displays, and photonic devices. Furthermore, PECVD is utilized in surface engineering applications to deposit protective and functional coatings onto various materials. Thin films of materials such as diamond-like carbon (DLC), silicon carbide (SiC), and titanium nitride (TiN) are deposited using PECVD to enhance wear resistance, corrosion resistance, and biocompatibility. These coatings find applications in automotive components, cutting tools, biomedical implants, and aerospace components, among others. Plasma-Enhanced Chemical Vapor Deposition (PECVD) stands as a versatile and indispensable technique in the realm of thin-film deposition. Its ability to operate at lower temperatures, offer superior control over film properties, and facilitate deposition on a wide range of substrates makes it well-suited for diverse applications across industries. As research and development efforts continue to advance, PECVD remains poised to play a pivotal role in enabling innovative solutions and driving progress in materials science and technology.

iii. Atomic Layer Deposition (ALD):

Atomic layer deposition (ALD) is a precise and conformal thin-film deposition technique that relies on sequential, self-limiting surface reactions to build up thin films layer by layer. In ALD, precursor gases are introduced into a reaction chamber alternately with a purge gas, such as nitrogen or argon. Each precursor pulse reacts with the substrate surface to form a monolayer of material, followed by purging to remove unreacted gases and byproducts. This cycle is repeated multiple times to achieve the desired film thickness with atomic-level precision and uniformity. Atomic Layer Deposition (ALD) represents a cutting-edge thin-film deposition technique characterized by its unparalleled precision, uniformity, and controllability. Rooted in sequential, self-limiting surface reactions, ALD enables the precise growth of thin films layer by layer with atomic-level accuracy. This comprehensive technique has found wide-ranging applications across industries, from semiconductor manufacturing to energy storage, catalysis, and beyond. At the heart of ALD lies its unique mechanism, which relies on sequential, self-limiting surface reactions to achieve precise control over film thickness and composition. The ALD process unfolds in a controlled environment typically within a vacuum chamber, where precursor gases are introduced alternately with a purge gas. Each precursor pulse reacts with the substrate surface to form a monolayer of material, followed by purging to remove unreacted gases and byproducts. This cycle is repeated iteratively until the desired film thickness is achieved, resulting in the deposition of thin films with atomic-level precision and uniformity.

One of the key advantages of ALD is its ability to deposit conformal coatings with uniform thickness and coverage, even on complex, three-dimensional substrates. The self-limiting nature of ALD ensures that each precursor pulse reacts only with available surface sites, preventing overgrowth and enabling precise control over film thickness and composition. This capability makes ALD well-suited for coating high-aspect-ratio structures, nanoporous materials, and other challenging substrates. Moreover, ALD offers superior control over film properties and characteristics, including thickness, composition, and crystallinity. By selecting appropriate precursor gases and process parameters, practitioners can tailor the properties of deposited thin films to meet specific application requirements. This level of control allows for the deposition of high-quality films with tunable properties such as electrical conductivity, optical transparency, and mechanical strength. ALD finds widespread applications across various industries, owing to its versatility and efficacy in depositing thin films with tailored properties. In semiconductor manufacturing, ALD is commonly used to deposit high-k dielectric materials such as hafnium oxide (HfO2) and aluminum oxide (Al_2O_3) onto silicon wafers. These thin films serve critical functions in the fabrication of advanced transistors, providing electrical insulation, gate capacitance, and reliability enhancement. In the field of energy storage, ALD is employed to deposit thin films of materials such as lithium oxide (Li_2O) and aluminum oxide (Al_2O_3) onto electrodes in lithium-ion batteries. These coatings serve to stabilize electrode materials, prevent side reactions, and improve battery performance, including cycle life, energy density, and safety. Furthermore, ALD is utilized in catalysis applications to deposit thin films of catalytic materials onto support substrates. Thin films of materials such as platinum (Pt), palladium (Pd), and ruthenium (Ru) are deposited using ALD to enhance catalytic activity, selectivity, and stability. These catalyst coatings find applications in automotive catalytic converters, fuel cells, and chemical reactors, among others. In the realm of surface engineering, ALD is employed to deposit protective and functional coatings onto various materials. Thin films of materials such as titanium nitride (TiN), aluminum oxide (Al_2O_3), and silicon dioxide (SiO_2) are deposited using ALD to enhance wear resistance, corrosion resistance, and biocompatibility. These coatings find applications in automotive components, cutting tools, biomedical implants, and aerospace components, among others. Atomic Layer Deposition (ALD) stands as a versatile and indispensable technique in the realm of thin-film deposition. Its unparalleled precision, uniformity, and controllability make it well-suited for diverse applications across industries, from semiconductor manufacturing to energy storage, catalysis, and surface engineering. As research and development efforts continue to advance, ALD remains poised to play a pivotal role in enabling innovative solutions and driving progress in materials science and technology. These CVD techniques are summarized in Fig. 2.5.

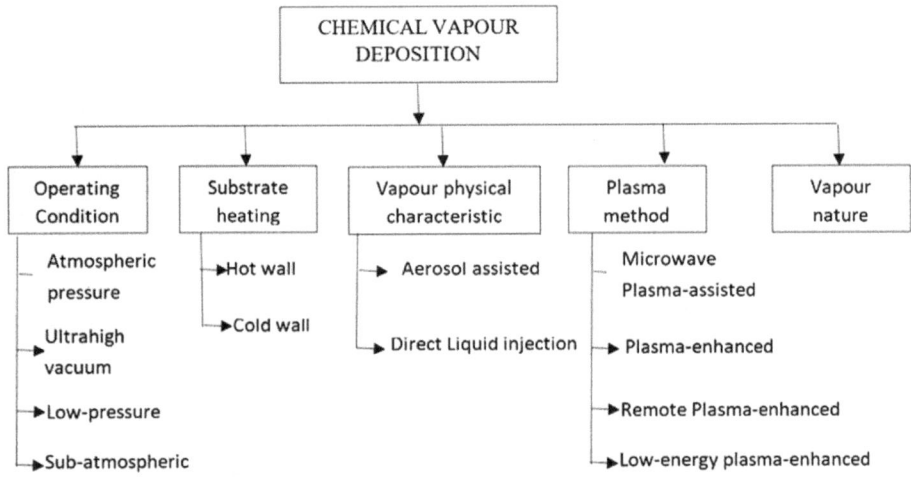

Fig. 2.5 Chemical vapour deposition classification

Applications of Chemical Vapor Deposition:

Chemical vapor deposition (CVD) finds numerous applications across various industries, where thin films are used to enhance surface properties, modify material properties, and create functional coatings with specific characteristics. Some common applications of CVD include:

i. Semiconductor Manufacturing: In the semiconductor industry, Chemical Vapor Deposition (CVD) plays a pivotal role in the fabrication of integrated circuits and microelectronics by depositing thin films of essential materials onto silicon wafers. These thin films serve diverse functions, including insulating layers, diffusion barriers, and conductive paths, crucial for the functionality and performance of semiconductor devices.

One of the primary materials deposited using CVD in the semiconductor industry is silicon dioxide (SiO_2). Silicon dioxide serves as an insulating layer, known as a dielectric, that electrically isolates different components of semiconductor devices. By depositing a thin film of silicon dioxide onto silicon wafers, CVD creates an insulating barrier that prevents electrical interference between adjacent components, ensuring the proper functioning of integrated circuits. Additionally, CVD is employed to deposit thin films of silicon nitride (Si_3N_4) in the semiconductor industry. Silicon nitride serves as a versatile material with various applications, including diffusion barriers and passivation layers. As a diffusion barrier, silicon nitride prevents the migration of impurities or dopants from one semiconductor layer to another, crucial for controlling the electrical properties of semiconductor devices. Furthermore, silicon nitride can serve as a passivation layer, protecting the surface of semiconductor devices from external contaminants and enhancing device reliability and stability. Furthermore, polysilicon (poly-Si) is another material deposited using CVD in the semiconductor industry. Polysilicon is a form of crystalline silicon with superior conductivity compared to its amorphous counterpart. By depositing thin films of polysilicon onto silicon wafers, CVD creates conductive paths that connect different components of integrated circuits, facilitating the flow of electrical currents within semiconductor devices. The utilization of CVD for depositing thin films of silicon dioxide, silicon nitride, and polysilicon enables the precise control over film thickness, composition, and quality, crucial for the fabrication of high-performance semiconductor devices. The versatility of CVD allows for the deposition of these materials in a controlled environment, ensuring uniformity and consistency across large batches of silicon wafers. Chemical Vapor Deposition (CVD) serves as a fundamental technique in the semiconductor industry for depositing thin films of essential materials onto silicon wafers. By depositing materials such as silicon dioxide, silicon nitride, and polysilicon, CVD enables the creation of insulating layers, diffusion barriers, and conductive paths vital for the fabrication of integrated circuits and microelectronics. The precise control and versatility of CVD make it an indispensable tool for achieving high-performance semiconductor devices with superior functionality and reliability.

ii. Optical Coatings: In the realm of optics, Chemical Vapor Deposition (CVD) serves as a foundational technique for depositing thin films of diverse materials onto optical components such as lenses, mirrors, and filters. These thin films, composed of oxides, nitrides, and metals, play a crucial role in controlling the transmission, reflection, and absorption of light, thus enhancing the optical performance of the components. One of the primary materials deposited using CVD in optical applications is oxides, including materials such as silicon dioxide (SiO_2) and titanium dioxide (TiO_2). These oxide thin films are employed in various optical coatings, including anti-reflection coatings. By depositing precise thicknesses of oxide layers onto optical surfaces, CVD coatings can minimize reflections and increase light transmission, thereby improving the efficiency and clarity of optical systems. Additionally, CVD is utilized to deposit thin films of nitrides onto optical components. Silicon nitride (Si3N4) and aluminum nitride (AlN) are examples of nitride materials commonly used in optical coatings. These thin films can serve multiple functions, including providing protective layers and enhancing the mechanical and thermal stability of optical components. Moreover, CVD enables the deposition of thin films of metals onto optical surfaces, such as aluminum (Al), gold (Au), and silver (Ag). Metal coatings play a crucial role in mirror coatings, where they provide high reflectivity across a broad range of wavelengths. By depositing metal thin films with precise control over thickness and uniformity, CVD coatings can achieve high-quality mirrors suitable for various optical applications.

The utilization of CVD for depositing thin films onto optical components enables the fabrication of optical coatings tailored to specific requirements. These coatings find widespread applications in cameras, telescopes, spectrophotometers, and other optical devices. For instance, anti-reflection coatings deposited using CVD reduce glare and unwanted reflections in camera lenses, improving image quality and contrast. Furthermore, mirror coatings created through CVD enable the construction of high-performance mirrors for telescopes and optical instruments. These mirrors exhibit exceptional reflectivity and durability, essential for capturing and focusing light in astronomical observations and scientific experiments. Chemical Vapor Deposition (CVD) plays a pivotal role in the fabrication of optical coatings for lenses, mirrors, and filters. By depositing thin films of materials such as oxides, nitrides, and metals, CVD enables precise control over light transmission, reflection, and absorption properties, enhancing the optical performance of components. These optical coatings find applications in various optical devices, including cameras, telescopes, and spectrophotometers, where they improve image quality, contrast, and sensitivity.

iii. *Hard Coatings*: In diverse industrial sectors, including automotive, aerospace, and manufacturing, Chemical Vapor Deposition (CVD) stands as a pivotal technique for enhancing the durability and performance of cutting tools, molds, and components. This process involves depositing hard, wear-resistant coatings onto various substrates to extend tool life, minimize friction, and bolster wear resistance. One of the primary functions of CVD coatings in these industries is to protect cutting tools from wear and abrasion during machining operations. By depositing thin films of materials such as diamond-like carbon (DLC), titanium nitride (TiN), and chromium nitride (CrN) onto cutting tool surfaces, CVD enhances their hardness and wear resistance, enabling them to withstand harsh operating conditions encountered in metal cutting, milling, and drilling processes. Moreover, CVD coatings play a crucial role in enhancing the performance of molds used in injection molding and die casting processes. By depositing wear-resistant coatings onto mold surfaces, CVD reduces friction and minimizes material adhesion, thus prolonging mold life and improving product quality and consistency. In the automotive industry, CVD coatings are extensively used to enhance the performance of engine components, such as piston rings, cylinder liners, and valve seats. These components operate in demanding environments characterized by high temperatures, pressures, and frictional forces. By depositing hard coatings such as DLC and TiN onto these components, CVD improves their wear resistance, reducing friction and minimizing component wear and degradation over time. Similarly, in the aerospace industry, CVD coatings are applied to critical components such as turbine blades, bearings, and gears to improve their performance and longevity. These components are subjected to extreme temperatures, pressures, and mechanical stresses during flight operations. By depositing hard coatings such as CrN and TiN onto aerospace components, CVD enhances their resistance to wear, corrosion, and fatigue, ensuring reliable performance in harsh operating environments. Furthermore, in the manufacturing sector, CVD coatings find applications in a wide range of components, including machine parts, dies, and tooling. By depositing hard coatings onto these components, CVD improves their durability, reduces maintenance requirements, and enhances overall productivity and efficiency in manufacturing processes. Chemical Vapor Deposition (CVD) plays a critical role in enhancing the performance and longevity of components in the automotive, aerospace, and manufacturing industries. By depositing hard, wear-resistant coatings onto cutting tools, molds, and components, CVD improves tool life, reduces friction, and enhances wear resistance, thereby contributing to increased productivity, reliability, and cost-effectiveness in industrial operations. Materials such as diamond-like carbon (DLC), titanium nitride (TiN), and chromium nitride (CrN) are commonly used for hard coatings in CVD applications, offering superior hardness, wear resistance, and thermal stability.

iv. Barrier Coatings: CVD is used to deposit thin films of materials such as silicon nitride and aluminum oxide onto flexible packaging materials, such as plastic films and metal foils, to create barrier coatings that prevent moisture, oxygen, and other gases from permeating through the packaging. Barrier coatings help extend the shelf life of packaged products and protect sensitive materials from degradation. CVD, or Chemical Vapor Deposition, serves as a crucial technique in the packaging industry, particularly for the creation of barrier coatings aimed at preserving the integrity of packaged products. This process involves the deposition of thin films of materials like silicon nitride and aluminum oxide onto flexible packaging materials such as plastic films and metal foils. The primary objective of employing CVD in this context is to develop barrier coatings that effectively block the permeation of moisture, oxygen, and other gases through the packaging material. By forming a protective layer, these coatings create a barrier that shields the packaged products from external environmental factors that could lead to degradation or spoilage. One of the significant benefits of utilizing CVD for this purpose is its ability to achieve uniform and precise coatings over large areas of flexible substrates. This uniformity ensures consistent protection across the entirety of the packaging material, leaving no weak spots vulnerable to gas permeation. The barrier coatings produced through CVD play a pivotal role in extending the shelf life of packaged products. By impeding the ingress of moisture and oxygen, these coatings help maintain the freshness, quality, and safety of the contents within. Additionally, they safeguard sensitive materials from degradation, ensuring that the packaged products reach consumers in optimal condition. Moreover, the versatility of CVD allows for the deposition of barrier coatings onto various types of flexible packaging materials, including plastic films and metal foils. This versatility enables manufacturers to adapt the barrier coatings to suit the specific requirements of different products and packaging formats, catering to a wide range of applications across industries. CVD serves as a vital tool in the packaging industry for the deposition of barrier coatings that protect packaged products from moisture, oxygen, and other gases. By extending the shelf life of products and safeguarding sensitive materials, these coatings contribute to ensuring product quality and consumer satisfaction. The precision, uniformity, and versatility of CVD make it an indispensable technique for creating effective barrier coatings on flexible packaging materials.

v. Biomedical Applications: CVD coatings play a critical role in enhancing the performance and biocompatibility of biomedical devices and implants, contributing to improved patient outcomes and longevity. By depositing thin films of materials such as titanium nitride, titanium oxide, and hydroxyapatite onto orthopedic implants, dental implants, and surgical instruments, CVD coatings offer a range of benefits including improved wear resistance, corrosion resistance, and biocompatibility. In the field of orthopedics, CVD coatings are extensively utilized to enhance the functionality and longevity of implants used in joint replacement surgeries. Titanium nitride (TiN) coatings, for example, are deposited onto orthopedic implants to improve wear resistance and reduce friction between the implant and surrounding tissues. These coatings promote smooth articulation, reduce the risk of implant failure due to wear-related issues, and contribute to the overall success of joint replacement procedures.

Similarly, CVD coatings find applications in dental implants to improve biocompatibility and tissue integration. Thin films of titanium oxide ($TiO2$) and hydroxyapatite, deposited using CVD, promote osseointegration—the process by which the implant fuses with surrounding bone tissue. These coatings enhance the stability and longevity of dental implants, reducing the risk of implant failure and improving patient satisfaction with dental restoration procedures. Moreover, in surgical instruments, CVD coatings are employed to improve wear resistance and reduce the risk of contamination and infection. Titanium nitride (TiN) coatings, for instance, are deposited onto surgical instruments such as scalpels and forceps to enhance their durability and longevity. These coatings reduce friction during surgical procedures, ensuring smooth operation and minimizing tissue trauma. Additionally, CVD coatings can prevent bacterial adhesion on the surface of surgical instruments, reducing the risk of post-operative infections and promoting better patient outcomes. The use of CVD coatings in biomedical devices and implants underscores their importance in improving performance, biocompatibility, and patient safety. By depositing thin films of materials such as titanium nitride, titanium oxide, and hydroxyapatite, CVD coatings offer a range of benefits including enhanced wear resistance, corrosion resistance, and tissue integration. These coatings play a vital role in promoting the success of orthopedic and dental implant procedures, as well as improving the functionality and safety of surgical instruments used in medical practice. As research and technology continue to advance, CVD coatings are expected to play an increasingly significant role in advancing the field of biomedical engineering and improving patient care.

Advancements in Chemical Vapor Deposition:

Over the years, advancements in chemical vapor deposition (CVD) technology have led to improvements in deposition rates, film quality, and process control, expanding the range of applications and capabilities of CVD coatings. Some notable advancements in CVD include:

i. High-Speed Deposition: Continuous advancements in Chemical Vapor Deposition (CVD) technology have revolutionized thin-film deposition processes, leading to higher deposition rates, increased productivity, and enhanced capabilities. These advancements, driven by improvements in equipment design, precursor chemistry, and process optimization, have propelled CVD techniques to new heights, enabling rapid and precise coating of large-area substrates across various industries. One of the significant breakthroughs in CVD technology is the development of high-speed deposition techniques such as Plasma-Enhanced CVD (PECVD) and Atomic Layer Deposition (ALD). PECVD utilizes plasma to enhance chemical reactions and deposition rates, enabling the rapid coating of large-area substrates with thin films of materials such as silicon nitride and silicon dioxide. By introducing energy into the deposition process, PECVD achieves higher deposition rates while maintaining precise control over film thickness and composition, making it ideal for applications requiring high throughput and uniformity. Similarly, ALD has emerged as a powerful technique for achieving precise and conformal thin-film deposition with atomic-level control. Unlike traditional CVD methods, which rely on gas-phase reactions to deposit thin films, ALD operates through self-limiting surface reactions, allowing for precise control over film thickness and composition. By depositing thin films layer by layer, ALD offers unparalleled uniformity and conformality, making it well-suited for coating complex three-dimensional structures and high-aspect-ratio substrates. Advancements in precursor chemistry have also played a significant role in enhancing CVD processes. By developing novel precursor molecules with improved reactivity and stability, researchers have expanded the range of materials that can be deposited using CVD techniques. These advancements have paved the way for the deposition of advanced materials such as metal oxides, nitrides, and carbides, opening up new possibilities for applications in electronics, energy storage, and biomedical devices. Furthermore, process optimization strategies have been instrumental in maximizing the efficiency and effectiveness of CVD processes. By fine-tuning process parameters such as temperature, pressure, and gas flow rates, researchers can optimize deposition conditions to achieve higher deposition rates, improved film quality, and enhanced control over film properties. Additionally, advances in equipment design, such as the development of multi-chamber deposition systems and in-situ monitoring techniques, have enabled greater automation, reliability, and scalability in CVD processes. Advances in CVD equipment design, precursor chemistry, and process optimization have revolutionized thin-film deposition processes, enabling higher deposition rates, increased productivity, and enhanced capabilities. Techniques such as PECVD and ALD offer rapid coating of large-area substrates with precise control over film thickness and composition, while advancements in precursor chemistry and process optimization have expanded the range of materials that can be deposited using CVD techniques. These advancements hold promise for a wide range of applications across industries, from electronics and energy storage to biomedical devices and beyond.

ii. Conformal Coatings: Conformal Chemical Vapor Deposition (CVD) coatings represent a significant advancement in thin-film deposition technology, offering precise coverage and adhesion to complex substrate surfaces with irregular geometries. These coatings, which adhere tightly to the substrate surface and conform to intricate features such as trenches, vias, and three-dimensional structures, play a critical role in various industries, including microelectronics, optics, and biomedical engineering. The development of advanced CVD techniques, such as Remote Plasma-Enhanced CVD (RPECVD) and Atomic Layer Deposition (ALD), has greatly expanded the capabilities for depositing conformal coatings on challenging substrates. RPECVD utilizes remote plasma to enhance precursor gas activation and surface reactivity, enabling the deposition of uniform and conformal coatings even on highly topographic surfaces. By dissociating precursor gases remotely from the substrate, RPECVD minimizes substrate damage and contamination, resulting in high-quality conformal coatings suitable for a wide range of applications. Similarly, Atomic Layer Deposition (ALD) has emerged as a powerful technique for achieving precise and conformal thin-film deposition with atomic-level control. Unlike conventional CVD methods, which rely on gas-phase reactions, ALD operates through self-limiting surface reactions, allowing for precise control over film thickness and conformality. By depositing thin films layer by layer, ALD ensures uniform coverage and adhesion to complex geometries, making it ideal for coating three-dimensional structures and high-aspect-ratio substrates. Conformal CVD coatings find numerous applications across industries, owing to their ability to provide uniform coverage and protection to intricate surfaces. In the microelectronics industry, conformal coatings are used to insulate and passivate components such as integrated circuits, microelectromechanical systems (MEMS), and interconnects. By depositing thin films of materials such as silicon dioxide and silicon nitride using ALD or RPECVD, manufacturers can ensure reliable performance and longevity of electronic devices, even in harsh operating environments. Moreover, conformal coatings play a crucial role in the fabrication of optical components such as lenses, mirrors, and photonic devices. By depositing thin films of materials such as titanium oxide and aluminum oxide using ALD or RPECVD, manufacturers can enhance the optical properties and durability of these components, ensuring consistent performance and reliability in optical systems. In the field of biomedical engineering, conformal coatings are used to improve the biocompatibility and functionality of medical implants and devices. Thin films of materials such as hydroxyapatite and titanium nitride, deposited using ALD or RPECVD, promote tissue integration and reduce the risk of infection and rejection. These coatings enable the fabrication of implantable devices with enhanced performance and biocompatibility, contributing to better patient outcomes and quality of life. Conformal Chemical Vapor Deposition (CVD) coatings represent a significant advancement in thin-film deposition technology, offering precise coverage and adhesion to complex substrate surfaces with irregular geometries. Advances in CVD techniques such as Remote Plasma-Enhanced CVD (RPECVD) and Atomic Layer Deposition (ALD) have enabled the deposition of conformal coatings with uniform thickness and coverage, even on irregular or highly topographic surfaces. These coatings find widespread applications across industries, including microelectronics, optics, and biomedical engineering, where they play a crucial role in enhancing performance, durability, and functionality of components and devices.

iii. Functional Coatings: Chemical Vapor Deposition (CVD) coatings offer a versatile platform for tailoring specific functionalities to meet the varied requirements of different applications. By engineering these coatings, manufacturers can impart desired properties such as electrical conductivity, optical transparency, or thermal insulation, thereby enhancing the performance, reliability, and efficiency of diverse devices and systems. In the realm of flexible electronics, CVD coatings play a crucial role in enabling the fabrication of lightweight, bendable electronic devices with enhanced functionality. Functional coatings engineered for electrical conductivity are applied to flexible substrates such as polymer films or textiles, allowing for the integration of electronic components into wearable devices, sensors, and displays. By providing a conductive surface, these coatings facilitate efficient charge transport and enable the seamless operation of flexible electronic systems in various applications, including healthcare, sports, and consumer electronics. Moreover, CVD coatings find extensive use in the fabrication of photovoltaic cells, where they contribute to improving the efficiency and reliability of solar energy harvesting devices. Functional coatings engineered for optical transparency and electrical conductivity are deposited onto solar cell substrates, enhancing light absorption, charge separation, and electron transport within the device. By optimizing these coatings, manufacturers can increase the efficiency of photovoltaic cells, leading to higher energy conversion rates and improved performance in diverse environmental conditions. Furthermore, in the field of biomedical engineering, CVD coatings are employed to enhance the functionality and biocompatibility of medical devices and implants. Functional coatings engineered for biocompatibility and surface modification are applied to implantable materials such as metals, ceramics, and polymers, promoting tissue integration, reducing inflammation, and preventing bacterial adhesion. These coatings enable the fabrication of medical implants with improved biocompatibility and longevity, leading to better patient outcomes and reduced risk of complications. In addition to these applications, CVD coatings find use in a wide range of other industries and technologies, including automotive, aerospace, and energy storage. Functional coatings engineered for thermal insulation, corrosion resistance, or catalytic activity are applied to various components and structures, enhancing their performance, durability, and efficiency in demanding environments. Chemical Vapor Deposition (CVD) coatings offer a versatile platform for engineering specific functionalities tailored to the requirements of different applications. By providing properties such as electrical conductivity, optical transparency, or thermal insulation, these coatings enhance the performance, reliability, and efficiency of diverse devices and systems, ranging from flexible electronics and photovoltaic cells to biomedical devices and beyond. As research and development efforts continue to advance, CVD coatings are expected to play an increasingly significant role in enabling innovative solutions across industries and technologies.

iv. Multicomponent Coatings: Recent advancements in Chemical Vapor Deposition (CVD) technology have ushered in a new era of coating capabilities, enabling the deposition of multicomponent coatings with precisely tailored properties and functionalities. These coatings, composed of two or more materials, offer synergistic effects that result in enhanced mechanical strength, chemical resistance, thermal stability, and other desirable properties. As a result, multicomponent coatings have emerged as indispensable solutions for demanding applications across industries such as aerospace, automotive, and energy. The development of innovative CVD chemistry has played a pivotal role in expanding the range of materials that can be incorporated into multicomponent coatings. By carefully selecting precursor gases and optimizing reaction conditions, researchers can deposit thin films composed of different materials with controlled composition and morphology. This precise control over material deposition enables the engineering of multicomponent coatings with tailored properties to meet specific application requirements. Furthermore, advancements in process control techniques have contributed to the successful deposition of multicomponent coatings with uniform thickness and coverage. Sophisticated process monitoring and control systems allow for real-time adjustment of deposition parameters, ensuring reproducibility and consistency in coating quality. By optimizing process parameters such as temperature, pressure, and precursor flow rates, researchers can achieve the desired composition and structure of multicomponent coatings, thereby maximizing their performance and reliability. Multicomponent coatings find wide-ranging applications in industries where superior material properties are critical for success. In the aerospace sector, for example, multicomponent coatings are used to protect aircraft components from harsh environmental conditions such as high temperatures, corrosion, and erosion. By combining materials with complementary properties, such as ceramic and metallic phases, multicomponent coatings offer enhanced durability and performance, ensuring the longevity and safety of aerospace structures and systems. Similarly, in the automotive industry, multicomponent coatings are applied to engine components, exhaust systems, and other critical parts to improve performance and longevity. These coatings provide enhanced wear resistance, thermal insulation, and corrosion protection, enabling automotive components to withstand the rigors of daily operation and prolonged exposure to harsh operating conditions. In the energy sector, multicomponent coatings are utilized in various applications, including fuel cells, batteries, and solar panels, to enhance efficiency and reliability. By incorporating materials with tailored properties, such as enhanced conductivity or catalytic activity, multicomponent coatings improve the performance and longevity of energy conversion and storage devices, contributing to the advancement of renewable energy technologies. Recent developments in CVD chemistry and process control have unlocked new possibilities for the deposition of multicomponent coatings with tailored properties and functionalities. These coatings offer synergistic effects that enhance mechanical strength, chemical resistance, thermal stability, and other key properties, making them indispensable solutions for demanding applications in aerospace, automotive, and energy industries. As research and development efforts continue to advance, multicomponent coatings are poised to play an increasingly significant role in enabling innovation and driving progress across diverse industrial sectors.

v. Environmentally Friendly Processes: Recent advancements in Chemical Vapor Deposition (CVD) technology have not only expanded the capabilities of thin-film deposition but also prioritized environmental sustainability by reducing energy consumption, waste generation, and emissions. These developments have led to the emergence of environmentally friendly CVD processes that offer enhanced efficiency, material utilization, and sustainability compared to traditional methods. Techniques such as Remote Plasma-Enhanced CVD (RPECVD), Atomic Layer Deposition (ALD), and Microwave Plasma-Enhanced CVD (MWPECVD) represent notable examples of these advancements. Remote Plasma-Enhanced CVD (RPECVD) stands out as a significant advancement in CVD technology, offering improved process efficiency and environmental sustainability. In RPECVD, plasma is generated remotely from the substrate, allowing for better control over the deposition process while minimizing substrate damage and contamination. This approach reduces energy consumption and waste generation, leading to more sustainable thin-film deposition processes. Atomic Layer Deposition (ALD) is another environmentally friendly CVD technique that has gained widespread adoption due to its precise control over film thickness and composition. ALD operates through self-limiting surface reactions, enabling atomic-level control over thin-film deposition. By depositing materials layer by layer, ALD minimizes material waste and ensures uniformity, leading to higher material utilization and reduced environmental impact compared to traditional CVD methods. Microwave Plasma-Enhanced CVD (MWPECVD) represents a further advancement in CVD technology, offering enhanced process efficiency and sustainability through the use of microwave-generated plasma. By utilizing microwave energy to generate plasma, MWPECVD achieves higher plasma densities and lower operating temperatures compared to conventional methods. This results in reduced energy consumption, shorter processing times, and lower emissions, contributing to improved environmental sustainability in thin-film deposition processes. These environmentally friendly

CVD techniques have broad applications across various industries, including electronics, optics, and renewable energy. In the electronics industry, for example, RPECVD, ALD, and MWPECVD are used to deposit thin films for semiconductor devices, displays, and photovoltaic cells, enabling the production of high-performance electronic components with minimal environmental impact. Advances in CVD technology have led to the development of environmentally friendly processes that prioritize energy efficiency, waste reduction, and emissions mitigation. Techniques such as Remote Plasma-Enhanced CVD (RPECVD), Atomic Layer Deposition (ALD), and Microwave Plasma-Enhanced CVD (MWPECVD) offer improved process efficiency, material utilization, and environmental sustainability compared to traditional methods. These advancements have significant implications for a wide range of industries, enabling the production of high-performance thin-film coatings with minimal environmental impact.

Chemical Vapor Deposition (CVD) is a highly versatile and widely used thin-film deposition technique that enables the creation of functional coatings with tailored properties and functionalities for a wide range of applications. From semiconductor manufacturing and optical coatings to hard coatings and biomedical implants, CVD plays a critical role in enhancing the performance, durability, and reliability of components and products across industries. With ongoing advancements in CVD technology, materials, and processes, the future holds exciting possibilities for innovative coatings with enhanced functionality, sustainability, and performance.

Sol–Gel Deposition: Sol–gel deposition involves the formation of thin films from a colloidal solution (sol) containing metal alkoxides or metal salts. In the sol–gel process, the precursor solution undergoes hydrolysis and condensation reactions to form a networked gel, which can then be deposited onto a substrate using techniques such as dip coating, spin coating, or spray coating. Sol–gel deposition offers versatility in film composition and processing conditions and is used in applications such as optical coatings, sensor materials, and biomaterials.

Sol–gel deposition is a versatile and widely-used technique for fabricating thin films and coatings with diverse applications in various industries, including electronics, optics, energy, biomedical engineering, and environmental remediation. The sol–gel process involves the conversion of a sol or solution into a gel-like material, followed by subsequent drying and densification steps to form a solid film. This method offers numerous advantages, including simplicity, low processing temperatures, tunable properties, and compatibility with a wide range of substrates. In this comprehensive exploration, we examine the principles, methods, applications, and advancements of sol–gel deposition, elucidating its significance and potential across different fields of science and technology. Sol–gel deposition relies on the controlled hydrolysis and condensation reactions of precursor molecules to form a sol, which is a colloidal suspension of nanoscale particles in a liquid medium. These precursor molecules typically consist of metal alkoxides, such as tetraethyl orthosilicate (TEOS) for silica-based coatings or titanium isopropoxide for titanium dioxide coatings, dissolved in a solvent, often alcohol or water. The hydrolysis of alkoxide groups (–OR) produces hydroxyl groups (–OH), while the condensation reactions lead to the formation of metal-oxygen-metal (M-O-M) linkages, resulting in the formation of a three-dimensional network structure within the sol. The sol is then subjected to gelation, a process where the sol transforms into a gel by forming a continuous network of interconnected particles. Gelation can be induced by various methods, including thermal treatment, chemical additives, or pH adjustments. During gelation, the sol undergoes a phase transition from a liquid-like state to a solid-like state, forming a three-dimensional network structure that retains the shape of the original sol. Following gelation, the gel undergoes drying or aging to remove the solvent and enhance the mechanical properties of the material. During drying, the solvent is evaporated, leaving behind a porous network of interconnected particles. Subsequent heat treatment or calcination is often employed to remove residual organic components, densify the material, and enhance its mechanical, thermal, and chemical stability.

One of the key advantages of sol–gel deposition is its versatility and tunability, allowing for precise control over the composition, structure, and properties of the resulting films. By adjusting parameters such as precursor concentration, solvent composition, pH, temperature, and processing time, researchers can tailor the properties of sol–gel coatings to meet specific application requirements. This flexibility enables the fabrication of thin films with a wide range of functionalities, including optical transparency, electrical conductivity, catalytic activity, and mechanical strength. Sol–gel deposition finds widespread applications across various industries due to its unique combination of properties and advantages. In the electronics industry, sol–gel coatings are used for fabricating thin-film transistors, capacitors, insulating layers, and dielectric materials. These coatings offer excellent dielectric properties, thermal stability, and compatibility with flexible substrates, making them ideal for flexible and wearable electronics. Moreover, sol–gel coatings are extensively employed in optical applications, such as anti-reflective coatings, optical filters, waveguides, and photonic devices. These coatings exhibit high optical transparency, low surface roughness, and precise control over refractive index, enabling the fabrication of high-performance optical components with tailored optical properties.

In the energy sector, sol–gel coatings play a critical role in various renewable energy technologies, including solar cells, fuel cells, batteries, and photocatalysts. These coatings are utilized for surface passivation, light trapping, charge transport, and catalytic enhancement, leading to improved energy conversion efficiency, durability, and performance of energy devices. Additionally, sol–gel deposition is widely applied in the biomedical field for fabricating biocompatible coatings, drug delivery systems, biosensors, and tissue engineering scaffolds. These coatings offer excellent biocompatibility, controlled release kinetics, and surface functionalization capabilities, enabling the development of advanced medical devices and therapies for diagnostics, drug delivery, and regenerative medicine. Furthermore, sol–gel coatings find applications in environmental remediation, including water purification, air filtration, and pollutant degradation. These coatings can be functionalized with specific molecules or nanoparticles to adsorb, degrade, or neutralize contaminants, providing effective solutions for addressing environmental pollution and improving public health. sol–gel deposition is a versatile and powerful technique for fabricating thin films and coatings with tailored properties and functionalities for a wide range of applications in electronics, optics, energy, biomedical engineering, and environmental remediation. With its simplicity, tunability, and compatibility with various substrates, sol–gel deposition offers numerous advantages and opportunities for innovation and advancement in science and technology. As research and development efforts continue to progress, sol–gel deposition is poised to play an increasingly significant role in addressing key challenges and driving progress across different fields of science and engineering.

Spin Coating: Spin coating is a simple and versatile technique used to deposit thin films onto substrates with high uniformity and control over thickness. In spin coating, a liquid precursor solution is dispensed onto a substrate, which is then spun at high speeds to spread the solution evenly across the surface. The centrifugal force causes the solvent to evaporate, leaving behind a thin film deposited onto the substrate. Spin coating is widely used in microelectronics, photovoltaics, and surface science research due to its simplicity, scalability, and ability to deposit uniform films over large areas. Spin coating is a widely used technique in the field of thin film deposition, offering a simple yet effective method for producing uniform coatings on flat substrates. This comprehensive exploration delves into the principles, methods, applications, and advancements of spin coating, elucidating its significance and potential across various disciplines, including microelectronics, optics, materials science, and biomedical engineering. Spin coating operates on the principle of centripetal force, where a liquid precursor solution is dispensed onto a rotating substrate, spreading into a thin film due to the centrifugal force generated by rotation. The rotational speed, viscosity of the precursor solution, and duration of spinning determine the thickness and uniformity of the resulting coating. By controlling these parameters, researchers can achieve precise control over the properties and characteristics of the deposited thin films. The spin coating process begins with the preparation of a precursor solution, typically consisting of a solvent and dissolved or dispersed functional materials, such as polymers, nanoparticles, or organic compounds. The precursor solution is then dispensed onto the center of a clean, flat substrate, which is mounted on a spin coater platform. The substrate is then rapidly spun at a controlled rotational speed, causing the precursor solution to spread outwards due to the centrifugal force. As the substrate spins, the solvent evaporates, leaving behind a thin film of the functional material on the substrate surface. The spinning process continues until the desired film thickness is achieved, at which point the rotation is gradually slowed down to prevent film deformation or cracking. One of the key advantages of spin coating is its ability to produce uniform coatings with precise control over thickness and morphology. The centrifugal force generated during spinning helps to spread the precursor solution evenly across the substrate surface, resulting in uniform film thicknesses even on large-area substrates. This uniformity is crucial for applications requiring high-quality thin films, such as microelectronics, optics, and sensor devices. Spin coating finds widespread applications across various industries and research fields, owing to its simplicity, versatility, and scalability. In the microelectronics industry, spin coating is used for the deposition of photoresist layers, dielectric materials, and insulating layers in semiconductor fabrication processes. These thin films play a crucial role in defining device structures, insulating electrical components, and protecting sensitive circuitry from environmental factors. Moreover, spin coating is extensively employed in the fabrication of optical coatings, including anti-reflection coatings, optical filters, and waveguides. These coatings are essential for controlling light transmission, reflection, and absorption properties in optical devices, such as lenses, mirrors, and photonic devices. By depositing thin films of materials with precise optical properties, spin coating enables the production of high-performance optical components with tailored functionalities. Additionally, spin coating is utilized in the field of materials science for the synthesis and deposition of functional materials, such as polymers, ceramics, and nanoparticles. These thin films are used for various purposes, including surface modification, functionalization, and encapsulation of materials for applications in sensors, actuators, and energy storage devices. In the biomedical engineering field, spin coating is employed for the fabrication of biocompatible coatings, drug delivery systems, and tissue engineering scaffolds. These coatings can be functionalized with bioactive molecules, growth factors, or drugs to promote cell adhesion, proliferation, and differentiation, facilitating applications in regenerative medicine, drug delivery, and biomedical diagnostics. Furthermore, spin coating is used in research laboratories and academic institutions for prototyping, proof-of-concept studies, and fundamental research in materials science, chemistry, and physics. Its simplicity, low cost, and accessibility make spin coating an attractive choice for researchers seeking to deposit thin films with precise control over properties and characteristics. Recent advancements in spin coating technology have focused on improving process efficiency, scalability, and control over thin film properties. Novel approaches, such as multi-step spin coating processes, dynamic control of rotational speed, and the use of alternative substrates and precursors, have enabled the fabrication of thin films with enhanced functionalities and performance characteristics. Spin coating is a versatile and widely-used technique for depositing uniform thin films on flat substrates, offering precise control over thickness, morphology, and properties. With applications spanning microelectronics, optics, materials science, biomedical engineering, and beyond, spin coating plays a crucial role in enabling innovation and advancement across various industries and research fields. As research and development efforts continue to progress, spin coating is poised to remain a key technology for the deposition of thin films and coatings with tailored functionalities and applications in diverse scientific and technological domains.

Spin coating thickness: In general, the thickness of a spin coated film is proportional to the inverse of the spin speed squared, as shown in the Eq. 2.1, where hf is the final film thickness and angular velocity/spin speed is the angular velocity/spin speed. The equation can also be used to calculate the spin curve.

$$h_f = \frac{1}{\sqrt{\omega}} \tag{2.1}$$

Attempts has been made to predict spin coating final film thickness without using experiment by different researchers. Equation 4 shows the Emslie, Bonner, and Peck Model used on an endless rotating disk, a non-volatile, viscous fluid. The model made various assumptions, including neglecting the effects of evaporation (which will vary depending on how volatile the solvent is) and ignoring the potential of non-Newtonian behaviour.

$$\frac{\partial h}{\partial t} + \frac{\rho \omega^2 r}{\eta} h^2 \frac{\partial h}{\partial r} = -\frac{2\rho \omega^2 h^3}{3\eta} \tag{2.2}$$

where t is the process start time, r is the distance from the center of rotation, ρ is the density, η is the viscosity, and h is the fluid layer thickness (rather than the dry thin film). Angular velocity is ω, rate of change in thickness is represented by $\partial h/\partial t$, and the rate of spreading is represented by $\partial h/\partial r$.

Equation 2.3 is obtained if the film is uniform, and the start time t becomes zero and there is no evaporation and hence is not used to compute the final dry film thickness.

$$h = \frac{h_0}{\left(1 + \frac{4\rho\omega^2}{3\eta} h_0^2 t\right)^{\frac{1}{2}}} \tag{2.3}$$

The modification of Emslie, Bonner, and Peck model to include evaporation rate of the solvent was done by Meyerhofer in 1978 and shown in Eq. 2.4.

$$\frac{dh}{dt} = -\frac{2\rho\omega^2 h^3}{3\eta} - E \tag{2.4}$$

where E stands for the uniform solvent evaporation rate, which is expressed in units of solvent volume evaporated per unit area per unit time.

Langmuir–Blodgett Technique: The Langmuir–Blodgett (LB) technique is a specialized method used to deposit thin organic films with precise control over thickness and molecular orientation. In the LB technique, organic molecules are spread as a monolayer on the surface of a liquid subphase, typically water or an organic solvent, in a Langmuir trough. By compressing or expanding the monolayer, the molecules can be transferred onto a solid substrate at the air–water interface, forming a thin film with a controlled number of molecular layers. The LB technique is used in applications such as molecular electronics, biomimetic materials, and surface science studies. The Langmuir–Blodgett (LB) technique is a versatile and powerful method for depositing well-ordered thin films with precise control over layer thickness, molecular orientation, and interfacial properties. Named after the scientists Irving Langmuir and Katharine Blodgett, who pioneered its development in the early twentieth century, the LB technique has found widespread applications in various fields, including surface science, materials research, nanotechnology, and biotechnology. This comprehensive exploration delves into the principles, methods, applications, and advancements of the Langmuir–Blodgett technique, elucidating its significance and potential across different disciplines.

Principles of Langmuir–Blodgett Technique
The Langmuir–Blodgett technique operates on the principles of molecular self-assembly and interfacial phenomena, leveraging the unique properties of amphiphilic molecules at the air–water interface. Amphiphilic molecules, such as fatty acids, phospholipids, and block copolymers, possess hydrophilic (water-attracting) and hydrophobic (water-repelling) regions within the same molecule. When spread onto the surface of a subphase (typically water), these molecules organize themselves into a monolayer with the hydrophilic heads facing the water and the hydrophobic tails oriented towards the air. The Langmuir–Blodgett technique involves the controlled compression of the monolayer at the air–water interface using a Langmuir trough, followed by the transfer of the monolayer onto a solid substrate. By controlling the surface pressure, temperature, and deposition speed, researchers can precisely manipulate the packing density and molecular orientation of the transferred monolayer, resulting in well-ordered thin films with tailored properties.

Methods of Langmuir–Blodgett Technique
The Langmuir–Blodgett technique encompasses several variations and modifications, each tailored to specific applications and requirements. The two primary modes of deposition are the Langmuir–Blodgett (LB) method and the Langmuir-Schaefer (LS) method, distinguished by the direction of deposition relative to the compression axis of the Langmuir trough. In the Langmuir–Blodgett (LB) method, the substrate is positioned horizontally beneath the monolayer at the air–water interface, and the monolayer is transferred onto the substrate by vertically lifting or dipping the substrate through the monolayer. This results in the deposition of a single layer of molecules onto the substrate, with precise control over layer thickness and molecular orientation. The LB method is well-suited for fabricating ultrathin films with nanometer-scale precision and excellent structural uniformity. In contrast, the Langmuir-Schaefer (LS) method involves the horizontal transfer of the monolayer onto the substrate by horizontally sliding the substrate across the compressed monolayer at the air–water interface. This results in the deposition of multiple monolayers in a layer-by-layer fashion, enabling the fabrication of thicker films with controlled thickness and stacking order. The LS method offers advantages for applications requiring thicker films or multilayer structures, such as optical coatings, sensors, and molecular electronics.

Applications of Langmuir–Blodgett Technique

The Langmuir–Blodgett technique finds diverse applications across various fields, owing to its ability to produce well-ordered thin films with tailored properties and functionalities. In the field of surface science, LB films are utilized for studying fundamental surface phenomena, such as molecular interactions, phase transitions, and surface patterning. These films serve as model systems for investigating surface properties and behaviors at the molecular level, providing insights into complex phenomena relevant to catalysis, adhesion, and wetting. In materials research and nanotechnology, LB films are employed for fabricating functional materials with tailored properties for applications in electronics, optics, and photonics. LB films can be engineered to exhibit unique electrical, optical, magnetic, and mechanical properties, making them attractive for a wide range of device applications, including organic semiconductors, photovoltaics, light-emitting diodes (LEDs), and sensors. By controlling the molecular structure and packing arrangement of the deposited molecules, researchers can tune the performance and functionality of LB-based devices for specific applications. Furthermore, in biotechnology and biomedical engineering, LB films are utilized for designing biomimetic interfaces, drug delivery systems, and biosensors. LB films can mimic the structure and function of biological membranes, enabling the study of biomolecular interactions, cell-surface interactions, and membrane transport processes. These films are also used for encapsulating and delivering therapeutic agents, such as drugs, proteins, and nucleic acids, for applications in drug delivery, gene therapy, and tissue engineering. Additionally, LB-based biosensors offer sensitive and selective detection of biomolecules and analytes, making them valuable tools for medical diagnostics, environmental monitoring, and food safety.

Advancements in Langmuir–Blodgett Technique

Recent advancements in the Langmuir–Blodgett technique have focused on enhancing film quality, scalability, and functionality through innovations in materials, instrumentation, and process control. Novel amphiphilic molecules, including dendrimers, peptides, and nanoparticles, have been developed for fabricating LB films with tailored properties and enhanced functionality. These molecules offer advantages such as precise control over molecular structure, surface chemistry, and intermolecular interactions, enabling the design of advanced materials for specific applications. Moreover, advancements in instrumentation and automation have improved the reproducibility, throughput, and scalability of LB film fabrication. Automated Langmuir troughs equipped with advanced sensors, actuators, and data acquisition systems enable precise control over experimental parameters, such as surface pressure, temperature, and deposition speed. These systems facilitate high-throughput fabrication of LB films with consistent quality and reproducibility, making them suitable for industrial-scale production and commercial applications. Furthermore, developments in surface patterning techniques, such as microcontact printing, photolithography, and soft lithography, have enabled the creation of complex and hierarchical structures in LB films. These patterning techniques offer precise control over

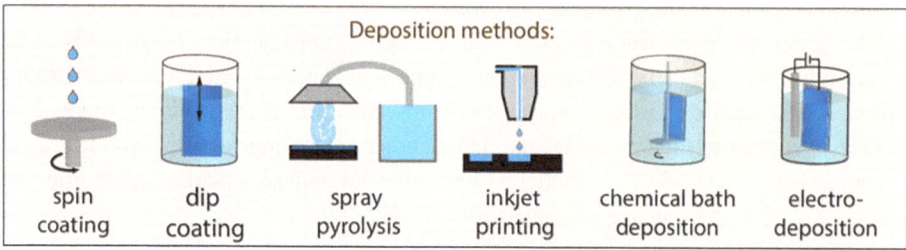

Fig. 2.6 Schematic summary of various deposition method

film morphology, topography, and functionality, allowing for the fabrication of nanostructured surfaces with tailored properties for applications in nanoelectronics, plasmonics, and bioengineering. The Langmuir–Blodgett technique is a versatile and powerful method for fabricating well-ordered thin films with precise control over thickness, molecular orientation, and interfacial properties. By leveraging the principles of molecular self-assembly and interfacial phenomena, the LB technique enables the deposition of functional materials onto solid substrates with tailored properties and functionalities. With applications spanning surface science, materials research, nanotechnology, biotechnology, and beyond, the Langmuir–Blodgett technique continues to drive innovation and advancement across various fields of science and technology. As research and development efforts continue to progress, the LB technique is poised to remain a key technology for the fabrication of advanced materials and devices with tailored properties and applications in diverse domains. A summary of the various deposition techniques is shown in Fig. 2.6.

Other Advanced Techniques: In addition to the techniques mentioned above, several advanced deposition techniques are used for thin film fabrication, including atomic layer deposition (ALD), molecular beam epitaxy (MBE), pulsed laser deposition (PLD), and chemical bath deposition (CBD) as shown in Table 2.1. These techniques offer precise control over film growth at the atomic or molecular scale and are used in specialized applications such as nanotechnology, quantum devices, and thin film epitaxy.

Pulsed Laser Deposition (PLD):

Pulsed Laser Deposition (PLD) is a versatile and powerful technique for depositing thin films of various materials with precise control over composition, structure, and properties. This comprehensive exploration delves into the principles, methods, applications, and advancements of Pulsed Laser Deposition, elucidating its significance and potential across different disciplines, including materials science, nanotechnology, electronics, optics, and energy research. Pulsed Laser Deposition (PLD) is a thin film deposition technique that utilizes high-energy pulsed lasers to ablate material from a target, forming a plasma plume, which is subsequently deposited onto a substrate to create thin films. PLD offers

Table 2.1   A summary of deposition method

Deposition type	Description	Merit	Parameters	Demerit
Solution based				
Chemical bath deposition	It is a thin-film deposition process that uses an aqueous precursor solution to produce solids from a solution or gas. Chemical Bath Deposition generates homogenous thin films of metal chalcogenides (primarily oxides, sulfides, and selenides)	CBD is generally cheap, convenient for large-area deposition, and has the capacity to tune thin film properties by modifying and controlling it	Temperature of solution, stirring time, precursor material concentration, and substrate type	There is waste of solution, and the substrate should be clean
Dip-coating	It is the process of slowly dipping a substrate into a solution. After the liquid component of this solution has dried, the substrate is left with a solid thin coating	It is affordable and thickness of deposited layer can finetuned easily	viscous drag, the forces of inertia, surface tension and gravitational force	The process is slow and screen blocking resulting in lower efficiency
Electrodeposition	The method of controlled deposition of material on conducting surfaces using electric current from a solution containing ionic species is known as electrodeposition or electroplating. it has been widely utilized to create thin films with no binder. Electrodeposition can be divided into three types: electroplating, electrophoretic deposition, and underpotential deposition	the capacity to wrap items with very thin metal layers to enhance their appearance	Content, current density and temperature. Step by step, the electrodeposition rate increases as the current density rises	Lack of thick shell

(continued)

Table 2.1 (continued)

Deposition type	Description	Merit	Parameters	Demerit
Screen printing	Screen printing (also known as serigraphy and serigraph printing) is a printing process in which ink (or dye) is transferred onto a substrate using a mesh, except for areas made impermeable to the ink by a blocking stencil. A blade or squeegee is dragged across the screen to fill the open mesh apertures with ink, and then a reverse stroke causes the screen to touch the substrate along a line of contact for a moment. As the screen springs back after the blade has passed, the ink wets the substrate and is drawn out through the mesh holes. Because each colour is printed separately, multiple screens can be utilized to create a multi-coloured image or design	Process is simple, costs are inexpensive, and the deposition area is large	The paste's viscosity, the amount of mesh screens, the snap-off distance between the screen and the substrate, and the pressure	Low output rate, poor ink mileage, and lengthy drying time are all issues that must be addressed
Sol–gel	The sol–gel method is a wet-chemical technique for making solid materials out of tiny molecules. The method entails converting monomers into a colloidal solution (sol), which serves as a precursor for forming an integrated network (or gel) of discrete particles or network polymers	Simplest, most homogeneous, least expensive, most reliable, most reproducible	amount of water, temperature, pH, solvent, and precursor are all factors to consider	Precursor costs are high, the process takes a long time, monolith synthesis is difficult, and the process chemistry is difficult to regulate and repeat

(continued)

Table 2.1 (continued)

Deposition type	Description	Merit	Parameters	Demerit
Spray pyrolysis	Spray pyrolysis is a method for coating large areas with films of very thin layers of uniform thickness. Spray coating is a process that involves forcing printing ink through a nozzle, resulting in a fine aerosol	Cost-effectiveness, versatility, thickness range, and processing speed		Non-uniformity of deposited films. Extra resources and cost are incurred to improve the deposited films
Physical vapour deposition				
Pulsed laser Deposition	A high-power pulsed laser beam is focused to impact a target of the desired composition in this physical vapor deposition technique. The material is then vaporized and deposited on a substrate facing the target as a thin film	The main benefit is that it is conceptually simple and versatile	maximum pulse energy, wavelength	Slow deposition rate, does not support large area deposition, target fragmentation in depositing some materials
Molecular beam epitaxy	MBE is used to create thin films of diverse materials layer by layer	Clean technique for obtaining films		The ultra-high vacuum (UHV) environment is demanding
Sputtering	Sputter deposition is a technique for depositing thin films that involves sputtering material from a 'target' onto a 'substrate.' Sputtering is a physical process in which atoms in a solid-state (target) are bombarded by energetic ions, causing them to be freed and flow into the gas phase (mainly noble gas ions)	Deposit a wide range of metal and metal oxide nanoparticles (NPs) and nanoclusters (NCs), as well as insulators, alloys, composites, and organic molecules		Sputtered films have an increased risk of contamination

(continued)

Table 2.1 (continued)

Deposition type	Description	Merit	Parameters	Demerit
Thermal evaporation	It's one of the most basic types of PVD, and it involves using a resistive heat source to evaporate a solid substance in a vacuum to make a thin film. In a high vacuum chamber, the material is heated until vapor pressure is achieved	Good purity because of low pressure	The main parameter is source material's vapor pressure at evaporation temperature	First, pollution from the boat's outgassing or evaporation, as well as the heating circuits used during the high-temperature evaporation process. Second, obtaining a suitable boat material is difficult, especially when evaporating refractory materials. Also, Poor step coverage, harder forming of alloys, and decreased throughput due to low vacuum
Sputtering	Sputtering is a process in which small particles of a solid material are ejected off its surface after being hit by intense plasma or gas particles. It can be an undesirable source of wear in precision components because it occurs naturally in space. It is, nevertheless, utilized in research and industry to execute accurate etching, carry out analytical methods, and deposit thin film layers in the creation of optical coatings, semiconductor devices, and nanotechnology products because of its ability to work on extremely small layers of material. It's a type of physical vapor deposition	Uniform covering across a large area with little contaminants	operating pressure, direct current power, and substrate temperature are all factors to consider	The bombardment's target region is too tiny, and the rate of deposition is often poor

several advantages, including the ability to deposit a wide range of materials, including oxides, metals, semiconductors, and polymers, with precise control over thickness, composition, and crystallinity. This technique is widely used in research and industrial applications for fabricating functional thin films for electronic, photonic, magnetic, and biomedical devices.

Principles of Pulsed Laser Deposition (PLD)
The principles of Pulsed Laser Deposition involve several key steps: laser ablation, plasma formation, and thin film deposition. Laser ablation is initiated by focusing a high-energy pulsed laser onto the surface of a target material. The intense laser pulse vaporizes and ionizes atoms or molecules from the target surface, creating a plasma plume consisting of energetic species, such as ions, atoms, and nanoparticles. The plasma plume expands away from the target and impinges onto a substrate placed in close proximity, where the deposited species condense to form a thin film. The film growth occurs through a combination of physical processes, including nucleation, condensation, and surface diffusion, leading to the formation of a dense, adherent thin film with desired properties.

Methods of Pulsed Laser Deposition (PLD)
Pulsed Laser Deposition can be performed using various configurations and setups, depending on the specific requirements of the deposition process and the desired properties of the thin films. The most common setup involves a vacuum chamber equipped with a target holder, a substrate holder, and a pulsed laser system. The target material, typically a solid pellet or thin film, is mounted on the target holder, while the substrate is placed on the substrate holder. The vacuum chamber is evacuated to high vacuum levels to minimize gas-phase reactions and contamination during deposition. A high-energy pulsed laser, such as an excimer laser or a Nd:YAG laser, is focused onto the target surface to initiate laser ablation and plasma formation. The substrate is positioned in the path of the plasma plume, where thin film deposition occurs. The deposition parameters, including laser fluence, pulse repetition rate, substrate temperature, and background gas pressure, can be adjusted to control the properties of the deposited thin films, such as thickness, composition, crystallinity, and morphology.

Applications of Pulsed Laser Deposition (PLD)
Pulsed Laser Deposition finds widespread applications across various fields, owing to its versatility, precision, and flexibility in depositing thin films of diverse materials with tailored properties. In the field of electronics, PLD is used for fabricating thin film transistors, capacitors, resistors, and interconnects for integrated circuits and microelectronic devices. These thin films exhibit high electrical conductivity, low resistivity, and excellent adhesion to substrates, making them suitable for applications in semiconductor manufacturing, flexible electronics, and display technologies. In the field of optics and photonics, PLD is employed for fabricating optical coatings, waveguides, photonic crystals, and

metamaterials with tailored optical properties, such as refractive index, absorption, and dispersion. These thin films exhibit high optical transparency, low optical losses, and precise control over spectral characteristics, enabling the development of advanced optical components for applications in lasers, sensors, telecommunications, and photonic integrated circuits. Moreover, in the field of energy research, PLD is utilized for fabricating thin film materials for energy harvesting, storage, and conversion devices, such as solar cells, fuel cells, batteries, and thermoelectric devices. These thin films exhibit enhanced energy conversion efficiency, electrochemical stability, and mechanical durability, enabling the development of high-performance energy technologies for sustainable energy production and storage. Furthermore, in the field of magnetic and spintronics, PLD is used for fabricating thin film materials with tailored magnetic properties, such as magnetization, coercivity, and magnetic anisotropy. These thin films exhibit unique magnetic behaviors, such as ferromagnetism, antiferromagnetism, and spin polarization, making them suitable for applications in magnetic data storage, magnetic sensors, and spintronic devices. Additionally, in the field of biomedical engineering, PLD is employed for fabricating biocompatible thin films for medical implants, drug delivery systems, and biosensors. These thin films exhibit excellent biocompatibility, low cytotoxicity, and controlled release kinetics, enabling the development of advanced medical devices and therapies for diagnostics, drug delivery, and tissue engineering.

Advancements in Pulsed Laser Deposition (PLD)
Recent advancements in Pulsed Laser Deposition have focused on enhancing film quality, deposition rate, and process efficiency through innovations in laser technology, target preparation, and substrate engineering. Advanced laser systems with higher pulse energies, shorter pulse durations, and higher repetition rates enable more efficient laser ablation and plasma generation, leading to higher deposition rates and improved film quality. Moreover, advancements in target fabrication techniques, such as laser machining, chemical etching, and sputtering, enable the production of high-purity, homogeneous targets with tailored microstructures and compositions, resulting in improved film properties and performance. Furthermore, innovations in substrate engineering, such as substrate heating, substrate biasing, and substrate rotation, enable better control over film growth kinetics, crystallographic orientation, and defect density, leading to the fabrication of thin films with enhanced structural, electrical, and optical properties. Pulsed Laser Deposition (PLD) is a versatile and powerful technique for depositing thin films of various materials with precise control over composition, structure, and properties. With applications spanning electronics, optics, energy, biomedical engineering, and beyond, PLD continues to drive innovation and advancement across different disciplines. As research and development efforts continue to progress, PLD is poised to remain a key technology for the fabrication of functional thin films and devices with tailored properties and applications in diverse scientific and technological domains.

Molecular beam epitaxy (MBE)

Molecular Beam Epitaxy (MBE) stands as a pivotal technique in the fabrication of semiconductor materials with exceptional precision and control. This comprehensive discourse ventures into the fundamentals, methodologies, applications, and advancements of Molecular Beam Epitaxy (MBE), unraveling its profound significance and expansive potential across various scientific and technological realms. Molecular Beam Epitaxy (MBE) represents a sophisticated thin film deposition technique renowned for its unparalleled precision in creating crystalline layers of semiconductor materials with atomic-scale accuracy. Originating in the 1960s, MBE has emerged as a cornerstone in materials science, offering a versatile platform for fabricating heterostructures, quantum wells, superlattices, and nanostructures with tailored properties and functionalities. By epitaxially growing atomically smooth layers of diverse materials, MBE enables the development of advanced electronic, optoelectronic, and photonic devices with unprecedented performance characteristics.

Principles of Molecular Beam Epitaxy (MBE)

At its core, Molecular Beam Epitaxy (MBE) operates on the principle of epitaxial growth, wherein thin films of crystalline materials are deposited on a substrate surface, maintaining the structural coherence and lattice alignment of the underlying substrate. In MBE, the epitaxial growth process occurs under ultrahigh vacuum conditions, ensuring precise control over film thickness, composition, and crystal quality. The technique relies on the flux of molecular or atomic beams directed towards the substrate surface, where they condense to form thin films through surface adsorption and incorporation processes. By precisely controlling the flux and energy of the incident beams, researchers can manipulate the growth kinetics, surface morphology, and crystal structure of the deposited layers, facilitating the fabrication of complex semiconductor structures with tailored properties.

Methods of Molecular Beam Epitaxy (MBE)

Molecular Beam Epitaxy (MBE) encompasses several variations and methodologies tailored to specific material systems and device applications. The two primary configurations are known as molecular beam epitaxy (MBE) and atomic layer molecular beam epitaxy (ALMBE), each offering distinct advantages and capabilities. In traditional MBE, molecular or atomic beams of the constituent elements are directed towards the substrate surface, where they react and incorporate into the growing film in a continuous fashion. This approach enables the epitaxial growth of multilayer structures with precise control over layer thickness, composition, and doping profiles. In contrast, atomic layer molecular beam epitaxy (ALMBE) employs a sequential, layer-by-layer deposition process, where individual atomic layers are grown sequentially by alternately exposing the substrate surface to precursor beams of the constituent elements. After each deposition step, the substrate surface is purged with an inert gas to remove excess reactants and reaction by-products, ensuring precise control over layer thickness and composition. ALMBE offers enhanced control over interface sharpness, interfacial roughness, and defect density, making it well-suited for fabricating high-quality heterostructures, quantum wells, and superlattices with atomically precise interfaces.

Applications of Molecular Beam Epitaxy (MBE)

Molecular Beam Epitaxy (MBE) finds diverse applications across various fields, including semiconductor electronics, optoelectronics, photonics, and quantum technologies. In the field of semiconductor electronics, MBE is utilized for fabricating high-performance transistors, diodes, and integrated circuits with superior electrical properties and device characteristics. By epitaxially growing ultrathin layers of semiconductor materials, such as silicon, gallium arsenide, and indium phosphide, MBE enables the realization of advanced device architectures, such as heterojunction bipolar transistors (HBTs), high-electron mobility transistors (HEMTs), and quantum cascade lasers (QCLs), with enhanced speed, power efficiency, and reliability. In the realm of optoelectronics and photonics, MBE plays a crucial role in fabricating semiconductor-based light-emitting diodes (LEDs), lasers, photodetectors, and photonic integrated circuits (PICs) for applications in telecommunications, sensing, and imaging. By epitaxially growing semiconductor heterostructures, such as quantum wells, quantum dots, and waveguides, MBE enables the development of advanced photonic devices with tailored optical properties, such as wavelength tunability, spectral purity, and coherence. Moreover, MBE is instrumental in the realization of emerging quantum technologies, such as quantum computing, quantum cryptography, and quantum sensing, by enabling the epitaxial growth of semiconductor-based quantum devices, such as quantum dots, spin qubits, and topological insulators, with precise control over spin and charge states. Furthermore, in the field of materials science and nanotechnology, MBE is employed for fabricating novel nanostructures, such as nanowires, nanorods, and nanotubes, with tailored dimensions, compositions, and properties. By epitaxially growing semiconductor nanostructures on predefined substrates, such as silicon, sapphire, and graphene, MBE enables the controlled synthesis of nanoscale materials with unique electronic, optical, and mechanical characteristics. These nanostructures hold promise for applications in nanoelectronics, nanophotonics, nanomagnetics, and nanomedicine, offering opportunities for developing next-generation devices and technologies with enhanced performance and functionality.

Advancements in Molecular Beam Epitaxy (MBE)

Recent advancements in Molecular Beam Epitaxy (MBE) have focused on enhancing film quality, growth rate, and process scalability through innovations in material sources, substrate engineering, and process control. Advanced molecular and atomic sources, such as solid-state effusion cells, gas-phase crackers, and plasma sources, enable the precise control of beam fluxes and compositions, leading to improved film stoichiometry, uniformity, and reproducibility. Moreover, innovations in substrate preparation techniques, such as surface cleaning, surface reconstruction, and surface passivation, enhance the nucleation and growth kinetics of epitaxial films, resulting in reduced defect densities, improved crystalline quality, and enhanced device performance. Furthermore, developments in in-situ monitoring and characterization techniques, such as reflection high-energy electron diffraction (RHEED), scanning tunneling microscopy (STM), and X-ray photoelectron spectroscopy (XPS), enable real-time monitoring of film growth kinetics, surface morphology, and chemical composition during epitaxial growth. These techniques provide valuable insights into the nucleation mechanisms, growth kinetics, and surface reactions involved in MBE processes, facilitating process optimization and defect control. Additionally, advancements in substrate engineering, such as substrate patterning, substrate heating, and substrate rotation, enable better control over film morphology, strain relaxation, and lattice mismatch, leading to the fabrication of heterostructures with tailored properties and functionalities.

Molecular Beam Epitaxy (MBE) stands as a transformative technique in materials science and semiconductor technology, offering unparalleled precision and control over the epitaxial growth of thin films with tailored properties and functionalities. With applications spanning electronics, optoelectronics, photonics, and quantum technologies, MBE continues to drive innovation and advancement across various scientific and technological domains. As research and development efforts continue to progress, MBE is poised to remain a key technology for fabricating advanced materials and devices with enhanced performance and functionality, paving the way for future breakthroughs in science, engineering, and technology.

Chemical Bath Deposition (CBD):

Chemical Bath Deposition (CBD) stands as a crucial technique in the fabrication of thin films and coatings, particularly for materials like semiconductors, oxides, and sulfides. This comprehensive exploration delves into the principles, methods, applications, and advancements of Chemical Bath Deposition, elucidating its significance and potential across various scientific and technological domains. Chemical Bath Deposition (CBD) is a widely utilized method for synthesizing thin films and coatings of various materials, including semiconductors, metal oxides, and metal sulfides. This technique involves the immersion of a substrate into a chemical solution, typically containing metal ions and complexing agents, under controlled temperature and pH conditions. Through chemical reactions occurring in the solution, thin films of the desired material are deposited onto the substrate surface, offering advantages such as simplicity, cost-effectiveness, and scalability. CBD finds applications in diverse fields, including photovoltaics, optoelectronics, sensors, and protective coatings, owing to its versatility and ease of implementation.

Principles of Chemical Bath Deposition (CBD)
The principles underlying Chemical Bath Deposition (CBD) stem from the complex chemical reactions that occur between precursor species in solution and the substrate surface. The process typically involves the following steps.

1. **Nucleation**: Initially, nucleation sites are formed on the substrate surface, where the deposition process begins. These nucleation sites serve as the initial points for the growth of the thin film.
2. **Adsorption**: Metal ions and complexing agents present in the solution are adsorbed onto the substrate surface, forming a precursor layer that serves as the building blocks for the thin film.
3. **Growth**: Through subsequent chemical reactions, the adsorbed precursor species undergo transformation, leading to the growth of the thin film on the substrate surface. This growth process can be controlled by adjusting parameters such as solution composition, temperature, pH, and deposition time.
4. **Nucleation and growth continue**: The deposition process continues until the desired thickness and properties of the thin film are achieved. Additional layers can be deposited by repeating the deposition process or by introducing additional precursor solutions.

Methods of Chemical Bath Deposition (CBD)

Chemical Bath Deposition (CBD) encompasses various methodologies and strategies tailored to specific material systems and deposition requirements. The choice of method depends on factors such as the desired film properties, substrate characteristics, and intended applications. Common variants of CBD.

1. **Single-step deposition**: In this approach, the entire deposition process, from nucleation to film growth, occurs in a single chemical bath solution. This method offers simplicity and ease of implementation but may lack precise control over film properties.
2. **Sequential deposition**: This method involves the sequential deposition of different layers or compositions onto the substrate surface by immersing the substrate in multiple chemical bath solutions consecutively. Sequential deposition allows for precise control over film composition, thickness, and properties, enabling the fabrication of complex multilayer structures.
3. **Template-assisted deposition**: In this approach, templates or substrates with predefined patterns or surface features are used to guide the nucleation and growth of the thin film. Template-assisted deposition allows for the fabrication of nanostructured or patterned thin films with controlled morphology and spatial arrangement.

Applications of Chemical Bath Deposition (CBD)

Chemical Bath Deposition (CBD) finds diverse applications across various fields, owing to its versatility, cost-effectiveness, and scalability. In the field of photovoltaics, CBD is utilized for fabricating thin film solar cells based on materials such as cadmium telluride (CdTe), copper indium gallium selenide (CIGS), and zinc oxide (ZnO). These thin film solar cells offer advantages such as low-cost manufacturing, lightweight, and flexibility, making them suitable for applications in building-integrated photovoltaics (BIPV), portable electronics, and off-grid power generation. In the field of optoelectronics, CBD is employed for fabricating transparent conducting oxides (TCOs), such as indium tin oxide (ITO) and fluorine-doped tin oxide (FTO), used in displays, touchscreens, and solar panels. CBD enables the deposition of TCO thin films with high optical transparency, low electrical resistivity, and good adhesion to substrates, making them ideal for applications requiring transparent and conductive coatings. Furthermore, in the field of sensors and detectors, CBD is utilized for fabricating thin film materials for gas sensors, chemical sensors, and biosensors. These thin films exhibit high surface area, chemical selectivity, and sensitivity to specific analytes, enabling the detection and quantification of gases, volatile organic compounds (VOCs), and biomolecules in environmental, medical, and industrial settings. Moreover, in the field of protective coatings and corrosion resistance, CBD is employed for depositing thin films of materials such as metal oxides, metal sulfides, and polymers onto substrates to enhance their mechanical, chemical, and thermal properties. These thin films provide barrier protection against moisture, corrosion, and abrasion, extending the lifespan and durability of coated materials in harsh environments.

Advancements in Chemical Bath Deposition (CBD)
Recent advancements in Chemical Bath Deposition (CBD) have focused on improving deposition efficiency, film quality, and process scalability through innovations in precursor chemistry, solution engineering, and process control. Advanced precursor materials, such as metal complexes, nanoparticle dispersions, and organic–inorganic hybrids, offer enhanced stability, reactivity, and solubility in solution, enabling the deposition of thin films with tailored properties and functionalities. Moreover, developments in solution engineering, such as pH adjustment, temperature control, and additive incorporation, optimize the deposition conditions and kinetics, leading to improved film uniformity, adhesion, and crystallinity. Furthermore, innovations in process control, such as automation, monitoring, and feedback control, enable real-time adjustment of deposition parameters, ensuring reproducibility and reliability in large-scale production. Additionally, advancements in post-deposition treatments, such as annealing, sintering, and surface modification, enhance the properties and performance of deposited thin films, enabling the fabrication of functional materials for advanced applications. These advancements pave the way for the continued growth and diversification of Chemical Bath Deposition (CBD) as a key technology for thin film fabrication in various scientific and technological domains.

Chemical Bath Deposition (CBD) stands as a versatile and cost-effective technique for depositing thin films and coatings of diverse materials with tailored properties and functionalities. With applications spanning photovoltaics, optoelectronics, sensors, protective coatings, and beyond, CBD continues to drive innovation and advancement across various scientific and technological domains. As research and development efforts continue to progress, CBD is poised to remain a key technology for thin film fabrication, offering opportunities for developing advanced materials and devices for diverse applications.

In summary, deposition techniques for thin films play a crucial role in controlling film properties and enabling a wide range of applications across various industries. By understanding the principles and capabilities of different deposition techniques, researchers can tailor thin films to meet specific requirements and drive innovation in materials science and engineering.

2.5   Emerging Trends in Thin Films

Thin films have been at the forefront of technological innovation for decades, finding applications in diverse fields ranging from electronics and optics to energy and healthcare. As technology continues to evolve, new trends are emerging in thin film research and development, driving advancements in materials science and engineering.

2.5.1   Nanotechnology Integration

One of the most significant emerging trends in thin films is the integration of nanotechnology, which involves the manipulation of materials at the nanoscale. Nanotechnology offers unprecedented control over material properties, allowing researchers to design thin films with enhanced functionalities and performance. For example, nanoscale thin films can exhibit unique optical, electronic, and mechanical properties that differ from their bulk counterparts. This opens up opportunities for applications such as nanoelectronics, nanophotonics, and nanomedicine, where precise control over material properties is essential.

One of the most significant emerging trends in thin films is the integration of nanotechnology, which involves the manipulation of materials at the nanoscale. This convergence of thin films and nanotechnology opens up new possibilities for developing advanced materials with tailored properties, enhanced functionalities, and novel applications across various industries.

1. **Enhanced Material Properties**: Integrating nanotechnology into thin films enables precise control over material properties at the nanoscale, resulting in enhanced mechanical, electrical, optical, and thermal characteristics. For example, nanomaterials such as nanoparticles, nanowires, and nanotubes can be incorporated into thin film coatings to improve hardness, strength, conductivity, transparency, and thermal stability. This enhancement in material properties opens up opportunities for developing high-performance coatings, functional surfaces, and advanced composites with superior mechanical and functional properties.

2. **Tailored Functionalities**: Nanotechnology allows for the engineering of thin film materials with tailored functionalities to meet specific application requirements. Functional nanomaterials, such as quantum dots, nanocomposites, and nanophosphors, can be integrated into thin films to impart unique properties such as luminescence, catalytic activity, self-cleaning, or anti-fouling capabilities. Thin film coatings with tailored functionalities find applications in areas such as optics, electronics, energy storage, biomedical devices, and environmental remediation, enabling innovative solutions for diverse industry challenges.

3. **Miniaturization and Nanoscale Devices**: The integration of nanotechnology into thin films enables the development of miniaturized devices and components with nanoscale dimensions. Nanoscale thin films, such as thin film transistors (TFTs), nanosensors, and nanoelectromechanical systems (NEMS), can be fabricated using nanolithography, atomic layer deposition (ALD), or molecular beam epitaxy (MBE) techniques. These nanoscale devices offer advantages such as high sensitivity, fast response times, and low power consumption, making them suitable for applications in electronics, sensing, healthcare, and information technology.

4. **Nanocoatings for Surface Engineering**: Nanotechnology-based thin film coatings offer advanced surface engineering solutions for enhancing the performance, durability, and functionality of materials and surfaces. Nanocoatings can be applied to substrates such as metals, ceramics, polymers, and textiles to provide properties such as scratch resistance, corrosion protection, water repellency, and antimicrobial activity. These nanocoatings find applications in automotive, aerospace, electronics, healthcare, and consumer goods industries, where surface properties play a critical role in product performance and longevity.

5. **Nanoparticle Thin Films for Energy Applications**: Nanoparticle-based thin films hold promise for energy-related applications such as solar cells, batteries, fuel cells, and energy storage devices. Nanoparticle thin films can be deposited onto substrates to form active layers for solar energy conversion, electrocatalytic electrodes for fuel cells, or electrode materials for batteries and supercapacitors. Nanotechnology enables the engineering of thin film materials with optimized morphologies, compositions, and interfaces to enhance energy conversion efficiencies, charge storage capacities, and device lifetimes, contributing to the advancement of renewable energy technologies.

6. **Nanofabrication Techniques for Thin Films**: Nanotechnology-driven fabrication techniques enable precise control over thin film deposition processes, enabling the fabrication of nanoscale structures and patterns with high resolution and fidelity. Techniques such as electron beam lithography, focused ion beam milling, and nanoimprint lithography allow for the creation of nanoscale features on thin film surfaces with sub-micron and sub-10-nm resolution. These nanofabrication techniques enable the development of next-generation electronic devices, sensors, photonic devices, and microfluidic systems with unprecedented precision and performance.

In summary, the integration of nanotechnology into thin films represents a significant emerging trend with wide-ranging implications for materials science, engineering, and technology innovation. This convergence enables the development of advanced materials, devices, and coatings with tailored properties, enhanced functionalities, and novel applications across diverse industries, driving forward the advancement of nanotechnology-enabled solutions in the twenty-first century.

2.5.2 Flexible Electronics

Another emerging trend in thin films is the development of flexible electronics, which involves the fabrication of electronic devices on flexible substrates such as plastics, polymers, and paper. Thin film technology plays a crucial role in enabling the flexibility and conformability of electronic components, allowing for the creation of bendable, stretchable, and wearable devices. Flexible electronics have applications in areas such as healthcare monitoring, wearable computing, and smart textiles, where traditional rigid electronics are impractical or cumbersome. Thin film transistors, sensors, and interconnects are key components of flexible electronic devices, offering high performance in a lightweight and flexible form factor.

Another emerging trend in thin films is the development of flexible electronics, which involves the fabrication of electronic devices on flexible substrates such as plastics, polymers, and paper. This trend is driven by the growing demand for lightweight, portable, and conformable electronic devices with bendable, stretchable, and foldable form factors. Flexible electronics offer numerous advantages over traditional rigid electronics, including enhanced mechanical flexibility, durability, and versatility, making them ideal for applications in wearable technology, healthcare monitoring, smart packaging, and flexible displays.

1. **Flexible Substrates and Thin Film Deposition**: Flexible electronics rely on thin film deposition techniques to fabricate electronic components and devices on flexible substrates. Thin film materials such as organic semiconductors, metal oxides, and conductive polymers can be deposited onto flexible substrates using techniques such as vacuum evaporation, sputtering, chemical vapor deposition (CVD), or inkjet printing. These thin film materials form the active layers of electronic devices such as transistors, sensors, and displays, enabling flexibility and conformability without sacrificing performance.
2. **Wearable Technology and Health Monitoring**: Flexible electronics are driving innovation in wearable technology and health monitoring devices, enabling the development of lightweight, comfortable, and unobtrusive solutions for continuous health monitoring and fitness tracking. Flexible sensors, such as strain sensors, temperature sensors, and bioelectronic sensors, can be integrated into wearable garments, patches,

or accessories to monitor vital signs, detect physiological parameters, and track physical activity in real-time. Thin film electronics enable the fabrication of flexible sensor arrays and circuits that conform to the contours of the body, ensuring comfort and reliability during wear.

3. **Smart Packaging and Internet of Things (IoT)**: Flexible electronics are revolutionizing the packaging industry by enabling the integration of smart functionalities into flexible and conformable packaging materials. Thin film sensors, RFID tags, and printed electronics can be embedded into flexible packaging films and labels to monitor product freshness, track shipment routes, and provide interactive consumer experiences. Flexible electronics also play a key role in the Internet of Things (IoT) ecosystem by enabling the development of connected sensors, smart tags, and wearable devices that seamlessly integrate into everyday objects and environments.

4. **Foldable and Rollable Displays**: Thin film technologies enable the fabrication of foldable and rollable displays for next-generation electronic devices such as smartphones, tablets, and e-readers. Flexible display technologies, such as organic light-emitting diodes (OLEDs) and electronic paper (e-paper), can be deposited onto flexible substrates to create lightweight, bendable, and durable display panels that can be folded, rolled, or curved without damage. These flexible displays offer advantages such as improved portability, ruggedness, and energy efficiency, paving the way for innovative form factors and user experiences in consumer electronics.

5. **Stretchable and Conformable Electronics**: Thin film electronics enable the development of stretchable and conformable electronic devices that can bend, stretch, and conform to irregular surfaces. Stretchable electronics use elastomeric substrates and stretchable interconnects to accommodate mechanical deformations without compromising electronic performance. These stretchable devices find applications in wearable electronics, medical implants, and soft robotics, where flexibility and conformability are essential for integration into complex and dynamic environments.

6. **Printed and Disposable Electronics**: Thin film printing techniques, such as inkjet printing, screen printing, and aerosol jet printing, enable the rapid and cost-effective fabrication of printed and disposable electronic devices on flexible substrates. Printed electronics allow for the mass production of electronic circuits, sensors, and displays on roll-to-roll or sheet-to-sheet printing platforms, reducing manufacturing costs and enabling scalable production. These printed electronic devices find applications in disposable medical sensors, electronic labels, and low-cost consumer electronics, democratizing access to electronic technology and reducing electronic waste.

In summary, the development of flexible electronics represents a transformative trend in thin film technology, enabling the fabrication of lightweight, portable, and conformable electronic devices with diverse applications across industries. As research and development efforts continue to advance, flexible electronics hold promise for revolutionizing the way we interact with technology, opening up new opportunities for innovation, connectivity, and user experience in the digital age.

2.5.3 Energy Harvesting and Storage

Thin films are also driving advancements in energy harvesting and storage technologies, offering efficient solutions for generating and storing renewable energy. Thin film solar cells, for example, provide lightweight and cost-effective alternatives to traditional silicon-based photovoltaic devices. Materials such as cadmium telluride (CdTe), copper indium gallium selenide (CIGS), and perovskite thin films are being investigated for their potential in thin film solar cells, offering high efficiency and scalability. Additionally, thin film batteries and supercapacitors are being developed for energy storage applications, offering high energy density, fast charging rates, and long cycle life. These advancements in thin film energy technologies hold promise for addressing global energy challenges and accelerating the transition to sustainable energy sources.

Thin films are indeed driving advancements in energy harvesting and storage technologies, offering efficient solutions for generating and storing renewable energy. This application of thin films plays a crucial role in addressing global energy challenges, reducing dependence on fossil fuels, and promoting sustainable development.

1. **Photovoltaic Thin Films**: Photovoltaic (PV) thin films, such as amorphous silicon (a-Si), cadmium telluride (CdTe), and copper indium gallium selenide (CIGS), are used to convert sunlight into electricity. These thin film solar cells offer advantages such as lightweight, flexibility, and scalability, making them suitable for a wide range of applications, including building-integrated photovoltaics (BIPV), portable electronics, and off-grid power systems. Thin film solar panels can be deposited onto flexible substrates using roll-to-roll manufacturing techniques, enabling low-cost, high-throughput production and deployment.

2. **Thin Film Batteries and Energy Storage**: Thin film batteries offer compact, lightweight, and high-performance energy storage solutions for various applications, including consumer electronics, medical devices, and electric vehicles. Thin film battery technologies, such as lithium-ion batteries (LIBs) and solid-state batteries, utilize thin film electrodes, electrolytes, and separators to store and deliver electrical energy. These thin film batteries offer advantages such as fast charging, high energy density, and long cycle life, making them ideal for portable and wearable electronics, as well as emerging applications in electric vehicles and grid-scale energy storage.

3. **Thin Film Supercapacitors and Energy Harvesters**: Thin film supercapacitors and energy harvesters offer efficient energy storage and harvesting solutions for applications requiring rapid energy release or capture. Thin film supercapacitors use porous electrodes and electrolytes to store electrical charge electrostatically, offering high

power density, rapid charge–discharge rates, and long cycle life. Thin film energy harvesters utilize piezoelectric, thermoelectric, or photovoltaic materials to convert mechanical, thermal, or solar energy into electrical energy, providing sustainable power sources for wireless sensors, IoT devices, and wearable electronics.

4. **Hybrid Thin Film Energy Systems**: Hybrid thin film energy systems combine multiple energy harvesting and storage technologies to maximize energy capture, storage, and utilization efficiency. For example, thin film solar panels can be integrated with thin film batteries and supercapacitors to create self-powered systems for remote monitoring, environmental sensing, and IoT applications. These hybrid energy systems leverage complementary energy sources and storage technologies to provide reliable, autonomous operation in off-grid or intermittent power supply environments.

5. **Thin Film Energy Conversion and Efficiency**: Thin films are also utilized in energy conversion and efficiency enhancement technologies, such as thermoelectric generators, transparent conductive coatings, and spectral selective absorbers. Thin film thermoelectric materials can convert waste heat into electricity, enabling energy recovery and efficiency improvements in industrial processes, automotive exhaust systems, and electronic devices. Transparent conductive thin films, such as indium tin oxide (ITO) or graphene, are used in solar cells, displays, and smart windows to improve light transmission and electrical conductivity, enhancing device performance and energy efficiency.

6. **Emerging Thin Film Energy Technologies**: Ongoing research and development efforts are focused on advancing emerging thin film energy technologies, such as perovskite solar cells, solid-state batteries, and thin film fuel cells. Perovskite thin film solar cells offer high efficiency, low-cost alternatives to traditional silicon-based solar cells, with potential applications in building-integrated photovoltaics and portable electronics. Solid-state thin film batteries use thin film electrolytes and electrodes to overcome safety and performance limitations of conventional liquid electrolyte batteries, enabling safer, longer-lasting energy storage solutions. Thin film fuel cells utilize electrochemical reactions to convert chemical energy into electrical energy, offering clean, efficient power generation for portable electronics, transportation, and distributed energy systems.

In summary, thin films play a vital role in driving advancements in energy harvesting and storage technologies, offering efficient solutions for generating, storing, and utilizing renewable energy. By leveraging the unique properties and capabilities of thin film materials, researchers and engineers continue to innovate and develop next-generation energy technologies that contribute to a more sustainable and resilient energy future.

2.5.4 Advancements in 2D Materials

The discovery and characterization of two-dimensional (2D) materials, such as graphene, transition metal dichalcogenides (TMDs), and black phosphorus, have opened up new avenues for thin film research and development. 2D materials exhibit unique electronic, optical, and mechanical properties due to their atomically thin nature, making them promising candidates for thin film applications. Thin film heterostructures composed of 2D materials offer opportunities for engineering novel electronic devices, sensors, and optoelectronic systems with unprecedented performance. Additionally, 2D materials are being explored for applications in flexible electronics, quantum computing, and nanophotonics, where their unique properties enable innovative functionalities and applications.

The discovery and characterization of two-dimensional (2D) materials have indeed opened up new avenues for thin film research and development, offering exciting opportunities for creating novel materials, devices, and applications with unique properties and functionalities. Graphene, transition metal dichalcogenides (TMDs), and black phosphorus are among the most notable 2D materials that have garnered significant attention from researchers and engineers worldwide.

1. **Graphene**: Graphene, a single layer of carbon atoms arranged in a two-dimensional honeycomb lattice, exhibits exceptional mechanical, electrical, thermal, and optical properties. Graphene's high carrier mobility, transparency, and mechanical flexibility make it an ideal candidate for a wide range of thin film applications, including transparent conductive coatings, flexible electronics, energy storage devices, and sensors. Graphene thin films can be synthesized using techniques such as chemical vapor deposition (CVD), epitaxial growth, or liquid-phase exfoliation, enabling scalable production and integration into various devices and systems.
2. **Transition Metal Dichalcogenides (TMDs)**: Transition metal dichalcogenides, such as molybdenum disulfide (MoS_2), tungsten diselenide (WSe_2), and tantalum disulfide (TaS_2), are layered materials composed of transition metal atoms sandwiched between chalcogen atoms in a hexagonal lattice structure. TMDs exhibit unique electronic, optical, and mechanical properties, including tunable bandgaps, strong light-matter interactions, and mechanical flexibility. TMD thin films find applications in optoelectronics, photovoltaics, catalysis, and flexible electronics, where their 2D nature and layer-dependent properties enable precise control over device performance and functionality.
3. **Black Phosphorus**: Black phosphorus, also known as phosphorene, is a layered material consisting of phosphorus atoms arranged in a puckered honeycomb lattice similar to graphene. Black phosphorus exhibits anisotropic electrical and optical properties, with a direct bandgap that can be tuned by varying the number of layers. Black phosphorus thin films offer advantages such as high carrier mobility, broadband light

absorption, and environmental stability, making them promising candidates for applications in transistors, photodetectors, photovoltaics, and flexible electronics. Thin film deposition techniques, such as mechanical exfoliation, liquid-phase exfoliation, and chemical vapor deposition, enable the fabrication of black phosphorus thin films with precise control over thickness, morphology, and crystal structure.

4. **Emerging 2D Materials**: In addition to graphene, TMDs, and black phosphorus, researchers are exploring a wide range of emerging 2D materials with unique properties and functionalities. These include materials such as hexagonal boron nitride (h-BN), transition metal carbides and nitrides (MXenes), silicene, germanene, and stanene. Each of these 2D materials offers distinct advantages and potential applications in areas such as electronics, photonics, spintronics, catalysis, and energy storage. By combining different 2D materials into heterostructures and hybrid materials, researchers can create novel thin film architectures with tailored properties and enhanced performance for specific applications.

5. **Characterization and Integration**: The characterization and integration of 2D materials into thin film structures require advanced techniques and methodologies to assess their structural, electronic, and optical properties at the nanoscale. Techniques such as scanning probe microscopy (SPM), transmission electron microscopy (TEM), Raman spectroscopy, and X-ray diffraction (XRD) provide valuable insights into the atomic structure, band structure, and defects of 2D materials. Integration of 2D materials into thin film devices often involves layer-by-layer assembly, chemical functionalization, or van der Waals epitaxy techniques to control the orientation, alignment, and interface properties of thin film structures.

In summary, the discovery and characterization of 2D materials have revolutionized thin film research and development, offering unprecedented control over material properties, device performance, and functionality. Graphene, TMDs, black phosphorus, and emerging 2D materials hold promise for creating next-generation thin film devices and systems with enhanced capabilities and applications in diverse fields, from electronics and photonics to energy and healthcare. Continued research and innovation in 2D materials are expected to drive further advancements in thin film technologies and unlock new opportunities for scientific discovery and technological innovation in the years to come.

2.5.5 Additive Manufacturing Applications

Additive manufacturing, also known as 3D printing, is emerging as a disruptive technology for thin film fabrication, offering new possibilities for rapid prototyping and customization. Additive manufacturing techniques such as inkjet printing, aerosol deposition, and laser-induced forward transfer (LIFT) enable the direct deposition of functional materials onto substrates, eliminating the need for traditional vacuum deposition processes. Additive manufacturing allows for the fabrication of complex thin film structures with precise control over composition, morphology, and thickness. This flexibility opens up opportunities for applications such as printed electronics, flexible sensors, and biomedical implants, where customized thin film devices can be produced on-demand and at low cost.

Additive manufacturing, commonly known as 3D printing, is indeed emerging as a disruptive technology for thin film fabrication, opening up new possibilities for rapid prototyping, customization, and innovation. While traditional thin film deposition techniques such as physical vapor deposition (PVD) and chemical vapor deposition (CVD) offer high precision and uniformity, they often involve complex processes, expensive equipment, and limited flexibility in terms of design and material selection. 3D printing, on the other hand, enables layer-by-layer fabrication of complex three-dimensional structures directly from digital designs, offering advantages such as design freedom, rapid iteration, and customization.

1. **Rapid Prototyping and Iterative Design**: Additive manufacturing accelerates the thin film fabrication process by enabling rapid prototyping and iterative design cycles. Designers and engineers can quickly translate digital concepts into physical prototypes using 3D printing, allowing for faster evaluation of design concepts, material properties, and performance characteristics. This iterative approach reduces time-to-market, minimizes development costs, and fosters innovation in thin film fabrication, particularly in research and development settings where rapid iteration and experimentation are critical.

2. **Complex Geometries and Customization**: Additive manufacturing enables the fabrication of thin film structures with complex geometries, intricate features, and customized designs that are difficult or impossible to achieve using traditional deposition techniques. 3D printing allows for the creation of highly customized thin film components tailored to specific applications, requirements, and user preferences. Designers can optimize thin film structures for mechanical strength, thermal conductivity, or optical properties by adjusting parameters such as layer thickness, infill density, and material composition, offering unparalleled design flexibility and customization capabilities.

3. **Multi-material Integration**: Additive manufacturing techniques such as multi-material 3D printing enable the integration of multiple materials into thin film structures, expanding the range of functionalities and applications. By selectively depositing different materials layer by layer, designers can create thin film composites, gradients, or hybrid structures with tailored properties and performance characteristics. This multi-material integration capability enables the fabrication of functional thin film devices, sensors, and actuators with diverse functionalities, such as flexible electronics, biomedical implants, and smart textiles.

4. **In-Situ Deposition and Integration**: Additive manufacturing platforms can be augmented with in-situ thin film deposition capabilities, allowing for the direct integration of thin films into 3D printed structures during fabrication. Techniques such as aerosol jet printing, inkjet printing, or direct-write lithography can be combined with 3D printing to deposit functional thin film coatings, conductive traces, or electronic components onto 3D printed substrates in a single manufacturing step. This in-situ deposition approach streamlines the fabrication process, reduces assembly complexity, and enables the integration of thin film functionalities directly into complex 3D structures.

5. **On-Demand Manufacturing and Mass Customization**: Additive manufacturing facilitates on-demand manufacturing and mass customization of thin film products, eliminating the need for expensive tooling, molds, and inventory stockpiles. With 3D printing, thin film components can be produced on-demand in small batches or individually tailored to specific customer requirements, preferences, and specifications. This on-demand manufacturing model reduces lead times, minimizes waste, and enhances supply chain efficiency, particularly in industries such as consumer electronics, healthcare, and aerospace where customization and rapid delivery are critical.

6. **Sustainable Manufacturing Practices**: Additive manufacturing promotes sustainable manufacturing practices by minimizing material waste, energy consumption, and environmental impact compared to traditional manufacturing methods. 3D printing techniques such as powder bed fusion, material extrusion, and vat photopolymerization use only the amount of material needed to build the desired object, reducing material waste and raw material consumption. Additionally, additive manufacturing enables the use of recycled or biodegradable materials for thin film fabrication, further reducing the environmental footprint of manufacturing processes.

In summary, additive manufacturing is revolutionizing thin film fabrication by offering new possibilities for rapid prototyping, customization, and innovation. By leveraging the capabilities of 3D printing, designers and engineers can create complex, customized thin film structures with enhanced functionalities and performance characteristics, driving forward the advancement of thin film technology in diverse industries and applications.

2.5.6 Advanced Deposition Techniques

Advancements in thin film deposition techniques are driving innovation in materials science and engineering, enabling the fabrication of thin films with tailored properties and complex structures. Techniques such as atomic layer deposition (ALD), molecular beam epitaxy (MBE), and pulsed laser deposition (PLD) offer precise control over thin film growth at the atomic or molecular scale. These advanced deposition techniques enable the deposition of thin films on diverse substrates, including flexible and curved surfaces, opening up new possibilities for device integration and manufacturing. Additionally, novel deposition methods such as initiated chemical vapor deposition (iCVD) and layer-by-layer assembly offer opportunities for bottom-up synthesis of thin films with controlled nanostructures and functionalities. Advancements in thin film deposition techniques are indeed driving innovation in materials science and engineering, empowering researchers and engineers to fabricate thin films with tailored properties and complex structures. Thin film deposition techniques play a crucial role in various industries, including electronics, optics, energy, healthcare, and aerospace, where precise control over material composition, morphology, and structure is essential for achieving desired performance characteristics.

1. **Physical Vapor Deposition (PVD)**:

   Physical vapor deposition techniques, such as sputtering and evaporation, are widely used for depositing thin films onto substrates by evaporating or sputtering material atoms from a solid source onto a substrate surface. Advancements in PVD technology, such as magnetron sputtering, ion beam sputtering, and pulsed laser deposition (PLD), enable precise control over thin film thickness, uniformity, and composition. These techniques offer advantages such as high deposition rates, scalability, and compatibility with a wide range of materials, making them suitable for applications in electronics, optics, and coatings. Physical Vapor Deposition (PVD) is a versatile and widely used thin-film deposition technique that plays a crucial role in various industries, including electronics, optics, automotive, aerospace, and biomedical. PVD involves the deposition of thin films of materials onto substrates through physical processes such as evaporation or sputtering, without involving chemical reactions. This comprehensive discourse explores the principles, techniques, applications, and advancements in Physical Vapor Deposition, highlighting its significance in modern manufacturing and technology.

2. **Chemical Vapor Deposition (CVD)**:

   Chemical vapor deposition techniques involve the chemical reaction of precursor gases to deposit thin films onto substrates by forming chemical bonds between reactive species and substrate surfaces. Advancements in CVD technology, such as plasma-enhanced CVD (PECVD), atomic layer deposition (ALD), and metal–organic CVD (MOCVD), enable the deposition of thin films with precise control over thickness, stoichiometry, and crystallinity. These techniques offer advantages such as high uniformity, conformality, and purity, making them ideal for applications in semiconductor manufacturing, photovoltaics, and catalysis.

3. **Molecular Beam Epitaxy (MBE)**: Molecular beam epitaxy is a precision thin film deposition technique that involves the deposition of material atoms onto a substrate surface under ultra-high vacuum conditions to create epitaxial layers with atomic-level precision. Advancements in MBE technology enable the growth of epitaxial thin

films with tailored crystal structures, interfaces, and doping profiles, offering unparalleled control over material properties and device performance. MBE is widely used in semiconductor research, quantum computing, and materials discovery, where precise control over thin film growth is essential for exploring novel electronic and optical phenomena.

4. **Atomic Layer Deposition (ALD)**: Atomic layer deposition is a thin film deposition technique that enables the sequential, self-limiting adsorption of precursor molecules onto a substrate surface, allowing for precise control over thin film thickness and composition at the atomic level. Advancements in ALD technology enable the deposition of ultrathin films with sub-nanometer precision, making it ideal for applications such as semiconductor device fabrication, nanoelectronics, and surface engineering. ALD offers advantages such as high uniformity, conformality, and scalability, making it suitable for coating complex 3D structures and nanoscale devices.

5. **Solution-Based Deposition Techniques**: Solution-based deposition techniques, such as spin coating, dip coating, and inkjet printing, offer cost-effective and scalable methods for depositing thin films from liquid precursors onto substrates. Advancements in solution-based deposition technologies enable the fabrication of thin films with tailored properties, including composition, morphology, and functionality. These techniques find applications in organic electronics, photovoltaics, biomedical coatings, and flexible electronics, where solution-processable materials offer advantages such as flexibility, low-cost, and compatibility with large-area substrates.

6. **Hybrid and Combinatorial Deposition Approaches**: Hybrid and combinatorial deposition approaches combine multiple thin film deposition techniques to achieve complex thin film structures and compositions with enhanced functionalities. For example, hybrid approaches such as sputtering-assisted CVD (SACVD) or plasma-enhanced atomic layer deposition (PEALD) combine aspects of PVD and CVD to deposit thin films with improved properties and performance. Combinatorial deposition techniques, such as combinatorial ALD or combinatorial sputtering, enable high-throughput screening of thin film libraries to identify optimal compositions and processing conditions for specific applications.

Advancements in thin film deposition techniques are driving innovation in materials science and engineering, enabling the fabrication of thin films with tailored properties and complex structures. By leveraging the capabilities of these deposition techniques, researchers and engineers can develop thin film materials and devices with enhanced functionalities and performance characteristics, contributing to advancements in electronics, optics, energy conversion, biomedical technology, and beyond.

In summary, emerging trends in thin films are driving advancements in materials science and engineering, offering new opportunities for innovation and technology development. From nanotechnology integration and flexible electronics to energy harvesting and 2D materials, thin films are enabling transformative applications across diverse industries. By leveraging advanced deposition techniques and materials design strategies, researchers can continue to push the boundaries of thin film technology and unlock new possibilities for future applications.

2.5.7   Artificial Intelligence and Thin Film

Artificial intelligence (AI) and thin film technologies represent two distinct but increasingly intertwined fields that hold immense potential for driving innovation and advancement across various industries. In recent years, the integration of AI techniques and algorithms into thin film research, development, and manufacturing processes has revolutionized the way thin films are designed, fabricated, and optimized, leading to enhanced performance, efficiency, and functionality. This discussion explores the intersection of artificial intelligence and thin film technologies, highlighting the synergies between these disciplines and the transformative impact they have on scientific research, industrial applications, and technological innovation.

Integration of AI in Thin Film Design and Optimization
One of the primary areas where artificial intelligence is making significant contributions to thin film technology is in the design and optimization of thin film materials and structures. Traditional methods for thin film design often rely on empirical approaches, trial-and-error experimentation, and intuition, which can be time-consuming, labor-intensive, and costly. AI-based approaches, such as machine learning algorithms, neural networks, and genetic algorithms, offer more efficient and systematic methods for exploring the vast design space of thin films, identifying optimal material compositions, layer structures, and deposition parameters. Machine learning algorithms can analyze large datasets of experimental and computational thin film properties, including thickness, composition, morphology, and performance characteristics, to identify patterns, correlations, and trends that may not be apparent through conventional analysis methods. By leveraging these insights, researchers can develop predictive models and algorithms that can accurately predict the properties and performance of thin films based on their composition, structure, and processing conditions. These predictive models can then be used to guide the design and optimization of thin film materials for specific applications, such as photovoltaics, optoelectronics, sensors, and coatings.

AI-Driven Process Control and Optimization
In addition to thin film design, artificial intelligence is also being utilized to enhance process control and optimization in thin film fabrication processes. Thin film deposition techniques, such as chemical vapor deposition (CVD), physical vapor deposition (PVD), and atomic layer deposition (ALD), involve complex interactions between precursor gases, substrate surfaces, and deposition conditions, which can impact film quality, uniformity,

and properties. AI-based control systems can monitor and analyze real-time process data, such as gas flow rates, temperature profiles, and deposition rates, to optimize process parameters and ensure consistent and reproducible thin film deposition. Furthermore, AI algorithms can employ feedback control mechanisms to adjust process parameters dynamically in response to variations in environmental conditions, substrate properties, or deposition kinetics, enabling precise control over film thickness, composition, and morphology. By integrating AI-driven process control systems into thin film deposition equipment, manufacturers can improve production yields, reduce material waste, and enhance the overall efficiency and reliability of thin film fabrication processes.

AI-Enabled Materials Discovery and Development
Another area where artificial intelligence is revolutionizing thin film technology is in materials discovery and development. Thin film materials with novel properties and functionalities are essential for advancing technologies in areas such as electronics, photonics, energy storage, and biomedicine. Traditional methods for materials discovery often rely on manual experimentation and screening of a limited number of candidate materials, which can be time-consuming and resource-intensive. AI-driven approaches, such as high-throughput screening, virtual screening, and materials informatics, enable rapid exploration of vast materials spaces to identify promising candidates with desired properties for thin film applications. Machine learning algorithms can analyze large databases of materials properties, synthesis parameters, and performance metrics to identify correlations and relationships between material composition, structure, and properties. By leveraging these insights, researchers can prioritize candidate materials for synthesis and characterization, accelerating the discovery of new thin film materials with tailored properties and functionalities. Moreover, AI-driven materials informatics platforms can facilitate collaborative research and data sharing among researchers, enabling faster dissemination of knowledge and accelerating the pace of materials discovery and development.

Challenges and Opportunities
While the integration of artificial intelligence and thin film technologies offers tremendous opportunities for innovation and advancement, it also poses several challenges and considerations. One of the primary challenges is the need for high-quality data for training machine learning algorithms and predictive models. Obtaining reliable and comprehensive datasets of thin film properties, deposition parameters, and performance metrics can be challenging, particularly for emerging materials and niche applications. Furthermore, the interpretability and transparency of AI-driven models and algorithms in thin film research and development are essential for gaining insights into the underlying physical and chemical processes governing thin film properties and behavior. Ensuring the robustness, reliability, and generalizability of AI-driven predictions and recommendations is crucial for building trust and confidence in AI-enabled thin film technologies among researchers, engineers, and stakeholders. Despite these challenges, the integration

of artificial intelligence and thin film technologies holds immense promise for accelerating materials discovery, optimizing fabrication processes, and advancing thin film applications across various industries. By harnessing the power of AI-driven approaches, researchers and manufacturers can unlock new opportunities for innovation and create thin film materials and devices with unprecedented performance, functionality, and efficiency. As AI continues to evolve and mature, its impact on thin film technology is expected to grow, ushering in a new era of materials science and engineering.

2.6   Future Prospects and Challenges in Thin Films

Thin films have emerged as a key enabling technology with vast potential for revolutionizing various industries and driving technological innovation. As researchers continue to push the boundaries of materials science and engineering, the future prospects for thin films appear promising. However, several challenges must be addressed to fully realize their potential.

Thin films offer a wide range of potential applications across diverse industries, spanning electronics, optics, energy, healthcare, and beyond. In the electronics industry, thin films are essential for the fabrication of integrated circuits, displays, sensors, and memory devices, enabling smaller, faster, and more efficient electronic devices. In the optics and photonics field, thin films are used in optical coatings, photonic devices, and laser systems, providing enhanced functionality and performance. Thin films also play a crucial role in energy technologies, including solar cells, batteries, and fuel cells, contributing to the development of sustainable energy solutions. Furthermore, thin films have applications in healthcare, such as drug delivery systems, medical implants, and biosensors, where their unique properties enable innovative biomedical devices. The potential applications of thin films are vast and continue to expand as researchers explore new materials, fabrication techniques, and integration strategies.

2.6.1   Addressing Challenges in Thin Film Production and Integration

Despite their immense potential, thin films face several challenges in production, processing, and integration. One of the primary challenges is achieving high throughput and scalability in thin film fabrication processes. Many thin film deposition techniques, such as physical vapor deposition (PVD) and chemical vapor deposition (CVD), are limited by their low deposition rates and batch processing nature, hindering mass production. Addressing this challenge requires the development of advanced deposition techniques that offer high throughput, uniformity, and cost-effectiveness. Additionally, integrating thin films into functional devices and systems poses challenges related to materials compatibility, interface engineering, and reliability. Thin film integration techniques must ensure robust adhesion, low defect density, and stable performance under operating conditions. Furthermore, the scalability and sustainability of thin film materials and processes must be considered to minimize environmental impact and resource consumption. Overcoming these challenges will require interdisciplinary collaboration and innovation in materials science, engineering, and manufacturing.

2.6.2 Role in Driving Technological Innovation

Despite the challenges, thin films are poised to play a pivotal role in driving technological innovation in the coming years. Advances in materials design, fabrication techniques, and device integration are unlocking new possibilities for thin film-based technologies. For example, the emergence of two-dimensional (2D) materials such as graphene and transition metal dichalcogenides (TMDs) has opened up new avenues for thin film research and development. These atomically thin materials offer unique electronic, optical, and mechanical properties, paving the way for next-generation electronic and optoelectronic devices. Furthermore, the integration of thin films with emerging technologies such as artificial intelligence (AI), Internet of Things (IoT), and quantum computing is enabling novel applications in smart systems, sensing networks, and quantum information processing. Thin films are also driving innovation in areas such as flexible electronics, wearable technology, and bioelectronics, where their lightweight, flexible, and biocompatible nature offers new opportunities for device design and functionality. By harnessing the potential of thin films and overcoming technical challenges, researchers can drive technological innovation and create transformative solutions to address global challenges in areas such as healthcare, energy, and environmental sustainability.

In summary, thin films hold immense promise for shaping the future of technology and driving innovation across various industries. While challenges in production, integration, and scalability remain, ongoing research and development efforts are poised to overcome these obstacles and unlock the full potential of thin film-based technologies. By leveraging advanced materials, fabrication techniques, and interdisciplinary collaboration, researchers can realize the vision of thin films as a versatile and transformative technology platform for the twenty-first century.

2.7   Thin Film for Sustainable Development

2.7.1   Introduction to Thin Film Technology in Sustainable Development

Thin film technology represents a critical frontier in the pursuit of sustainable development. With its diverse applications across various industries, thin films offer innovative solutions that contribute to environmental preservation, resource efficiency, and socio-economic progress. This introductory discourse aims to examine the essence of thin film technology, exploring its fundamental principles, applications, and significance in fostering sustainability across different sectors. One of the most prominent areas where thin film technology plays a pivotal role is in renewable energy systems. Thin film solar cells, for instance, offer a lightweight, flexible alternative to traditional silicon-based photovoltaic cells. By leveraging materials like amorphous silicon, cadmium telluride, and copper indium gallium selenide, thin film solar cells can be manufactured cost-effectively and integrated into diverse settings, including building facades, vehicles, and wearable devices. This scalability and versatility contribute significantly to the adoption of solar energy as a sustainable power source.

Moreover, thin film coatings are instrumental in enhancing the energy efficiency of buildings and infrastructure. Low-emissivity (Low-E) coatings, applied to windows and glass facades, mitigate heat transfer, thereby reducing the need for heating and cooling. Similarly, smart coatings incorporating nanomaterials can regulate light transmission and reflectivity based on environmental conditions, optimizing indoor comfort while minimizing energy consumption. These advancements underscore the role of thin film technology in promoting sustainable architecture and urban development. Beyond energy, thin films find applications in environmental monitoring and remediation. Thin film sensors equipped with nanomaterial-based transducers offer real-time detection of pollutants, pathogens, and toxins in air, water, and soil. Furthermore, thin film membranes enable efficient separation and purification processes in wastewater treatment, desalination, and gas filtration. By facilitating resource recovery and minimizing environmental contamination, these technologies contribute to the preservation of ecosystems and public health.

In the realm of electronics and communication, thin film technology drives innovation in miniaturization, performance, and connectivity. Thin film transistors (TFTs) are fundamental building blocks in modern displays, sensors, and wearable electronics. Flexible and stretchable thin film circuits enable the development of conformal devices that can adapt to complex surfaces and body movements, heralding a new era of personalized and ubiquitous computing. Additionally, thin film antennas and metamaterials enable high-speed data transmission and wireless connectivity, supporting the infrastructure for the Internet of Things (IoT) and smart cities. The agricultural sector also stands to benefit from the integration of thin film technology. Thin film coatings can be tailored to provide controlled release of nutrients, pesticides, and growth regulators, enhancing crop productivity while minimizing environmental impact. Moreover, thin film sensors and imaging technologies enable precision agriculture practices, allowing farmers to monitor soil moisture, nutrient levels, and crop health with unprecedented accuracy. By optimizing resource utilization and reducing chemical inputs, these innovations contribute to sustainable food production and rural livelihoods.

In summary, thin film technology holds immense promise as a driver of sustainable development across various domains. Whether in renewable energy, green buildings, environmental monitoring, electronics, or agriculture, thin films offer solutions that prioritize efficiency, resource conservation, and environmental stewardship. As research and development efforts continue to advance the capabilities and affordability of thin film materials and processes, their integration into mainstream practices will play a pivotal role in shaping a more sustainable and resilient future for humanity.

2.7.2 Applications of Thin Film Technology in Renewable Energy Systems

Thin film technology has emerged as a key player in the renewable energy sector, offering versatile solutions to harness solar energy efficiently and sustainably. With advancements in materials science and manufacturing processes, thin film solar cells have become increasingly competitive with traditional silicon-based photovoltaics. Moreover, thin film coatings and materials find applications beyond solar energy, contributing to the efficiency and sustainability of various renewable energy systems.

Thin film solar cells, characterized by their lightweight and flexible nature, represent a significant advancement in photovoltaic technology. Unlike rigid silicon panels, thin film solar cells can be integrated into a variety of surfaces, including building facades, roofs, and even curved or irregular structures. This flexibility opens up new possibilities for solar energy deployment in urban environments, where space constraints and architectural considerations often limit the use of conventional solar panels. One of the most common types of thin film solar cells is based on amorphous silicon (a-Si), a non-crystalline form of silicon. Amorphous silicon thin film solar cells are deposited onto substrates using techniques such as plasma-enhanced chemical vapor deposition (PECVD) or sputtering. While they may have lower efficiencies compared to crystalline silicon cells, amorphous silicon thin film solar cells offer advantages in terms of manufacturing cost, scalability, and performance in low-light conditions. Another promising thin film material for solar cells is cadmium telluride (CdTe). Cadmium telluride thin film solar cells boast higher efficiencies compared to amorphous silicon cells and are manufactured using a simpler deposition process. As a result, CdTe thin film solar cells have gained commercial traction and are deployed in utility-scale solar power plants around the world. Their low-cost production and high efficiency make them a competitive option for large-scale solar energy generation.

Copper indium gallium selenide (CIGS) is another thin film material with promising applications in solar photovoltaics. CIGS thin film solar cells offer high efficiencies comparable to crystalline silicon cells while maintaining the advantages of thin film technology, such as flexibility and lightweight. Research efforts are focused on optimizing the performance and stability of CIGS thin film solar cells to enhance their competitiveness in the market. In addition to traditional thin film solar cells, emerging materials such as perovskites show great potential for revolutionizing the solar energy landscape. Perovskite solar cells, composed of organic–inorganic hybrid materials, have demonstrated remarkable efficiency improvements in recent years. Perovskite thin film solar cells can be fabricated using solution-based processes, which are more cost-effective and scalable compared to vacuum deposition techniques. However, challenges related to stability, scalability, and toxicity need to be addressed before perovskite solar cells can be deployed on a large scale. Apart from solar energy generation, thin film technology contributes to the efficiency and sustainability of other renewable energy systems. Thin film coatings play a crucial role in enhancing the performance of solar thermal collectors, which convert sunlight into heat for various applications, including space heating, water heating, and industrial processes. Selective absorber coatings applied to solar thermal collectors improve heat absorption and minimize heat losses, increasing overall efficiency and reducing energy consumption. Furthermore, thin film materials find applications in other renewable energy technologies, such as wind and hydroelectric power generation. Protective coatings applied to wind turbine blades enhance durability and weather resistance, reducing maintenance costs and extending the lifespan of wind turbines. Similarly, hydrophobic coatings applied to turbine surfaces in hydroelectric power plants prevent fouling and corrosion, ensuring optimal performance and efficiency.

Thin film technology plays a vital role in advancing renewable energy systems, particularly in the field of solar photovoltaics. From amorphous silicon and cadmium telluride to copper indium gallium selenide and perovskites, thin film materials offer diverse options for efficient and sustainable solar energy generation. Moreover, thin film coatings contribute to the performance and durability of various renewable energy technologies, including solar thermal collectors, wind turbines, and hydroelectric systems. As research and development efforts continue to drive innovation in thin film materials and processes, the integration of thin film technology will further accelerate the transition towards a cleaner, more sustainable energy future.

2.7.3 Thin Film Materials and Their Environmental Impact

Thin film materials have become indispensable in various industries due to their unique properties and diverse applications. From electronics and energy to healthcare and aerospace, thin films play a crucial role in advancing technology and driving innovation. However, the widespread use of thin film materials raises concerns about their environmental impact throughout their lifecycle, from raw material extraction and manufacturing to disposal and recycling. This comprehensive discourse aims to explore the environmental implications of thin film materials, highlighting both the challenges and opportunities for mitigating their environmental footprint.

Thin film materials encompass a broad range of substances, including metals, semiconductors, ceramics, and polymers, deposited onto substrates to form thin layers with specific functionalities. Common deposition techniques such as physical vapor deposition (PVD), chemical vapor deposition (CVD), and sputtering enable precise control over film thickness, composition, and structure. While thin films offer advantages such as lightweight, flexibility, and scalability, their environmental impact stems from various factors, including raw material sourcing, energy consumption, waste generation, and chemical usage. One of the primary concerns regarding thin film materials is the extraction and processing of raw materials required for their production. Many thin film materials, such as indium, gallium, tellurium, and rare earth elements, are considered critical or strategic metals due to their limited availability and geopolitical significance. The extraction of these materials often involves environmentally damaging practices such as open-pit mining, which can lead to habitat destruction, soil erosion, water pollution, and deforestation. Additionally, the processing of raw materials into thin film components may require large amounts of energy and water, further exacerbating their environmental footprint. Furthermore, the manufacturing processes involved in thin film deposition contribute to environmental pollution and resource depletion. Traditional deposition techniques, such as PVD and CVD, rely on vacuum chambers, high temperatures, and chemical precursors, which can generate hazardous emissions, greenhouse gases, and toxic byproducts. For example, the production of thin film semiconductors often involves the use of toxic gases such as arsine, phosphine, and silane, posing risks to human health and the environment. Moreover, the energy-intensive nature of thin film manufacturing contributes to carbon emissions and exacerbates climate change, further underscoring the need for sustainable alternatives. In addition to the environmental impact of thin film production, concerns arise regarding the disposal and recycling of thin film materials at the end of their lifecycle. Thin film devices, such as solar panels, displays, and electronic components, contain valuable materials that can be recovered and reused through recycling processes. However, the complexity of thin film structures and the presence of contaminants pose challenges to efficient recycling. Current recycling technologies for thin film materials are limited in their effectiveness and scalability, leading to significant amounts of electronic waste ending up in landfills or being incinerated, releasing hazardous pollutants into the environment.

Despite these challenges, efforts are underway to address the environmental impact of thin film materials and promote more sustainable practices throughout their lifecycle. One approach is the development of eco-friendly thin film materials derived from renewable resources or recycled sources. For example, researchers are exploring alternative thin film materials based on biodegradable polymers, cellulose nanocrystals, and organic compounds derived from biomass. These bio-based thin films offer advantages such as biodegradability, renewability, and reduced environmental toxicity, making them attractive options for various applications, including packaging, coatings, and electronics. Moreover, advancements in thin film deposition technologies are driving improvements in energy efficiency, material utilization, and waste reduction. Novel deposition techniques such as atomic layer deposition (ALD), molecular beam epitaxy (MBE), and solution processing enable precise control over thin film growth and composition while minimizing resource consumption and emissions. By optimizing process parameters and materials usage, manufacturers can reduce environmental impacts and enhance the sustainability of thin film production.

In parallel, initiatives focused on circular economy principles aim to promote the reuse, remanufacturing, and recycling of thin film materials to minimize waste and conserve resources. Closed-loop recycling systems for thin film devices, coupled with innovative separation and purification technologies, offer opportunities to recover valuable materials and reduce the demand for virgin resources. Additionally, product stewardship programs and extended producer responsibility (EPR) frameworks encourage manufacturers to take responsibility for the environmental impacts of their products throughout their lifecycle, incentivizing eco-design, resource recovery, and waste management strategies. Furthermore, lifecycle assessment (LCA) methodologies provide valuable insights into the environmental performance of thin film materials and enable informed decision-making to minimize environmental impacts. By quantifying the energy, resource, and emissions profiles of thin film products from cradle to grave, LCAs help identify hotspots, prioritize improvement opportunities, and inform sustainability strategies. Incorporating environmental considerations into product design, procurement, and end-of-life management processes ensures that thin film materials are produced and used in a manner that minimizes adverse environmental effects and maximizes long-term sustainability.

Thin film materials play a vital role in modern technology, offering diverse functionalities and enabling innovative applications across various industries. However, their widespread use raises significant environmental concerns related to raw material extraction, manufacturing processes, and end-of-life disposal. Addressing these challenges requires a concerted effort from stakeholders across the value chain, including researchers, manufacturers, policymakers, and consumers. By embracing sustainable practices, eco-friendly materials, and circular economy principles, the environmental impact of thin film materials can be minimized, paving the way for a more sustainable and resource-efficient future.

2.7.4 Advancements in Thin Film Technology for Green Building Solutions

Green building solutions have become increasingly essential in the construction industry as society seeks to address environmental concerns, improve energy efficiency, and promote sustainable development. Thin film technology has emerged as a key enabler of green building solutions, offering innovative materials and coatings that enhance building performance, reduce energy consumption, and minimize environmental impact. This comprehensive discourse explores the latest advancements in thin film technology for green building applications, highlighting their benefits, challenges, and potential for transforming the built environment.

Thin film coatings play a crucial role in improving the energy efficiency of buildings by controlling heat transfer, managing light transmission, and enhancing durability. Low-emissivity (Low-E) coatings, applied to windows and glass facades, reduce heat loss during the winter and heat gain during the summer, thereby reducing the need for heating, cooling, and artificial lighting. These coatings typically consist of thin layers of metallic or metal oxide materials deposited onto glass substrates using techniques such as sputtering or chemical vapor deposition (CVD). By reflecting infrared radiation while allowing visible light to pass through, Low-E coatings help maintain comfortable indoor temperatures, improve occupant comfort, and lower energy bills.

In addition to Low-E coatings, spectrally selective coatings offer selective control over the transmission and absorption of solar radiation, allowing buildings to optimize daylighting and passive solar heating while minimizing solar heat gain. These coatings can be tailored to specific climate conditions and building orientations, maximizing energy savings and reducing reliance on mechanical HVAC systems. By integrating spectrally selective coatings into building envelopes, architects and designers can create energy-efficient and sustainable buildings that prioritize occupant comfort and environmental stewardship. Furthermore, smart coatings incorporating thin film materials enable dynamic control of light transmission, transparency, and thermal properties based on environmental conditions and user preferences. Electrochromic and thermochromic coatings, for example, respond to changes in temperature, sunlight, or electrical voltage, allowing windows and glass facades to adjust their tint, opacity, and thermal insulation properties in real-time. These smart coatings offer opportunities for passive climate control, daylight harvesting, and glare reduction, enhancing building performance while reducing energy consumption and carbon emissions.

Moreover, anti-reflective coatings applied to solar panels and building-integrated photovoltaic (BIPV) systems improve light capture efficiency and energy generation by minimizing surface reflections and maximizing incident sunlight absorption. These coatings typically consist of thin layers of dielectric materials with tailored refractive indices, deposited onto solar cell surfaces using techniques such as sputtering or atomic layer deposition (ALD). By reducing optical losses and enhancing light transmission, anti-reflective coatings increase the overall efficiency and output of solar energy systems, contributing to the sustainability of green buildings and renewable energy integration. In addition to coatings, thin film materials find applications in other green building solutions, including energy-efficient lighting, sensors, and building-integrated electronics. Organic light-emitting diodes (OLEDs) and thin film transistors (TFTs), for example, enable the development of flexible, lightweight, and energy-efficient lighting systems and displays that can be seamlessly integrated into building surfaces and interior spaces. These thin film devices offer advantages such as high brightness, wide viewing angles, and low power consumption, making them ideal for ambient lighting, signage, and decorative elements in green buildings. Furthermore, thin film sensors and actuators enable smart building systems for environmental monitoring, occupancy detection, and adaptive control of lighting, heating, and ventilation. These sensors can be integrated into building materials and surfaces to provide real-time data on indoor air quality, temperature, humidity, and occupancy levels, enabling automated responses to optimize energy efficiency and occupant comfort. By leveraging thin film technology, building managers can implement proactive maintenance strategies, reduce energy waste, and create healthier indoor environments that promote well-being and productivity.

In summary, advancements in thin film technology offer a myriad of opportunities for enhancing green building solutions and promoting sustainable development in the construction industry. From energy-efficient coatings and smart materials to integrated electronics and sensors, thin film innovations enable architects, designers, and building professionals to create high-performance buildings that minimize environmental impact, maximize energy efficiency, and enhance occupant comfort. By embracing thin film technology, the green building sector can accelerate the transition towards a more sustainable and resilient built environment, benefiting both present and future generations.

2.7.5 Thin Film Coatings for Sustainable Agriculture and Food Packaging

Thin film coatings play a crucial role in modern agriculture and food packaging, offering solutions to enhance crop productivity, prolong shelf life, and reduce food waste. With growing concerns about food security, environmental sustainability, and food safety, thin film technology offers innovative approaches to address these challenges while minimizing environmental impact. This comprehensive discourse explores the applications, benefits, and implications of thin film coatings in sustainable agriculture and food packaging, highlighting their potential to transform the food supply chain and promote a more sustainable future. Thin film coatings find numerous applications in agriculture, ranging from crop protection and soil enhancement to precision farming and post-harvest management. These coatings are designed to improve seed germination, nutrient uptake, water retention, and pest resistance, thereby enhancing crop yields and reducing the need for synthetic inputs such as fertilizers and pesticides. Moreover, thin film materials can be tailored to release nutrients and agrochemicals gradually, minimizing leaching and runoff while maximizing plant uptake and efficiency.

One of the key applications of thin film coatings in agriculture is seed coating, where seeds are coated with thin layers of protective materials to enhance germination, seedling vigor, and crop establishment. These coatings may contain beneficial microorganisms, growth-promoting substances, or biostimulants that improve soil health, root development, and plant growth. By encapsulating seeds in thin film coatings, farmers can optimize seed performance, reduce seedling mortality, and achieve higher yields with lower inputs, contributing to sustainable agriculture practices and resource conservation. Furthermore, thin film coatings find applications in soil stabilization and erosion control, where they are applied to soil surfaces to prevent erosion, retain moisture, and enhance nutrient availability. These coatings can be composed of biodegradable polymers, natural resins, or soil-binding agents that form protective barriers against wind, water, and microbial degradation. By stabilizing soil aggregates and minimizing erosion, thin film coatings help preserve soil fertility, mitigate nutrient runoff, and protect water quality, promoting sustainable land management practices and ecosystem resilience. In addition to agricultural applications, thin film coatings play a critical role in food packaging, where they provide barrier properties, antimicrobial protection, and freshness preservation to extend the shelf life of perishable foods and reduce food waste. Thin film materials such as biodegradable polymers, edible films, and nanocomposites offer alternatives to conventional plastic packaging, which is often non-biodegradable, non-renewable, and environmentally harmful.

Biodegradable thin film coatings derived from renewable resources such as starch, cellulose, and chitosan offer advantages such as compostability, biocompatibility, and reduced environmental impact. These coatings can be applied to food surfaces or used as wrappers and pouches to provide moisture resistance, gas permeability, and microbial barrier properties while ensuring product safety and quality. By replacing petroleum-based plastics with biodegradable alternatives, thin film coatings contribute to reducing plastic pollution, conserving natural resources, and promoting a circular economy. Moreover, thin film coatings incorporating antimicrobial agents, antioxidants, and oxygen scavengers help inhibit microbial growth, delay oxidation, and preserve sensory attributes in packaged foods. These coatings can be applied to food contact surfaces, packaging materials, or directly onto food products to extend shelf life, reduce spoilage, and maintain freshness throughout the supply chain. By minimizing food waste and spoilage, thin film coatings contribute to improving food security, resource efficiency, and economic viability in the food industry.

Nanocomposite thin film coatings, reinforced with nanoparticles such as silver, zinc oxide, or clay minerals, offer enhanced barrier properties, mechanical strength, and antimicrobial activity compared to conventional materials. These coatings can be applied to food packaging films, trays, and containers to provide multifunctional performance, including gas barrier, moisture barrier, UV protection, and antimicrobial effects. By incorporating nanotechnology into food packaging, thin film coatings enable safer, more sustainable packaging solutions that meet regulatory requirements and consumer preferences for environmentally friendly and safe food products. Furthermore, active packaging systems utilizing thin film coatings with intelligent functionalities, such as oxygen scavenging, ethylene absorption, and pH sensing, enable real-time monitoring and control of food quality and safety. These coatings can interact with food products and the surrounding environment to detect and respond to changes in temperature, humidity, or microbial activity, extending shelf life, and ensuring product integrity. By incorporating active packaging technologies, thin film coatings contribute to reducing food waste, improving supply chain efficiency, and enhancing consumer confidence in the safety and quality of packaged foods.

In summary, thin film coatings play a vital role in sustainable agriculture and food packaging by offering solutions to enhance crop productivity, preserve food freshness, and reduce environmental impact throughout the food supply chain. From seed coatings and soil amendments to biodegradable packaging and active coatings, thin film technology offers innovative approaches to address the challenges of food security, food safety, and food waste. By embracing thin film coatings, farmers, food producers, and packaging manufacturers can contribute to building a more sustainable and resilient food system that meets the needs of present and future generations.

2.7.6 Economic and Social Implications of Thin Film Innovation in Sustainable Development

Thin film innovation holds significant promise in driving sustainable development across various sectors, with far-reaching economic and social implications. As thin film technology advances, its applications in renewable energy, green buildings, agriculture, and healthcare are reshaping industries, creating new opportunities, and addressing global challenges such as climate change, resource scarcity, and social inequality. This comprehensive discourse explores the economic and social implications of thin film innovation in sustainable development, highlighting its potential to catalyze economic growth, enhance social welfare, and foster inclusive prosperity.

Thin film innovation stimulates economic growth by creating new industries, generating employment opportunities, and fostering technological innovation. As demand for thin film products and solutions increases, businesses invest in research and development, manufacturing facilities, and distribution networks, creating jobs across the value chain.

From scientists and engineers to technicians, manufacturers, and sales professionals, thin film innovation generates employment in diverse sectors, contributing to economic development and prosperity. Thin film innovation disrupts traditional industries and transforms business models, driving competitiveness and market differentiation. Industries such as solar energy, electronics, and packaging are undergoing rapid changes as thin film technologies enable higher efficiency, lower costs, and improved performance. Companies that embrace thin film innovation gain a competitive edge by offering innovative products, reducing environmental impact, and meeting evolving consumer preferences for sustainability and quality. Thin film innovation enhances export competitiveness and trade balances by providing differentiated products and services with high value-added content. Countries with strong capabilities in thin film research, manufacturing, and exports benefit from increased demand for sustainable technologies and solutions worldwide. Moreover, thin film products, such as solar panels, electronic devices, and advanced materials, contribute to export growth, job creation, and revenue generation, strengthening national economies and reducing dependency on fossil fuels and imported goods. Thin film innovation attracts investment and financing from public and private sources, fueling research, development, and commercialization efforts. Venture capital firms, private equity funds, and government agencies allocate funding to startups, research institutions, and companies engaged in thin film research and development. Additionally, financial instruments such as green bonds, impact investments, and project financing support thin film projects and initiatives with environmental and social benefits, accelerating their deployment and scale-up Thin film innovation drives cost reduction and affordability across various industries, making sustainable technologies more accessible to businesses and consumers. Advances in materials science, manufacturing processes, and scale economies enable manufacturers to produce thin film products at lower costs while maintaining or improving performance. This cost-effectiveness drives adoption and market penetration of thin film solutions, democratizing access to clean energy, efficient buildings, and eco-friendly products for individuals and communities worldwide.

Thin film innovation improves access to clean energy and basic services for underserved communities, empowering them to overcome energy poverty, improve living standards, and enhance quality of life. Off-grid and remote communities benefit from decentralized renewable energy solutions, such as solar panels and thin film batteries, which provide reliable electricity for lighting, heating, cooking, and communication. Moreover, thin film technologies enable water purification, healthcare delivery, and education services in resource-constrained settings, bridging the gap in access to essential services and promoting social inclusion. Thin film innovation contributes to environmental conservation and climate resilience by reducing carbon emissions, mitigating environmental degradation, and promoting sustainable resource management practices. Renewable energy systems powered by thin film solar panels replace fossil fuel-based energy sources, reducing greenhouse gas emissions and air pollution. Thin film coatings and materials protect natural habitats, conserve water resources, and enhance soil fertility, preserving biodiversity and ecosystem services. Furthermore, thin film technologies enable climate adaptation strategies, such as flood protection, drought resilience, and disaster prepared-ness, strengthening community resilience to climate change impacts. Thin film innovation improves health and well-being by enhancing access to clean water, nutritious food, and healthcare services, particularly in underserved communities. Thin film membranes and filtration systems provide safe drinking water by removing contaminants, pathogens, and pollutants from water sources. Additionally, thin film coatings and sensors enable food safety monitoring, disease detection, and medical diagnostics, empowering individuals to make informed decisions about their health and nutrition. By addressing health dis-parities and promoting preventive care, thin film technologies contribute to better health outcomes, reduced healthcare costs, and improved quality of life for individuals and com-munities. Thin film innovation fosters education and skills development by providing opportunities for learning, training, and capacity building in science, technology, engineer-ing, and mathematics (STEM) fields. Research institutions, universities, and vocational training centers offer programs and courses on thin film materials, deposition techniques, and applications, preparing students and professionals for careers in emerging industries. Moreover, collaborative research projects, internships, and knowledge exchange programs facilitate technology transfer and skill transfer, empowering individuals to contribute to innovation, entrepreneurship, and sustainable development in their communities. Thin film innovation promotes community engagement and empowerment by involving stakehold-ers in decision-making, planning, and implementation processes. Participatory approaches, community-based projects, and citizen science initiatives enable local communities to co-create and co-manage thin film technologies and solutions that address their specific needs and priorities. By fostering collaboration, inclusivity, and ownership, thin film innovation strengthens social cohesion, resilience, and adaptive capacity, building trust and solidarity among diverse stakeholders and fostering sustainable development from the ground up.

Thin film innovation has profound economic and social implications for sustainable development, driving economic growth, enhancing social welfare, and fostering inclusive prosperity. By unlocking new opportunities, addressing global challenges, and empow-ering communities, thin film technologies contribute to building a more sustainable, resilient, and equitable future for all. As governments, businesses, and civil society orga-nizations continue to invest in thin film research, development, and deployment, the economic and social benefits of thin film innovation are expected to accelerate, creating lasting positive impacts for generations to come.

2.8   Summary

Thin films represent a pivotal area of materials science and engineering, offering a diverse array of applications across multiple industries. Their history is deeply intertwined with scientific advancements and technological innovations, dating back to the early developments in vacuum deposition techniques and optical coatings. Over time, thin films have played a crucial role in the semiconductor industry, contributing to the miniaturization of electronic devices and the advancement of integrated circuits. Furthermore, thin films have made significant contributions to optical coatings, enabling enhanced performance in optical devices and systems. Advancements in magnetic recording technology have also been propelled by thin film innovations, leading to higher data storage densities and improved magnetic storage media. As technology progresses, emerging trends in thin films are shaping the future of materials science and engineering. Nanotechnology integration offers precise control over material properties at the nanoscale, opening up new possibilities for enhanced functionality and performance. Flexible electronics are enabling the development of bendable, stretchable, and wearable devices, revolutionizing healthcare, consumer electronics, and beyond.

Energy harvesting and storage technologies are benefiting from thin film advancements, with thin film solar cells and batteries offering sustainable solutions for renewable energy generation and storage. The discovery and characterization of two-dimensional materials are driving innovations in thin film research, unlocking new opportunities for electronic, optoelectronic, and quantum devices. Additive manufacturing techniques are also revolutionizing thin film fabrication, providing rapid prototyping and customization capabilities for electronic and sensor applications. Advanced deposition techniques such as atomic layer deposition (ALD) and molecular beam epitaxy (MBE) offer precise control over thin film growth, enabling the development of novel materials and devices.

Despite the immense potential of thin films, challenges remain in production, scalability, and integration. Addressing these challenges requires interdisciplinary collaboration and innovation in materials science, engineering, and manufacturing. By leveraging emerging trends and overcoming technical obstacles, researchers can drive technological innovation and create transformative solutions for the future.

Applications of 3D Printing in the Fourth Industrial Revolution

3

3.1 Introduction

The Fourth Industrial Revolution, characterized by the fusion of technologies blurring the lines between physical, digital, and biological domains, has brought about unprecedented advancements across various sectors. One of the pivotal technologies driving this revolution is 3D printing, also known as additive manufacturing. This innovative approach to manufacturing is reshaping industries, offering new possibilities in design, production, and distribution. In this chapter, the transformative impact of 3D printing in the context of the Fourth Industrial Revolution is examined.

In the realm of modern manufacturing, few technologies have garnered as much attention and excitement as 3D printing. Also known as additive manufacturing, 3D printing has emerged as a transformative force, revolutionizing traditional manufacturing processes and unlocking new possibilities across a myriad of industries. At its core, 3D printing enables the creation of three-dimensional objects layer by layer from digital models, offering unparalleled flexibility, precision, and customization. The journey of 3D printing traces back to the 1980s, when visionary pioneers like Chuck Hull introduced the world to the concept of additive manufacturing. Since then, technology has evolved at a remarkable pace, driven by advancements in materials, printing techniques, and software capabilities. What began as a novel innovation has now blossomed into a multifaceted tool with applications ranging from rapid prototyping and custom manufacturing to medical device production and space exploration. In recent years, 3D printing has transcended its niche status to become a mainstream technology, reshaping industries and challenging traditional manufacturing paradigms. Its ability to democratize production, empower customization, and foster innovation has captured the imagination of entrepreneurs, engineers, and designers worldwide. From small-scale startups to multinational corporations, organizations of all sizes are harnessing the power of 3D printing to streamline processes, reduce costs, and unlock new opportunities for growth and creativity.

The Fourth Industrial Revolution, characterized by the convergence of digital, physical, and biological technologies, the role of 3D printing in shaping the future of manufacturing has never been more profound. With each passing day, new breakthroughs and applications emerge, pushing the boundaries of what is possible and propelling us towards a future where the impossible becomes routine. In this exploration of 3D printing, this chapter looks into its origins, evolution, and transformative impact on industries ranging from aerospace and automotive to healthcare and consumer goods. The challenges and opportunities it presents, from material limitations and scalability issues to sustainability considerations and ethical implications are considered. Moreover, we envision the future trajectory of 3D printing, envisioning a world where on-demand manufacturing, personalized products, and distributed production networks are the norm.

3.1.1   3D Printing (Additive Manufacturing)

3D printing, also known as additive manufacturing, is a revolutionary process of creating three-dimensional objects from a digital file. Unlike traditional subtractive manufacturing methods that involve cutting or drilling material away from a solid block, 3D printing builds objects layer by layer, adding material precisely where needed. This technology has transformed various industries, enabling rapid prototyping, customized production, and complex geometries that were previously impossible to achieve. An illustration of a 3D printing process is shown in Fig. 3.1.

Fig. 3.1 Schematic depiction of 3D printing

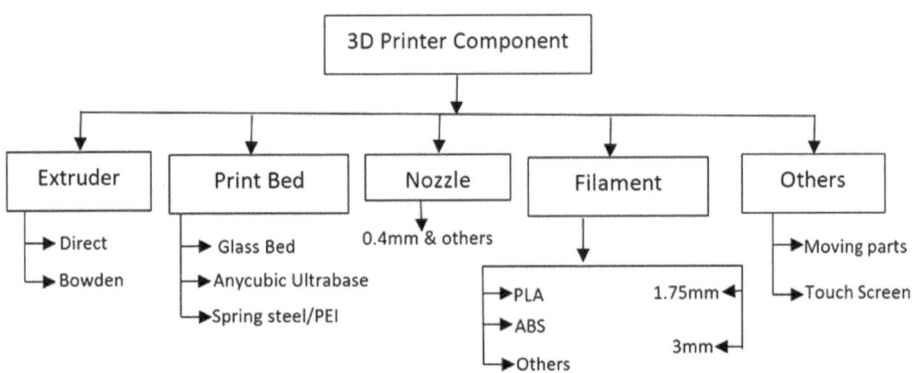

Fig. 3.2 Key components of 3D printer

Components of a 3D printer
A 3D printer is a complex machine comprised of various components, each playing a crucial role in the additive manufacturing process. These components work in harmony to transform digital designs into physical objects layer by layer. A detailed overview of the key components of a 3D printer is shown in Fig. 3.2.

1. **Frame**: The frame provides the structural support and stability necessary to hold all other components in place during the printing process. It's typically constructed from sturdy materials like metal or reinforced plastic to withstand the forces generated during printing.
2. **Extruder Assembly**: The extruder assembly is responsible for feeding filament (the raw material) into the printer and melting it to create the printed object. It consists of several sub-components, including:
   - **Hotend**: The hotend is the component responsible for heating the filament to its melting point. It typically includes a heating element (such as a resistor or cartridge heater) and a nozzle through which the molten filament is extruded.
   - **Filament Drive**: Also known as the filament feeder or extruder motor, the filament drive mechanism grips the filament and feeds it into the hotend at a controlled rate. It may utilize gears, rollers, or a hobbed bolt to ensure consistent filament extrusion.
3. **Build Platform**: The build platform is the surface where the printed object is formed. It may be stationary or movable, depending on the printer design. The build platform provides a stable foundation for the print and ensures proper adhesion during the printing process.

4. **Heated Bed**: Many 3D printers feature a heated bed, which helps maintain adhesion between the printed object and the build platform. Heating the bed to a specific temperature can prevent warping and improve the overall quality of the print, particularly with materials like ABS that are prone to shrinkage.

5. **Print Head/Nozzle**: The print head or nozzle is where the molten filament is deposited onto the build platform or previous layers to create the 3D object. Nozzles come in various sizes, allowing for different layer heights and levels of detail in the final print.

6. **Stepper Motors**: Stepper motors are used to control the movement of the printer's axes, including the X, Y, and Z axes, as well as other components like the extruder and build platform. These motors provide precise and controlled motion, enabling accurate positioning of the print head and build platform.

7. **Controller Board**: The controller board serves as the brain of the 3D printer, interpreting instructions from the slicing software and coordinating the movements of the printer components. It controls functions such as temperature regulation, motor control, and communication with the slicing software.

8. **Endstops and Limit Switches**: Endstops and limit switches are sensors located at various points on the printer's axes to define the boundaries of the print volume and ensure accurate positioning. They signal the controller board when the print head or build platform reaches its maximum or minimum position.

9. **Cooling Fans**: Cooling fans are used to regulate the temperature of components such as the hotend and stepper motors, preventing overheating and ensuring stable printing conditions. They may also be used to quickly cool printed layers, improving print quality and reducing warping.

10. **LCD Display and Control Interface**: Some 3D printers feature an LCD display and control interface, allowing users to interact with the printer directly. This interface typically provides options for selecting files, monitoring print progress, adjusting settings, and troubleshooting issues.

11. **Power Supply**: The power supply unit provides electrical power to the various components of the 3D printer, including the motors, heaters, and controller board. It converts mains voltage (AC) to the appropriate voltage and current required by the printer's electronics.

Each of these components plays a vital role in the 3D printing process, contributing to the overall functionality, reliability, and performance of the printer. Understanding how these components work together is essential for achieving successful prints and maximizing the capabilities of the 3D printing technology.

Principle of 3D Printing:

The principle of 3D printing involves slicing a digital model into thin horizontal layers using computer-aided design (CAD) software. These layers are then sequentially printed, with each layer fusing or solidifying to the layer beneath it, gradually building up the final object. Different 3D printing processes employ various methods to deposit or solidify material, such as extrusion, curing, sintering, or melting, depending on the type of printer and material being used. The principle of 3D printing, also known as additive manufacturing, revolves around the concept of building objects layer by layer from digital models. Unlike traditional subtractive manufacturing processes, where material is removed from a solid block to create the desired shape, 3D printing adds material incrementally to construct objects from the ground up. This additive approach offers significant advantages in terms of flexibility, customization, and complexity, making it a transformative technology across various industries.

A 3D printer, slicing software, and filament constitute the essential components needed to kickstart the 3D printing process. Among these, the role of computer-aided design (CAD) and slicing software is paramount. CAD software serves as the creative hub, allowing users to craft intricate 3D models from scratch or modify existing designs to suit their needs. Whether it's engineering parts, artistic sculptures, or architectural prototypes, CAD software provides the canvas for digital creation. For custom-designed models, CAD becomes indispensable, offering precise control over dimensions, shapes, and features. Once the model is finalized, the CAD software exports it into an STL (stereolithography) file format—a universal format compatible with most slicing software. This STL file essentially serves as the blueprint for the 3D object. The second critical software component is the slicing program. This software takes the STL file and translates it into a language the 3D printer can understand—typically G-code. The slicing software divides the model into thin horizontal layers, akin to slicing a loaf of bread, and generates instructions for the printer on how to construct each layer. This includes specifying the path the printer's nozzle or extruder should follow, the speed at which it should move, and the amount of filament to be deposited. G-code, the output of the slicing process, provides precise instructions to the printer's motors on how to execute the print job. It dictates the movement of the printer's axes (X, Y, and Z) and controls other parameters like extrusion temperature and fan speed. To initiate the printing process, the G-code file is transferred to the 3D printer. This can be done using various methods such as loading the file onto an SD card or transmitting it wirelessly via Wi-Fi. Once the printer receives the G-code instructions, it's ready to commence the print job.

The actual printing process can vary depending on factors like the type of printer, material being used, and complexity of the object. The fundamental principle of 3D printing involves key steps as shown in Fig. 3.3.

1. Digital Design: The process of 3D printing commences with the creation of a digital 3D model, often crafted using renowned computer-aided design (CAD) software such as AutoCAD, SolidWorks, or Fusion 360. These industry-standard software packages empower designers with a suite of powerful tools and functionalities to translate creative visions into tangible digital representations. Whether designing complex

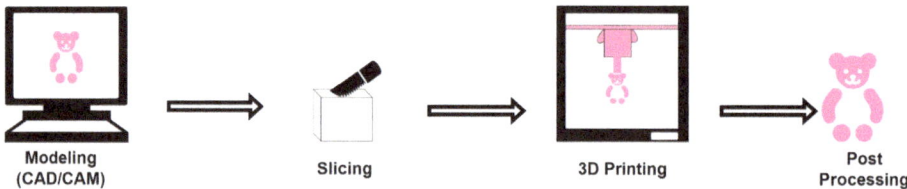

Fig. 3.3 The 3D printing Process

mechanical parts, architectural prototypes, or artistic sculptures, CAD software offers a versatile platform for creating intricate digital models with precision and accuracy. In addition to CAD software, designers may also utilize cutting-edge 3D scanning technologies pioneered by industry leaders such as Faro Technologies, Artec 3D, or Trimble. These companies specialize in developing advanced 3D scanners that utilize laser scanning, structured light scanning, or photogrammetry to capture detailed surface data and create accurate 3D representations of real-world objects. Such technologies are indispensable for reverse engineering, quality control, and digitization of physical prototypes or artifacts. Once the digital model is created or captured, it undergoes a series of iterative design processes using renowned simulation and analysis software like ANSYS, Siemens NX, or Autodesk Simulation. These software solutions enable designers to optimize designs for strength, durability, and performance through sophisticated simulation tools and analysis capabilities. By leveraging these tools, designers can validate concepts, test designs, and optimize geometries, thereby reducing the need for costly physical prototypes and iterations. The versatility and flexibility afforded by digital modeling in 3D printing revolutionize the design process, enabling designers to explore creative concepts, iterate rapidly, and customize objects to suit specific requirements and preferences. From intricate mechanical components to elaborate architectural structures, the possibilities are virtually limitless, allowing for innovation and experimentation across various industries. Moreover, digital modeling facilitates customization and personalization, allowing for the creation of bespoke products tailored to individual preferences and requirements. Companies like Stratasys, 3D Systems, and Formlabs are at the forefront of 3D printing technology, offering a wide range of additive manufacturing solutions that empower designers to bring their creative visions to life. Whether producing customized medical implants, personalized fashion accessories, or unique architectural elements, these industry leaders provide the tools and expertise to enable designers to create one-of-a-kind objects that resonate with end-users on a personal level. The process of 3D printing begins with the creation of a digital 3D model using renowned computer-aided design (CAD) software or 3D scanning technologies pioneered by industry leaders. These tools and technologies empower designers to translate creative visions into tangible digital representations, unleashing unparalleled flexibility and creativity in the design process. From concept to creation, digital modeling in 3D printing offers a transformative approach to design and manufacturing, enabling innovation, customization, and optimization across various industries.

2. *Slicing*: Once the digital model is created, a pivotal step in the 3D printing process involves slicing the model into thin cross-sectional layers. This crucial task is typically performed using specialized slicing software, which plays a fundamental role in preparing the digital model for printing. Each slice generated by the slicing software represents a horizontal section of the object to be printed, with the number of slices determined by factors such as the desired resolution and layer thickness. Slicing software, such as Ultimaker Cura, Simplify3D, or Slic3r, offers a comprehensive suite of tools and functionalities for precisely slicing digital models and optimizing the printing process. Designers can specify parameters such as layer height, infill density, print speed, and support structures to tailor the printing process to their specific requirements and preferences. The slicing process begins by importing the digital model into the slicing software, where it is analyzed and processed to generate a series of 2D slices, or layers, that represent the object's geometry at different heights. The software divides the digital model into discrete layers based on the specified layer thickness, creating a stack of slices that collectively define the object's shape and structure. Each slice is meticulously crafted to ensure optimal print quality and fidelity to the original digital model. The slicing software takes into account factors such as overhangs, bridges, and intricate details, generating support structures as needed to stabilize overhanging features during printing. Support structures are temporary structures added to the design to prevent sagging or deformation of unsupported regions during the printing process. Once the slicing process is complete, the slicing software generates a set of instructions, known as G-code, that serves as a roadmap for the 3D printer. G-code contains a series of commands that specify the exact movements, temperatures, and extrusion rates required to reproduce each layer of the object. These instructions guide the 3D printer's movements and operations throughout the printing process, ensuring precise deposition of material and accurate replication of the digital model in physical form. The G-code generated by the slicing software provides detailed instructions for every aspect of the printing process, including the movement of the print head, the extrusion of material, and the activation of auxiliary functions such as heating, cooling, and bed leveling. Each command is meticulously executed by the 3D printer's onboard controller, orchestrating the intricate dance of motors, heaters, and sensors required to bring the digital model to life. During the printing process, the 3D printer follows the instructions in the G-code, depositing material layer by layer to build up the object from the bottom up. The print head moves along the X, Y, and Z axes, precisely extruding material onto the build platform according to the specified path and parameters. As each layer is completed, the build platform is incrementally lowered to make room for the next layer, gradually building up the object's shape and structure.

The layer-by-layer deposition of material continues until the entire object is printed, with each layer seamlessly integrated to form a cohesive whole. Depending on the 3D printing technology used, the deposited material may be thermoplastic filament, liquid resin, powdered material, or droplets of binder or material, each offering unique properties and capabilities. Throughout the printing process, the slicing software and G-code serve as critical components, orchestrating the complex interplay of digital design data, printer settings, and material properties to ensure successful fabrication of the desired object. By meticulously slicing the digital model into precise layers and generating detailed instructions for the printer, slicing software enables designers to translate virtual designs into physical reality with unparalleled accuracy and precision. The slicing of digital models into thin cross-sectional layers is a pivotal step in the 3D printing process, facilitated by specialized slicing software. Each slice generated by the slicing software represents a horizontal section of the object to be printed, with the number of slices determined by factors such as the desired resolution and layer thickness. Slicing software generates a set of instructions, known as G-code, that guides the 3D printer in depositing material layer by layer to recreate the digital model in physical form. This intricate process ensures precise replication of the digital model and enables designers to bring their creative visions to life with unparalleled accuracy and fidelity.

3. *Printing Process*: With the sliced model and G-code prepared, the 3D printing process commences, marking the transformation of digital design data into physical reality. As the 3D printer springs to life, it embarks on a meticulous journey to fabricate the desired object layer by layer. This transformative process unfolds through a series of precise movements and material depositions orchestrated by the instructions encoded within the G-code. At the heart of the printing process lies the chosen printing material, carefully selected to suit the specific printing technique and application. Whether it be thermoplastics like PLA or ABS, metal powders, ceramics, or resins, the material serves as the building blocks for the object's creation. Each material possesses unique properties and characteristics that dictate its suitability for particular applications, ranging from prototyping and manufacturing to biomedical and aerospace applications. As the printing material is fed into the printer's extrusion system, it undergoes a transformation dictated by the printing technique employed. In extrusion-based printing, such as fused deposition modeling (FDM) or fused filament fabrication (FFF), thermoplastic filaments are heated to their melting point within the printer's extruder assembly. The molten material is then extruded through a heated nozzle, depositing precise layers onto the build platform below. Alternatively, in polymerization-based printing techniques like stereolithography (SLA) or digital light processing (DLP), liquid photopolymer resins are cured or solidified using UV light. The printer's build platform descends incrementally as each layer is exposed to UV light, causing the resin to solidify and adhere to the previous layer. This process continues until the entire object is formed, with each layer seamlessly integrated to create a cohesive whole. In powder-based

printing methods like selective laser sintering (SLS) or selective laser melting (SLM), a bed of powdered material is selectively fused together using a laser or electron beam. The printer's print head or laser system scans across the surface of the powder bed, selectively melting or sintering the material according to the instructions encoded within the G-code. Layer by layer, the object emerges from the powdered bed, with each pass of the laser contributing to its gradual formation. Regardless of the printing technique employed, the printer nozzle or print head moves with precision along the X, Y, and Z axes, depositing material according to the instructions provided by the G-code. The intricate dance of motors and actuators orchestrates the controlled movement of the print head, ensuring precise deposition of material and accurate replication of the digital model. Layer by layer, the object begins to take shape, with each successive layer adhering to the previous one to form a cohesive whole. The printer's intricate movements and material depositions result in the gradual emergence of intricate geometries, complex structures, and fine details, faithfully replicating the digital model in physical form. Throughout the printing process, the printer's parameters and settings are carefully monitored and adjusted to ensure optimal print quality and performance. Parameters such as print speed, nozzle temperature, layer thickness, and infill density may be fine-tuned to achieve desired outcomes and address specific requirements. As the printing progresses, the object gradually materializes before the eyes, evolving from a digital abstraction into a tangible reality. Whether producing prototypes, functional parts, or artistic creations, the 3D printing process empowers designers and manufacturers to bring their ideas to life with unprecedented speed, precision, and flexibility. The 3D printing process unfolds through a meticulously orchestrated series of movements and material depositions guided by the instructions encoded within the G-code. With the sliced model and G-code prepared, the printer embarks on the task of fabricating the desired object layer by layer. Whether heating, melting, or curing the chosen printing material, the printer executes precise movements along the X, Y, and Z axes, depositing material according to the instructions provided by the G-code. Layer by layer, the object gradually takes shape, with each successive layer adhering to the previous one to form a cohesive whole. Through this transformative process, the digital model is brought to life, transcending the virtual realm to become a tangible reality.

4. *Material Deposition and Solidification*: As the 3D printer executes its precise movements and material depositions guided by the G-code, each layer of material is strategically fused or solidified to gradually form a solid object. This crucial step in the 3D printing process is essential for achieving the desired properties and mechanical strength of the final object. Depending on the specific printing technique employed, this may involve heating, cooling, curing, or sintering the material to bring the digital model to life in physical form. In fused deposition modeling (FDM), one of the most widely used 3D printing techniques, thermoplastic filaments are fed into the printer's extruder assembly, where they are heated to their melting point and extruded through

a nozzle onto the build platform below. As the molten filament is deposited onto the platform, it quickly cools and solidifies, forming a solid layer that adheres to the previous layer. This process is repeated layer by layer until the entire object is formed, with each successive layer seamlessly integrated to create a cohesive whole. The key to the success of FDM lies in the precise control of the extrusion temperature, extrusion rate, and cooling conditions, which influence the quality and mechanical properties of the printed object. By carefully adjusting these parameters, designers can achieve optimal adhesion between layers, minimize warping and distortion, and produce parts with the desired strength and dimensional accuracy. In stereolithography (SLA) and digital light processing (DLP), polymerization-based 3D printing techniques, liquid photopolymer resins are cured or solidified using ultraviolet (UV) light. In SLA, a UV laser selectively cures layers of liquid resin on the surface of a resin tank, while in DLP, a digital light projector illuminates the entire surface of the resin with UV light, curing an entire layer simultaneously. As the UV light interacts with the photopolymer resin, it triggers a chemical reaction known as photopolymerization, causing the resin to solidify and harden. This process results in the formation of solid layers that gradually build up to create the final object. The printer's build platform descends incrementally after each layer is cured, making room for the next layer to be formed. The success of SLA and DLP lies in the precise control of the curing parameters, including exposure time, light intensity, and resin viscosity, which influence the resolution, surface finish, and mechanical properties of the printed object. By optimizing these parameters, designers can achieve high levels of detail, fine features, and smooth surface finishes, making SLA and DLP ideal for producing intricate prototypes, jewelry, and dental appliances. In selective laser sintering (SLS) and selective laser melting (SLM), powder-based 3D printing techniques commonly used for metal and ceramic materials, a high-powered laser selectively fuses powdered material together to form each layer of the object. In SLS, the laser selectively sinters powdered material, while in SLM, the laser fully melts the powder to create a dense, solid layer. As the laser scans across the surface of the powder bed, it selectively heats and fuses the particles together, forming a solid layer that adheres to the previous layer. This process is repeated layer by layer until the entire object is formed, with each layer fully integrated to create a dense, homogeneous structure. The success of SLS and SLM lies in the precise control of the laser parameters, including laser power, scanning speed, and powder bed temperature, which influence the density, porosity, and mechanical properties of the printed object. By optimizing these parameters, designers can achieve parts with excellent mechanical strength, dimensional accuracy, and surface finish, making SLS and SLM ideal for producing functional prototypes, end-use parts, and complex geometries. As the 3D printer deposits material layer by layer, each layer is strategically fused or solidified to create a solid object. Depending on the printing technique employed, this may involve heating, cooling, curing, or sintering the material to achieve the desired properties and mechanical strength. Whether using FDM, SLA, SLS, or SLM, the precise control of printing parameters is essential for achieving optimal print quality, dimensional accuracy, and mechanical performance. Through careful calibration and optimization, designers can harness the power of 3D printing to bring their creative visions to life with unparalleled precision and efficiency.

5. *Post-processing*: Once the 3D printing process reaches its culmination and the object is fully fabricated layer by layer, the journey is not yet complete. The printed object may undergo post-processing steps to refine its surface finish, enhance its accuracy, or improve its mechanical properties. These post-processing techniques play a crucial role in elevating the quality and aesthetics of the final product, ensuring that it meets the desired standards and specifications. One of the primary post-processing steps often encountered is the removal of support structures. During the printing process, especially in techniques like FDM or SLS, support structures are often generated to provide stability for overhanging features or intricate geometries. Once the printing is complete, these support structures need to be carefully removed using tools like pliers, wire cutters, or specialized support removal systems. This step requires precision to avoid damaging the printed object while ensuring that all support material is completely removed. Following support removal, the printed object may undergo sanding to smooth out rough surfaces and eliminate visible layer lines. Sanding is particularly beneficial for improving surface finish and achieving a uniform texture across the entire object. Depending on the desired outcome, various grits of sandpaper may be used, starting from coarse grits to remove larger imperfections and gradually progressing to finer grits for a smoother finish. Polishing is another common post-processing technique used to enhance the appearance of 3D printed objects. By applying abrasive compounds or polishing pastes to the surface of the printed object and buffing it with a cloth or polishing tool, a glossy finish can be achieved. This technique is especially effective for materials like resin or thermoplastics, where surface imperfections can be minimized to create a sleek, professional look. In addition to sanding and polishing, painting is often employed to add color, texture, or additional features to the printed object. Acrylic paints, spray paints, or airbrushing techniques can be used to customize the appearance of the object according to specific design requirements or aesthetic preferences. Painting can also be used to highlight details, create gradients, or add intricate patterns to the surface of the printed object, enhancing its visual appeal and overall attractiveness. For certain applications, applying additional coatings or treatments may be necessary to improve the mechanical properties or functional performance of the printed object. For example, applying a clear coat or protective sealant can provide added durability and resistance to environmental factors such as moisture, heat, or UV radiation. Similarly, surface treatments such as chemical vapor smoothing or flame polishing can be used to further refine surface finish and eliminate minor imperfections. The specific post-processing steps required may vary depending on factors such as the printing material, technique, and desired outcome. For instance,

materials like metals or ceramics may require additional heat treatment or annealing processes to enhance their mechanical properties and structural integrity. Meanwhile, materials like thermoplastics or resins may benefit from annealing, UV curing, or chemical treatments to improve their strength, flexibility, or chemical resistance. In some cases, post-processing steps may be minimal, especially for objects printed with high-resolution techniques like SLA or SLS that inherently produce smoother surface finishes and finer details. However, for objects with complex geometries, large overhangs, or rough surfaces, extensive post-processing may be necessary to achieve the desired level of quality and functionality. Ultimately, post-processing plays a critical role in enhancing the overall quality, appearance, and performance of 3D printed objects. By carefully selecting and executing appropriate post-processing techniques, designers and manufacturers can ensure that their creations meet the highest standards of excellence and fulfill the intended purpose with precision and reliability. Whether it's removing support structures, sanding, polishing, painting, or applying specialized coatings, each post-processing step contributes to the transformation of raw printed objects into refined, finished products ready for use or display.

6. Material Versatility: One of the defining features and strengths of 3D printing lies in its remarkable versatility and adaptability regarding the materials it can utilize to create objects with diverse properties and functionalities. Unlike traditional manufacturing methods limited by material constraints, 3D printers have the capability to work with an extensive range of materials, spanning from plastics and metals to ceramics, composites, and even biological materials. This versatility opens up endless possibilities for innovation and customization, empowering designers and engineers to create products tailored to meet specific performance criteria and application requirements. Plastics are among the most commonly used materials in 3D printing, owing to their affordability, ease of use, and wide availability. Thermoplastics like polylactic acid (PLA) and acrylonitrile butadiene styrene (ABS) are popular choices for desktop 3D printers due to their low cost, ease of printing, and versatility. PLA, derived from renewable resources such as corn starch or sugarcane, is biodegradable and environmentally friendly, making it a sustainable option for prototyping and consumer goods. ABS, known for its strength, durability, and impact resistance, is commonly used in industrial applications such as automotive parts, electronic enclosures, and functional prototypes. Furthermore, advanced engineering thermoplastics such as polyether ether ketone (PEEK), polyamide (PA), and polyethylene terephthalate glycol (PETG) offer enhanced mechanical properties, thermal stability, and chemical resistance, making them suitable for demanding applications in aerospace, medical, and automotive industries. These materials command higher prices due to their superior performance characteristics and specialized processing requirements but offer significant benefits in terms of strength, durability, and reliability. Metal 3D printing, also known as additive manufacturing (AM) or direct metal laser sintering (DMLS), has emerged

as a transformative technology for producing high-performance metal parts with complex geometries and superior mechanical properties. Metals such as stainless steel, aluminum, titanium, and nickel alloys can be processed using metal 3D printing techniques, offering exceptional strength, corrosion resistance, and thermal conductivity. The aerospace, automotive, and healthcare industries are among the primary beneficiaries of metal 3D printing, leveraging its capabilities to produce lightweight components, intricate structures, and patient-specific implants. Ceramic materials, prized for their high temperature resistance, electrical insulation, and biocompatibility, are gaining traction in 3D printing for applications in electronics, aerospace, and biomedical fields. Silicon carbide (SiC), alumina (Al_2O_3), and zirconia (ZrO_2) are commonly used ceramic materials in 3D printing, offering excellent mechanical properties, thermal stability, and chemical inertness. Ceramic 3D printing techniques such as binder jetting and stereolithography (SLA) enable the fabrication of complex ceramic parts with precise geometries and fine details, opening up new opportunities for advanced manufacturing and prototyping. Composites, composed of a combination of two or more materials with distinct properties, offer a unique blend of characteristics such as strength, stiffness, and lightweight. Carbon fiber-reinforced polymers (CFRP), glass fiber-reinforced polymers (GFRP), and metal matrix composites (MMC) are examples of composite materials commonly used in 3D printing. These materials are valued for their high strength-to-weight ratio, fatigue resistance, and thermal stability, making them ideal for aerospace, automotive, and sports equipment applications. Biological materials, including living cells, proteins, and biomaterials, hold tremendous promise for biomedical applications such as tissue engineering, regenerative medicine, and drug delivery. Bioprinting, a specialized form of 3D printing, enables the precise deposition of biological materials to create complex three-dimensional structures mimicking native tissues and organs. Materials such as hydrogels, alginate, and bioinks are commonly used in bioprinting, offering biocompatibility, cell adhesion, and tissue regeneration properties essential for biomedical research and clinical applications. The market for 3D printing materials is experiencing rapid growth, driven by technological advancements, expanding applications, and increasing demand for customized, high-performance products. According to a report by MarketsandMarkets, the global 3D printing materials market is projected to reach \$4.5 billion by 2025, with a compound annual growth rate (CAGR) of 23.5% from 2020 to 2025. This growth is fueled by the rising adoption of 3D printing across various industries, including aerospace, automotive, healthcare, and consumer goods, as well as ongoing research and development efforts to expand the range of printable materials and improve printing processes. The ability of 3D printing to utilize a wide range of materials is a key enabler of its versatility and applicability across diverse industries. From plastics and metals to ceramics, composites, and biological materials, each material offers unique properties and functionalities that can be harnessed to meet specific design requirements and performance criteria. As the 3D printing materials market continues to evolve and innovate, fueled by technological advancements and expanding applications, the possibilities for customized, high-performance products are virtually limitless.

7. Layer-by-Layer Construction: Central to the principle of 3D printing is the layer-by-layer construction of objects, which enables precise control over geometry, complexity, and internal structures. Unlike traditional manufacturing methods, which often involve multiple steps and assembly processes, 3D printing builds objects in a single, continuous process, layer by layer. This additive approach allows for the creation of highly intricate and detailed designs that would be difficult or impossible to achieve using conventional methods. 3D printing offers unprecedented design freedom, allowing designers to create complex shapes, geometries, and structures that were previously impractical or unattainable. With traditional manufacturing methods, designers are often constrained by the limitations of machining, casting, or molding processes. In contrast, 3D printing enables the fabrication of organic, asymmetrical, and highly customized designs without the need for specialized tooling or equipment. This design freedom encourages innovation, experimentation, and creativity in product development and design. Another fundamental principle of 3D printing is its ability to accelerate the product development cycle through rapid prototyping. By quickly transforming digital designs into physical prototypes, 3D printing allows designers and engineers to test and iterate their ideas in a fraction of the time required by traditional methods. This iterative approach enables faster validation of design concepts, identification of potential issues or improvements, and ultimately, shorter time-to-market for new products and innovations. 3D printing enables mass customization and personalization of products, allowing manufacturers to produce individualized items tailored to the unique preferences and requirements of each customer. Whether it's custom-fit prosthetics, personalized medical implants, or bespoke consumer goods, 3D printing makes it possible to create one-of-a-kind products at scale. This customization capability not only enhances customer satisfaction but also opens up new opportunities for niche markets, small-batch production, and personalized healthcare solutions.

Unlike traditional manufacturing methods, which often generate significant waste through material removal or excess inventory, 3D printing minimizes waste by only using the exact amount of material needed to build each object. This principle of additive manufacturing results in higher material efficiency, reduced scrap, and lower environmental impact compared to subtractive processes. Additionally, 3D printing supports sustainability initiatives by enabling recycling of materials, use of biodegradable polymers, and local production to reduce transportation emissions. 3D printing facilitates distributed manufacturing, allowing production to be decentralized and localized to meet demand closer to the point of consumption. This principle of additive manufacturing reduces reliance on centralized factories, supply chains, and global logistics networks, leading to shorter lead times, lower inventory costs, and increased resilience to disruptions. By enabling on-demand produc-tion of parts and products, 3D printing supports just-in-time manufacturing, customization, and agile response to changing market conditions.

In summary, the principle of 3D printing revolves around additive manufacturing processes that build objects layer by layer from digital designs. This approach offers numerous advantages, including material versatility, design freedom, rapid prototyping, customization, waste reduction, and distributed manufacturing. By harnessing the power of 3D printing, industries can innovate faster, produce more efficiently, and create products that are tailored to individual needs and preferences, ultimately leading to a more sustain-able and prosperous future. Overall, the principle of 3D printing enables the creation of complex, customized, and functional objects with unprecedented speed and efficiency. By building objects layer by layer from digital designs, 3D printing offers limitless possibili-ties for innovation, prototyping, and manufacturing across industries, from aerospace and automotive to healthcare and consumer goods. As the technology continues to advance and mature, it is poised to revolutionize the way we design, produce, and interact with objects in the digital age.

Types and Classifications of 3D Printing:

There are several types of 3D printing technologies, each with its unique process and characteristics. Some common types include:

1. *Fused Deposition Modeling (FDM) or Fused Filament Fabrication (FFF)*: This process involves extruding thermoplastic filaments through a heated nozzle, which deposits layers of molten material to build the object.Fused Deposition Modeling (FDM), also known as Fused Filament Fabrication (FFF), is one of the most widely used 3D printing technologies, particularly in desktop or hobbyist 3D printers. It was developed by Scott Crump in the late 1980s and commercialized by his company Stratasys in the early 1990s. FDM/FFF works by extruding thermoplastic filament through a heated nozzle, which melts the material, allowing it to be deposited layer by layer to build up the desired object.

The process begins with a digital model of the object, typically created using computer-aided design (CAD) software. This model is then sliced into thin horizontal layers, with each layer serving as a blueprint for the 3D printer. The printer's software interprets these layers and generates a toolpath that dictates the movement of the print head and the deposition of material. During printing, the filament is fed into the extruder, which heats it to its melting point. The molten material is then deposited onto the build platform or previous layers, where it quickly cools and solidifies, bonding to the underlying layers. This layer-by-layer approach continues until the entire object is formed. FDM/FFF offers several advantages that have contributed to its widespread adoption. Firstly, it is relatively simple and cost-effective compared to other 3D printing technologies, making it accessible to a wide range of users, from hobbyists to professionals. Additionally, FDM/FFF printers are known for their ease of use and low maintenance requirements, making them suitable for both beginners and experienced users alike. Another advantage of FDM/FFF is its versatility in terms of materials. A variety of thermoplastic filaments are available for use with FDM/FFF printers, including PLA, ABS, PETG, TPU, and more. Each filament type offers different properties such as strength, flexibility, temperature resistance, and surface finish, allowing users to choose the most suitable material for their specific application. Furthermore, FDM/FFF excels in producing functional prototypes, concept models, and low-volume production parts with reasonable accuracy and surface finish. While not as precise as some other 3D printing technologies, such as SLA or SLS, FDM/FFF can achieve tolerances within a few tenths of a millimeter, depending on factors such as nozzle size, layer height, and printer calibration.

Despite its many advantages, FDM/FFF also has some limitations. One notable drawback is its relatively slow printing speed compared to other 3D printing technologies, especially when printing objects with intricate geometries or fine details. Additionally, FDM/FFF parts may exhibit visible layer lines or surface imperfections, which can require post-processing techniques such as sanding, painting, or chemical smoothing to improve the overall appearance. Fused Deposition Modeling (FDM) or Fused Filament Fabrication (FFF) is a popular and accessible 3D printing technology that has found widespread use in various industries and applications. With its simplicity, versatility, and cost-effectiveness, FDM/FFF continues to be a valuable tool for prototyping, production, and innovation in the ever-expanding world of additive manufacturing.

2. Stereolithography (SLA): SLA utilizes a vat of liquid photopolymer resin and a UV laser to selectively cure the resin layer by layer, solidifying it into the desired shape. Stereolithography (SLA) is a pioneering 3D printing technology that revolutionized the additive manufacturing industry when it was invented by Chuck Hull in the late 1980s. It was the world's first commercial 3D printing process and laid the foundation for subsequent advancements in additive manufacturing. SLA operates on the principle of photopolymerization, where liquid photopolymer resin is selectively cured by a UV laser to create solid objects layer by layer. The SLA process begins with a digital 3D model of the object, typically designed using computer-aided design (CAD) software. This digital model is sliced into thin cross-sectional layers, each representing a horizontal section of the final object. The SLA printer's software then uses these slices to generate a series of two-dimensional patterns, known as cross-sections or layers, which will be traced by the UV laser during printing. During printing, a platform submerged in a vat of liquid photopolymer resin is lowered incrementally, allowing the UV laser to trace the first layer of the object's geometry on the surface of the resin. Wherever the laser beam strikes the resin, it causes a chemical reaction that solidifies the material, forming a thin layer of cured resin. Once the first layer is completed, the platform moves slightly downward, and the process repeats, with the laser tracing the next layer

on top of the previous one. This layer-by-layer approach continues until the entire object is formed, with each layer fusing seamlessly to the layers below it. SLA offers several advantages that have contributed to its popularity and widespread adoption in various industries. One of the key advantages is its ability to produce highly detailed, intricate, and accurate parts with smooth surface finishes. The fine resolution and layer thickness achievable with SLA make it ideal for applications requiring tight tolerances, such as prototyping, jewelry design, and dental models. Additionally, SLA is capable of producing parts with a wide range of mechanical properties by using different types of photopolymer resins. These resins can be formulated to exhibit properties such as high strength, flexibility, temperature resistance, and biocompatibility, making SLA suitable for a diverse array of applications across industries, from aerospace and automotive to healthcare and consumer goods. Another advantage of SLA is its speed and efficiency compared to other 3D printing technologies. With its ability to cure entire layers of resin simultaneously using a single UV laser beam, SLA can achieve faster printing speeds and higher throughput, making it well-suited for rapid prototyping and small-batch production.

Despite its many advantages, SLA also has some limitations and considerations. One notable limitation is the necessity of post-processing to remove excess resin and support structures and to cure the printed parts fully. Additionally, the cost of SLA printers and materials may be higher than other 3D printing technologies, particularly for industrial-grade systems and specialized resins. Stereolithography (SLA) is a pioneering 3D printing technology that has played a significant role in advancing additive manufacturing. With its ability to produce highly detailed, accurate, and functional parts with a wide range of mechanical properties, SLA continues to be a valuable tool for prototyping, production, and innovation in various industries.

3. Selective Laser Sintering (SLS): SLS uses a high-powered laser to selectively sinter powdered material, such as plastics, metals, or ceramics, into a solid layer, with each layer fused to the previous one. Selective Laser Sintering (SLS) is a powerful additive manufacturing technology that has gained prominence in various industries for its ability to produce high-quality, functional parts with a wide range of materials. Developed in the 1980s by Carl Deckard and Joseph Beaman at the University of Texas at Austin, SLS has since become a cornerstone of the additive manufacturing landscape, offering unparalleled versatility, accuracy, and efficiency. At its core, SLS works by selectively fusing powdered material using a high-powered laser to create three-dimensional objects layer by layer. Unlike other 3D printing processes that rely on liquid resin or filament feedstock, SLS utilizes powdered materials such as plastics, metals, ceramics, and composites. This powder is spread evenly across the build platform, and a laser selectively sinters, or fuses, the powder particles together based on

a digital model of the object being printed. The SLS process begins with the preparation of the build chamber, where a thin layer of powdered material is spread evenly across the build platform using a recoating mechanism. A high-powered laser then scans the surface of the powder bed, selectively sintering the particles together based on the cross-sectional shape of the object being printed. Once a layer is completed, the build platform is lowered by a small increment, and the process is repeated, with each subsequent layer fusing to the layer below it.

One of the key advantages of SLS is its ability to produce parts with complex geometries and intricate internal features without the need for support structures. Unlike other 3D printing processes that require supports to hold up overhanging features during printing, SLS relies on unsintered powder surrounding the part to provide support, making it ideal for producing parts with internal channels, cavities, and lattice structures. Moreover, SLS offers a wide range of materials with varying properties and characteristics, allowing users to choose the most suitable material for their specific application requirements. Common materials used in SLS include nylon, polyamide, polystyrene, metals such as aluminum, titanium, and stainless steel, as well as ceramics and composites. These materials exhibit properties such as high strength, temperature resistance, chemical resistance, and biocompatibility, making SLS suitable for a diverse array of applications across industries, from aerospace and automotive to healthcare and consumer goods. Another advantage of SLS is its scalability and cost-effectiveness for small to medium batch production. Since SLS does not require support structures and can print multiple parts simultaneously within the same build volume, it offers significant time and cost savings compared to traditional manufacturing methods such as machining or casting. Additionally, SLS enables rapid iteration and design optimization, allowing users to quickly test and refine prototypes or iterate on designs without the need for expensive tooling or molds. Despite its many advantages, SLS also has some limitations and considerations. One notable limitation is the surface finish of SLS parts, which may exhibit a rough texture or grainy appearance due to the nature of the sintering process. Post-processing techniques such as sanding, polishing, or media blasting may be required to achieve smoother surface finishes for certain applications.

Furthermore, the cost of SLS equipment and materials may be prohibitive for some users, particularly for industrial-grade systems and specialized materials. Additionally, the handling and disposal of unsintered powder can present challenges in terms of safety, cleanliness, and environmental impact, requiring proper ventilation, filtration, and waste management protocols. Selective Laser Sintering (SLS) is a versatile and powerful additive manufacturing technology that offers numerous advantages for producing high-quality, functional parts with a wide range of materials. With its ability to produce complex geometries, diverse material options, and cost-effective batch production capabilities, SLS continues to be a valuable tool for prototyping, production, and innovation in various industries. As technology advances and materials evolve, SLS is expected to play an increasingly important role in shaping the future of additive manufacturing.

4. *Digital Light Processing (DLP):* Similar to SLA, DLP employs a digital light projector to cure liquid resin into a solid layer, but it uses a different method of light projection. Digital Light Processing (DLP) is an advanced 3D printing technology that utilizes digital light projection to cure liquid photopolymer resin and create three-dimensional objects layer by layer. Developed in the 1980s by Larry Hornbeck at Texas Instruments, DLP was initially used in projection systems for televisions and projectors before being adapted for additive manufacturing applications.

At its core, DLP works by projecting a digital image onto a vat of liquid photopolymer resin, selectively curing the resin layer by layer to build up the desired object. Unlike other 3D printing technologies that use lasers or extrusion methods, DLP employs a digital micromirror device (DMD) chip, consisting of millions of microscopic mirrors arranged in a grid pattern. Each mirror can be individually tilted to reflect light either towards or away from the resin vat, creating a pattern of light and shadow that corresponds to the cross-sectional shape of the object being printed. The DLP printing process begins with the preparation of the build platform, which is submerged in a vat of liquid photopolymer resin. A light source, typically a high-intensity LED or UV lamp, illuminates the DMD chip, projecting a two-dimensional image onto the surface of the resin. The pattern of light and shadow generated by the DMD chip selectively cures the resin, solidifying it into a thin layer of the desired shape. Once a layer is completed, the build platform is lowered by a small increment, and the process is repeated, with each subsequent layer fusing to the layer below it. This layer-by-layer approach continues until the entire object is formed, with each layer solidifying in the desired shape. After printing is complete, the printed object is removed from the vat, cleaned of excess resin, and cured under UV light to fully harden the material.

One of the key advantages of DLP is its ability to produce highly detailed, accurate, and smooth parts with fine features and intricate geometries. The digital micromirror array allows for precise control over the exposure of each pixel, enabling high-resolution printing with minimal pixelation or stair-stepping artifacts. Additionally, DLP is capable of producing parts with consistent mechanical properties and isotropic strength, making it suitable for functional prototypes, jewelry, dental models, and other applications requiring high precision and surface finish. Moreover, DLP offers a wide range of photopolymer resins with varying properties and characteristics, including standard resins, engineering resins, flexible resins, and biocompatible resins. These resins can be formulated to exhibit properties such as high strength, temperature resistance, flexibility, transparency, and biocompatibility, allowing users to choose the most suitable material for their specific application requirements. Another advantage of DLP is its speed and efficiency compared to other 3D printing technologies. With its ability to cure entire layers of resin simultaneously using a digital light projection system, DLP can achieve faster printing speeds and higher throughput, making it well-suited for rapid prototyping, small-batch production, and on-demand manufacturing. Despite its many advantages, DLP also has some limitations and considerations. One notable limitation is the necessity of support structures to hold up overhanging features during printing, as well as the need for post-processing to remove excess resin and support structures and to cure the printed parts fully. Additionally, the cost of DLP printers and materials may be higher than other 3D printing technologies, particularly for industrial-grade systems and specialized resins. Digital Light Processing (DLP) is a sophisticated and versatile 3D printing technology that offers numerous advantages for producing high-quality, detailed, and functional parts with a wide range of materials. With its ability to achieve high-resolution printing, consistent mechanical properties, and fast printing speeds, DLP continues to be a valuable tool for prototyping, production, and innovation in various industries. As technology advances and materials evolve, DLP is expected to play an increasingly important role in shaping the future of additive manufacturing.

3D printer Technology

3D printing, also known as additive manufacturing, encompasses a diverse range of technologies that enable the layer-by-layer fabrication of three-dimensional objects from digital design data. These technologies can be broadly classified into four main categories: extrusion printing, polymerization, powder-based printing, and droplet printing as summarized in Fig. 3.4. Each category utilizes distinct processes and materials to build objects layer by layer, catering to a wide range of applications and industries.

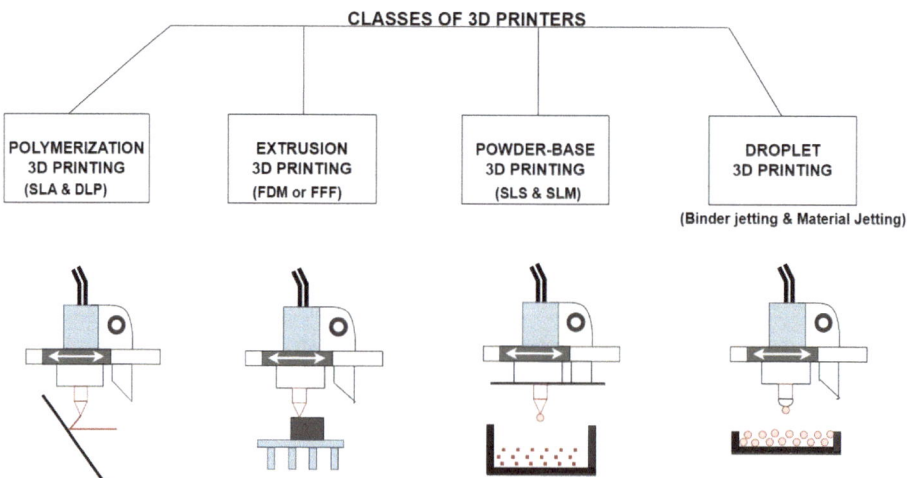

Fig. 3.4 Classification of 3D printer technology

Extrusion printing, also referred to as fused deposition modeling (FDM) or fused filament fabrication (FFF), is one of the most widely used 3D printing technologies. In extrusion printing, a thermoplastic filament is heated to its melting point and extruded through a nozzle onto a build platform, where it solidifies to form each layer of the object. The build platform is lowered by a fraction of a millimeter after each layer is deposited, allowing subsequent layers to adhere to the previous ones. Extrusion printing offers several advantages, including affordability, simplicity, and versatility. FDM printers are relatively inexpensive and easy to use, making them accessible to hobbyists, educators, and small businesses. Additionally, a wide range of thermoplastic materials can be used in extrusion printing, including PLA, ABS, PETG, and TPU, allowing for diverse applications across various industries. However, extrusion printing also has limitations, including limited resolution and surface finish compared to other 3D printing technologies. The layer-by-layer deposition process can result in visible layer lines and surface imperfections, which may require post-processing to achieve smooth, high-quality finishes. Additionally, the mechanical properties of parts produced through extrusion printing can be anisotropic, meaning they may vary depending on the direction of printing.

Polymerization-based 3D printing, also known as stereolithography (SLA) and digital light processing (DLP), utilizes photopolymerization to create objects from liquid resin materials. In SLA printing, a UV laser or light source is used to selectively cure layers of liquid resin to create the desired object. In DLP printing, a digital light projector is used to project UV light onto the entire surface of the resin, curing an entire layer simultaneously. Polymerization-based printing offers several advantages, including high resolution, accuracy, and surface finish. SLA and DLP printers are capable of producing intricate details and fine features with excellent dimensional accuracy, making them well-suited for applications requiring high precision and aesthetics. Additionally, a wide range of photopolymer resins are available, including standard, engineering, and specialty materials with various properties such as flexibility, toughness, and biocompatibility. However, polymerization-based printing also has limitations, including limited build volume and slower print speeds compared to extrusion-based technologies. SLA and DLP printers typically have smaller build volumes and longer print times due to the layer-by-layer curing process. Additionally, post-processing of printed parts may be required to remove uncured resin and achieve desired mechanical properties.

Powder-based 3D printing, also known as selective laser sintering (SLS) and selective laser melting (SLM), utilizes a bed of powdered material, such as plastic, metal, or ceramic, to build objects layer by layer. In SLS printing, a laser selectively fuses powdered material together to create each layer of the object, while in SLM printing, the powdered material is fully melted and fused to form each layer. Powder-based printing offers several advantages, including high resolution, complexity, and material versatility. SLS and SLM printers are capable of producing complex geometries and intricate designs with excellent resolution and surface finish. Additionally, a wide range of materials can be used in powder-based printing, including polymers, metals, and ceramics, offering diverse applications across various industries. However, powder-based printing also has limitations, including limited resolution and post-processing requirements. The layer-by-layer fusion process can result in visible layer lines and surface roughness, which may require post-processing such as sanding, polishing, or coating to achieve smooth, high-quality finishes. Additionally, powder-based printing typically requires additional equipment and safety precautions due to the handling of powdered materials and exposure to high temperatures.

Droplet-based 3D printing, also known as binder jetting and material jetting, utilizes droplets of liquid binder or material to selectively bond particles together to create each layer of the object. In binder jetting, a liquid binder is deposited onto a bed of powdered material, such as sand or metal, to bind the particles together and form each layer. In material jetting, liquid photopolymer resin is jetted onto a build platform and cured with UV light to create each layer of the object. Droplet-based printing offers several advantages, including speed, scalability, and material versatility. Binder jetting and material jetting printers are capable of producing parts quickly and cost-effectively, making them suitable for high-volume production applications. Additionally, a wide range of materials can be used in droplet-based printing, including metals, ceramics, plastics, and composites, offering diverse applications across various industries. However, droplet-based printing also has limitations, including limited resolution and material properties. The layer-by-layer deposition process can result in visible layer lines and surface roughness, which may require post-processing to achieve desired surface finishes. Additionally, printed parts may have anisotropic mechanical properties, meaning they may vary depending on the direction of printing.

3D printing encompasses a diverse range of technologies that enable the layer-by-layer fabrication of three-dimensional objects from digital design data. Each 3D printing technology has its own advantages and limitations, catering to specific applications and industries. Extrusion printing, polymerization, powder-based printing, and droplet-based printing offer unique capabilities and material options, allowing for a wide range of applications in manufacturing, healthcare, aerospace, automotive, and other industries. As 3D printing technology continues to evolve and mature, its role in manufacturing and production is expected to expand, driving innovation and transformation across various sectors.

Filaments Used in 3D Printing:

The choice of filament material significantly impacts the properties and applications of 3D-printed objects. An example of a filament is shown in Fig. 3.5.

Common filaments include:

1. *Thermoplastics*: PLA (Polylactic Acid), ABS (Acrylonitrile Butadiene Styrene), PETG (Polyethylene Terephthalate Glycol), and TPU (Thermoplastic Polyurethane) are

Fig. 3.5 A picture of filaments
used for 3D printing

widely used for their affordability, ease of use, and versatility. Thermoplastic filament is one of the most commonly used materials in 3D printing, offering versatility,
affordability, and ease of use. Thermoplastics are polymers that become soft and
moldable when heated and solidify when cooled, making them ideal for additive
manufacturing processes like Fused Deposition Modeling (FDM) or Fused Filament
Fabrication (FFF). These filaments come in various types, each with unique properties
and characteristics suited to different applications.

One of the most popular thermoplastic filaments used in 3D printing is Polylactic Acid
(PLA). PLA is a biodegradable and environmentally friendly thermoplastic derived from
renewable resources such as corn starch or sugarcane. It is known for its low melting
point, ease of printing, and vibrant color options. PLA produces prints with a glossy finish
and excellent dimensional accuracy, making it ideal for prototyping, hobbyist projects, and
educational purposes. Another widely used thermoplastic filament is Acrylonitrile Butadiene Styrene (ABS). ABS is a strong, impact-resistant material known for its durability,
heat resistance, and versatility. It has a higher melting point than PLA, making it suitable
for applications requiring higher temperature resistance, such as automotive parts, functional prototypes, and mechanical components. ABS prints have a matte finish and can be
sanded, glued, and painted easily for post-processing. Polyethylene Terephthalate Glycol
(PETG) is another popular thermoplastic filament known for its durability, transparency,
and chemical resistance. PETG combines the strength and temperature resistance of ABS
with the ease of printing and low warping properties of PLA. It produces prints with
excellent layer adhesion, minimal shrinkage, and minimal odor during printing, making it
suitable for a wide range of applications, including medical devices, food containers, and
mechanical parts.

Thermoplastic elastomers (TPE) and Thermoplastic Polyurethane (TPU) filaments are flexible, rubber-like materials known for their elasticity, resilience, and soft-touch feel. These filaments are commonly used in applications requiring impact resistance, vibration dampening, and flexibility, such as gaskets, seals, and prosthetics. TPE and TPU filaments offer excellent layer adhesion and can be printed with varying degrees of flexibility and hardness, depending on the specific material formulation. Polyvinyl Alcohol (PVA) is a water-soluble thermoplastic filament used as a support material in dual-extrusion 3D printing. PVA dissolves in water, making it ideal for creating complex geometries and overhangs that would otherwise require cumbersome support structures. PVA supports are easy to remove by soaking the printed object in water, leaving behind a clean, smooth surface finish. PVA is commonly used in conjunction with PLA and other materials that do not adhere well to themselves for applications requiring intricate details and overhangs. Thermoplastic filaments are essential materials in 3D printing, offering a wide range of options for creating functional prototypes, end-use parts, and artistic creations. From biodegradable PLA to durable ABS, flexible TPU, and water-soluble PVA, thermoplastics provide users with the flexibility, versatility, and performance needed to bring their ideas to life. As technology advances and materials evolve, thermoplastic filaments are expected to continue playing a crucial role in the future of additive manufacturing, driving innovation, and expanding the possibilities of 3D printing.

2. *Metals*: Aluminum, titanium, stainless steel, and cobalt-chrome alloys are used in metal 3D printing processes like SLM and EBM for applications requiring high strength, durability, and heat resistance. Metals filaments for 3D printing represent a significant advancement in additive manufacturing technology, allowing users to create metal parts with complex geometries, high strength, and excellent mechanical properties. Metal filaments are used in processes such as Selective Laser Melting (SLM) and Electron Beam Melting (EBM), where metal powders are fused together layer by layer to build up the final object. These filaments come in various types of metals and metal alloys, each with unique properties and characteristics suited to different applications.

One of the most commonly used metals in 3D printing is stainless steel. Stainless steel filaments offer excellent corrosion resistance, strength, and durability, making them suitable for a wide range of industrial applications, including aerospace, automotive, and medical devices. Stainless steel prints can be post-processed through polishing, sandblasting, or electroplating to achieve a smooth surface finish and improve aesthetics. Another popular metal filament is aluminum. Aluminum is known for its lightweight, high strength-to-weight ratio, and excellent thermal conductivity, making it ideal for applications requiring parts with reduced weight and enhanced heat dissipation. Aluminum prints are commonly used in automotive, aerospace, and consumer electronics industries for components such as heat sinks, brackets, and housings. Titanium is a high-performance metal filament prized for its exceptional strength, biocompatibility, and corrosion resistance. Titanium prints are used in medical implants, aerospace components, and military applications due to their lightweight, durable, and biocompatible properties. Titanium prints can be post-processed through machining or surface finishing techniques to achieve precise dimensions and desired surface finishes.

Cobalt-chrome alloys are another popular choice for metal 3D printing filaments. Cobalt-chrome filaments offer high strength, wear resistance, and biocompatibility, making them suitable for medical implants, dental prosthetics, and aerospace components. Cobalt-chrome prints can be post-processed through polishing, grinding, or electroplating to achieve smooth surface finishes and precise tolerances. Precious metals such as gold, silver, and platinum are also available as 3D printing filaments, albeit at a higher cost. Precious metal prints are used in jewelry making, luxury goods, and high-end electronics for their aesthetic appeal, conductivity, and corrosion resistance. Precious metal prints can be post-processed through polishing, plating, or stone setting to achieve intricate designs and luxurious finishes. Inconel and other nickel-based alloys are used in applications requiring high temperature resistance, corrosion resistance, and mechanical strength. Inconel prints are commonly used in aerospace, automotive, and oil and gas industries for components such as turbine blades, exhaust systems, and heat exchangers. Inconel prints can be post-processed through heat treatment or surface coating to enhance their properties and performance. Metal filaments for 3D printing offer a wide range of options for creating high-performance, functional parts with excellent mechanical properties and surface finishes. From stainless steel and aluminum to titanium, cobalt-chrome, and precious metals, metal filaments provide users with the flexibility, versatility, and performance needed to meet the demands of diverse applications across industries. As technology advances and materials evolve, metal filaments are expected to continue playing a crucial role in the future of additive manufacturing, driving innovation, and expanding the possibilities of metal 3D printing.

3. Ceramics: Alumina, zirconia, and silicon carbide powders are employed in ceramic 3D printing processes for applications demanding high temperature resistance, electrical insulation, and chemical inertness. Ceramic filament for 3D printing represents a significant advancement in additive manufacturing technology, enabling the production of ceramic parts with intricate geometries, high precision, and excellent thermal and chemical properties. Ceramic filaments are used in processes such as Ceramic Stereolithography (CSL), Binder Jetting, and Selective Laser Sintering (SLS), where ceramic powders are fused together layer by layer to build up the final object. These filaments come in various types of ceramics, each with unique properties and characteristics suited to different applications.

One of the most commonly used ceramic materials in 3D printing is alumina (Al_2O_3). Alumina filaments offer excellent mechanical strength, high temperature resistance, and electrical insulation properties, making them suitable for a wide range of industrial applications, including aerospace, automotive, electronics, and medical devices. Alumina prints can be post-processed through polishing, glazing, or firing to achieve precise dimensions and desired surface finishes. Silicon carbide (SiC) is another popular ceramic filament known for its exceptional hardness, abrasion resistance, and thermal conductivity. Silicon carbide prints are used in applications requiring wear-resistant parts, such as cutting tools, nozzles, and bearings, as well as in high-temperature environments such as furnace components and heat exchangers. Silicon carbide prints can be post-processed through machining or surface finishing techniques to achieve precise dimensions and smooth surface finishes. Zirconia (ZrO_2) is a high-performance ceramic filament prized for its excellent mechanical properties, biocompatibility, and corrosion resistance. Zirconia prints are used in dental implants, medical devices, aerospace components, and industrial wear parts due to their strength, durability, and aesthetic appeal. Zirconia prints can be post-processed through sintering, polishing, or glazing to achieve precise dimensions and desired surface finishes.

Other ceramic materials commonly used in 3D printing include boron nitride (BN), aluminum nitride (AlN), and cordierite ($Mg_2Al_4Si_5O_{18}$). Boron nitride prints offer excellent thermal conductivity, lubricity, and dielectric properties, making them suitable for applications such as heat sinks, insulators, and electronic components. Aluminum nitride prints are prized for their high thermal conductivity, electrical insulation, and chemical stability, making them ideal for use in electronic substrates, heat sinks, and semiconductor packaging. Cordierite prints offer high thermal shock resistance, low thermal expansion, and excellent dimensional stability, making them suitable for applications such as kiln furniture, automotive catalyst substrates, and ceramic filters. Ceramic filament for 3D printing offers a wide range of options for creating high-performance, functional parts with excellent mechanical, thermal, and chemical properties. From alumina and silicon carbide to zirconia, boron nitride, and aluminum nitride, ceramic filaments provide users with the flexibility, versatility, and performance needed to meet the demands of diverse applications across industries. As technology advances and materials evolve, ceramic filaments are expected to continue playing a crucial role in the future of additive manufacturing, driving innovation, and expanding the possibilities of ceramic 3D printing.

4. *Composites Filaments*: Carbon fiber, fiberglass, and Kevlar reinforcements are combined with thermoplastic or resin matrices to produce high-strength, lightweight parts for aerospace, automotive, and sporting goods industries. Composite filaments for 3D printing represent a cutting-edge advancement in additive manufacturing technology, offering the ability to produce parts with enhanced mechanical properties, lightweight characteristics, and tailored performance. These filaments combine traditional thermoplastic polymers with reinforcing fibers, particles, or additives to create materials with superior strength, stiffness, and durability compared to conventional plastics.

One of the most common types of composite filament used in 3D printing is carbon fiber-reinforced filament. Carbon fiber filaments consist of thermoplastic polymers, such as PLA or ABS, infused with carbon fibers, resulting in materials with exceptional strength-to-weight ratios, high stiffness, and excellent dimensional stability. Carbon fiber-reinforced filaments are used in applications requiring lightweight, rigid parts with superior mechanical properties, such as aerospace components, automotive parts, and sporting goods. Another popular type of composite filament is glass fiber-reinforced filament. Glass fiber filaments combine thermoplastic polymers with glass fibers, offering improved strength, impact resistance, and heat resistance compared to standard plastics. Glass fiber-reinforced filaments are used in applications requiring durable, impact-resistant parts, such as machine components, tooling, and structural components. In addition to carbon fiber and glass fiber, composite filaments can incorporate a wide range of other reinforcing materials, including Kevlar, basalt, wood, metal powders, and ceramics. Each type of reinforcement imparts unique properties to the filament, allowing for the customization of material characteristics to meet specific application requirements.

Composite filaments offer several advantages over traditional plastics for 3D printing applications. By combining thermoplastic polymers with reinforcing materials, composite filaments can achieve higher strength, stiffness, and toughness while maintaining the ease of printing and post-processing associated with standard plastics. This allows for the production of parts with complex geometries, fine details, and superior mechanical performance. Moreover, composite filaments enable the creation of lightweight parts with excellent strength-to-weight ratios, making them ideal for applications where weight savings are critical, such as aerospace, automotive, and robotics. The enhanced properties of composite filaments also allow for the production of functional prototypes, end-use parts, and tooling with improved performance and durability compared to traditional plastics.

Composite filaments for 3D printing represent a significant advancement in additive manufacturing technology, offering the ability to produce parts with enhanced mechanical properties, lightweight characteristics, and tailored performance. By combining thermoplastic polymers with reinforcing materials, composite filaments enable the creation of durable, high-performance parts for a wide range of applications across industries. As technology advances and materials evolve, composite filaments are expected to continue playing a crucial role in the future of additive manufacturing, driving innovation and expanding the possibilities of 3D printing. A summary of key materials used in 3D printing is shown in Fig. 3.6.

Generations of 3D Printing:

The evolution of 3D printing technology can be categorized into several generations, each marked by significant advancements in materials, processes, capabilities, and applications. These generations represent key milestones in the development of additive manufacturing and have contributed to the widespread adoption and transformative impact of 3D printing across various industries as shown in Fig. 3.7.

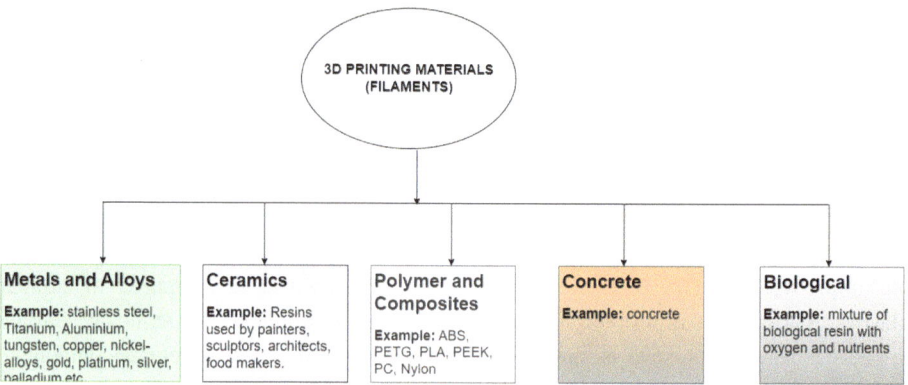

Fig. 3.6 The major 3D printing materials

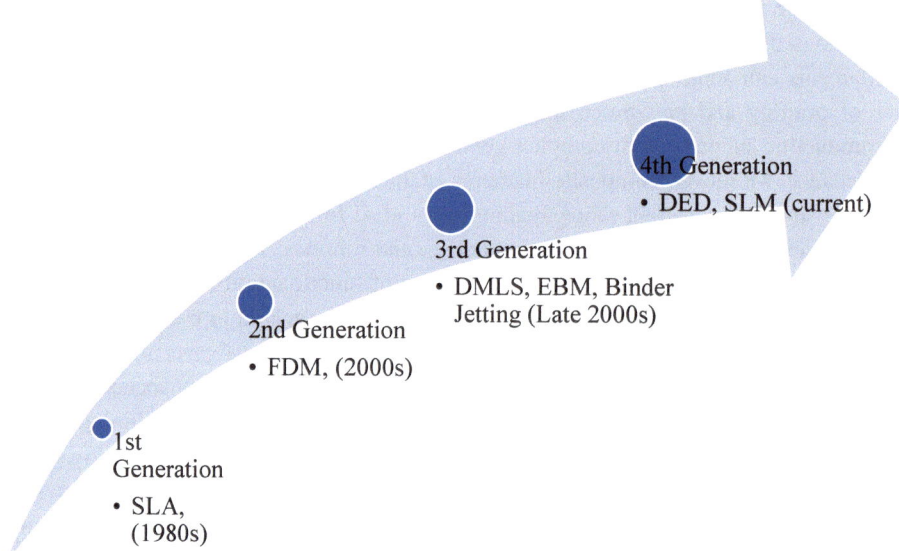

Fig. 3.7 The key generations of 3D printing

First Generation:

The advent of the first generation of 3D printing in the 1980s marked a significant mile-stone in manufacturing technology. Stereolithography (SLA), invented by Chuck Hull in 1983, introduced the concept of curing liquid resin with a UV laser to create solid objects layer by layer. Similarly, Selective Laser Sintering (SLS), developed by Carl Deckard and Joseph Beaman in the late 1980s, used a high-powered laser to selectively fuse powdered material, such as plastics, metals, or ceramics, into solid objects. These technologies were primarily used for rapid prototyping and tooling applications in industries like aerospace, automotive, and healthcare, where they offered advantages such as faster turnaround times, reduced costs, and increased design flexibility compared to traditional manufacturing methods. The emergence of the first generation of 3D printing in the 1980s represented a significant breakthrough in manufacturing technology, revolutionizing the way objects are designed, prototyped, and produced. Two pioneering additive manufacturing techniques, Stereolithography (SLA) and Selective Laser Sintering (SLS), paved the way for the widespread adoption of 3D printing across various industries.

Invented by Chuck Hull in 1983, Stereolithography (SLA) introduced a revolutionary approach to 3D printing by utilizing a process called photopolymerization. This technique involved curing liquid resin with a UV laser to create solid objects layer by layer. SLA printers consist of a vat of liquid resin, a platform that moves vertically, and a UV laser that selectively solidifies the resin according to the desired 3D model. As each layer is cured, the platform gradually moves downwards, allowing subsequent layers to be added until the entire object is formed. Developed by Carl Deckard and Joseph Beaman in the late 1980s, Selective Laser Sintering (SLS) introduced another innovative approach to 3D printing. SLS utilizes a high-powered laser to selectively fuse powdered material, such as plastics, metals, or ceramics, into solid objects layer by layer. During the printing process, a thin layer of powdered material is spread across the build platform, and the laser is precisely controlled to sinter (fuse) the particles together based on the digital design data. Once a layer is complete, the platform descends, and a new layer of powder is spread, repeating the process until the object is fully formed.

Applications and Advantages:

SLA and SLS technologies were initially employed for rapid prototyping and tooling applications across industries such as aerospace, automotive, and healthcare. These 3D printing techniques offered several advantages over traditional manufacturing methods. SLA and SLS enabled rapid prototyping and production of complex parts with shorter lead times compared to conventional manufacturing processes. This accelerated development cycles and allowed companies to bring products to market more quickly. By eliminating the need for expensive tooling and molds required in traditional manufacturing, SLA and SLS significantly reduced production costs, especially for low-volume or customized parts. 3D printing technologies offered unparalleled design freedom, allowing engineers and designers to create intricate geometries, internal structures, and customized features that were not feasible with traditional manufacturing methods. Materials Versatility: SLA and SLS could work with a wide range of materials, including plastics, metals, ceramics, and composites, providing flexibility in material selection to meet specific application requirements. Since their inception, SLA and SLS have continued to evolve and improve, with advancements in materials, resolution, and printing speed. The development of new resins, powders, and additive manufacturing techniques has expanded the capabilities of

3D printing, enabling applications in areas such as bioprinting, aerospace manufacturing, and consumer products. Furthermore, the integration of SLA and SLS with digital design tools, automation, and artificial intelligence (AI) has streamlined the 3D printing workflow, making it more accessible and efficient for a broader range of users. As 3D printing technologies continue to advance, they hold the potential to revolutionize manufacturing across industries, driving innovation, sustainability, and economic growth in the years to come.

ii. *Second Generation*:

The second generation of 3D printing emerged in the early 2000s with the development of desktop or hobbyist 3D printers, such as Fused Deposition Modeling (FDM) machines. These printers, pioneered by companies like Stratasys, MakerBot, and RepRap, brought additive manufacturing technology into homes, schools, and small businesses, democratizing access to 3D printing and sparking a wave of creativity and innovation. FDM printers extruded thermoplastic filaments, such as PLA and ABS, through a heated nozzle to build objects layer by layer, making them suitable for a wide range of applications, including prototyping, customization, and educational purposes. Fused Deposition Modeling (FDM) emerged as a leading 3D printing technology during this period. FDM printers function by extruding thermoplastic filaments, such as polylactic acid (PLA) and acrylonitrile butadiene styrene (ABS), through a heated nozzle. The nozzle deposits the molten filament onto a build platform, where it rapidly solidifies to form layers that gradually build up to create the desired 3D object. FDM printers offer a straightforward and versatile approach to additive manufacturing, making them suitable for a wide range of applications. The introduction of desktop 3D printers, particularly those utilizing FDM technology, democratized access to 3D printing by lowering barriers to entry in terms of cost, size, and complexity. Hobbyists, educators, and small businesses embraced these accessible machines as tools for experimentation, creativity, and innovation. With the ability to turn digital designs into physical objects quickly and affordably, individuals and organizations explored new avenues in product development, customization, and education.

FDM 3D printers found widespread applications across various domains. FDM technology revolutionized the prototyping process by enabling engineers, designers, and inventors to rapidly iterate on product designs and concepts. With FDM printers, prototypes could be produced quickly and cost-effectively, allowing for more efficient development cycles and faster time-to-market for new products. Customization: The ability to produce customized parts and products on-demand made FDM 3D printers invaluable in industries such as consumer goods, healthcare, and automotive. From personalized consumer products to bespoke medical devices and customized automotive components, FDM technology facilitated the creation of unique and tailored solutions to meet individual needs and preferences. FDM 3D printers became indispensable tools in educational settings, providing students and educators with hands-on learning experiences in design, engineering, and manufacturing. By bringing 3D printing capabilities into classrooms, schools, and makerspaces, FDM printers fostered creativity, problem-solving skills, and STEM education initiatives. As desktop 3D printing technology continues to evolve, advancements in materials, resolution, and functionality are expanding the capabilities and applications of FDM printers. The development of new materials, including engineering-grade thermoplastics and composite filaments, is broadening the range of industries and applications that can benefit from FDM technology. Furthermore, the integration of FDM printers with digital design software, cloud-based platforms, and IoT connectivity is enhancing workflow efficiency, collaboration, and remote access to 3D printing capabilities. As FDM technology becomes more accessible, affordable, and versatile, it holds the potential to revolutionize manufacturing, product development, and education on a global scale, empowering individuals and organizations to turn ideas into reality with unprecedented speed and ease.

iii. *Third Generation*:

The late 2000s marked the onset of the third generation of 3D printing, characterized by significant advancements in industrial-scale additive manufacturing technologies. During this period and continuing to the present day, industrial-grade 3D printers capable of producing high-quality, functional parts using a diverse array of materials have become increasingly prevalent. Technologies such as Direct Metal Laser Sintering (DMLS), Electron Beam Melting (EBM), and Binder Jetting have revolutionized various industries, including aerospace, automotive, healthcare, and consumer goods. These advancements have facilitated the production of complex, high-performance parts with superior mechanical properties and unparalleled design freedom. Moreover, the proliferation of industrial-scale 3D printing has led to the emergence of service bureaus and contract manufacturers specializing in additive manufacturing, offering on-demand production services to a diverse range of clients.

i. *Direct Metal Laser Sintering (DMLS)*: DMLS utilizes a high-powered laser to selectively fuse metal powder, layer by layer, to produce fully dense metal parts. This technology enables the fabrication of complex geometries and intricate structures with exceptional mechanical properties, making it ideal for aerospace, automotive, and medical applications that demand high strength-to-weight ratios and precise tolerances. Direct Metal Laser Sintering (DMLS) is an advanced additive manufacturing technology that enables the production of high-quality metal parts with intricate geometries and excellent mechanical properties. Also known as Selective Laser Melting (SLM), DMLS utilizes a high-powered laser to selectively fuse metal powder particles, layer by layer, to build up complex 3D objects. The DMLS process begins with the creation of a digital 3D model of the desired part using computer-aided design (CAD) software. The CAD model is then sliced into thin cross-sectional layers, and the data is transferred to the DMLS machine. Inside the DMLS machine, a thin layer of metal powder is evenly spread across the build platform. A high-powered laser is directed onto the surface of the powder bed, selectively melting and fusing the metal particles according to the cross-sectional shape of the desired part. Once a layer is complete, the build platform descends, and a new layer of powder is spread over the previous layer. This process is repeated layer by layer until the entire part is fabricated. DMLS is compatible with a wide range of metal materials, including stainless steel, aluminum, titanium, and nickel-based alloys. These materials offer excellent mechanical properties, high strength-to-weight ratios, and corrosion resistance, making them suitable for a variety of demanding applications across industries such as aerospace, automotive, and medical.

The key advantage is that DMLS enables the fabrication of highly complex parts with intricate geometries, internal features, and thin walls that are difficult or impossible to achieve using traditional manufacturing methods. The layer-by-layer nature of DMLS allows for complete design freedom, enabling engineers and designers to create optimized parts with minimal material waste and maximum functionality. DMLS is well-suited for rapid prototyping and iterative design processes, allowing for quick turnaround times and cost-effective testing of new product concepts and iterations. DMLS offers exceptional dimensional accuracy and repeatability, ensuring tight tolerances and precise geometries in the manufactured parts. Parts produced using DMLS exhibit excellent mechanical properties, including high strength, hardness, and fatigue resistance, comparable to those of conventionally manufactured metal components. DMLS enables customization and personalization of parts to meet specific customer requirements or design specifications, offering flexibility in manufacturing.

DMLS finds applications across a wide range of industries and sectors. In the aerospace industry, DMLS is used to produce lightweight, complex components such as turbine blades, engine components, and structural parts for aircraft and spacecraft. The ability to manufacture parts with reduced weight and improved performance is critical for achieving fuel efficiency and reducing emissions in aerospace applications. In the medical field, DMLS is employed to fabricate patient-specific implants, prosthetics, and surgical instruments. The customizable nature of DMLS allows for the creation of orthopedic implants tailored to individual patient anatomy, leading to better outcomes and faster recovery times. In the automotive industry, DMLS is utilized for prototyping, tooling, and production of components such as engine parts, brackets, and exhaust systems. The lightweight, high-strength properties of DMLS-produced parts contribute to improved fuel efficiency, performance, and durability in automotive applications. DMLS is used for the fabrication of injection molds, dies, and tooling inserts in manufacturing processes. The ability to produce complex tooling components with rapid turnaround times reduces lead times and costs associated with traditional tooling methods. DMLS is employed in the production of high-end consumer goods, including jewelry, watches, and luxury accessories. The ability to create intricate designs and customized products appeals to consumers seeking unique and personalized items.

As DMLS technology continues to advance, we can expect further improvements in process efficiency, material selection, and part quality. Continued research and development efforts are focused on expanding the range of compatible materials, enhancing surface finish and post-processing capabilities, and increasing the scalability and affordability of DMLS systems. With ongoing advancements, DMLS is poised to play an increasingly prominent role in the manufacturing landscape, driving innovation, efficiency, and competitiveness across industries.

ii. *Electron Beam Melting (EBM)*: EBM operates by using an electron beam to selectively melt metal powder in a high-vacuum environment, resulting in the production of dense, near-net-shape metal parts. EBM offers advantages such as faster build rates and superior material properties compared to traditional manufacturing methods, making it well-suited for aerospace components, orthopedic implants, and tooling applications.

Electron Beam Melting (EBM) is a cutting-edge additive manufacturing technology that enables the production of complex metal parts with exceptional mechanical properties and dimensional accuracy. Developed by Arcam AB (now a part of GE Additive) in the 1990s, EBM utilizes an electron beam to selectively melt and fuse metal powder particles, layer by layer, to build up 3D objects. The EBM process begins with the creation of a digital 3D model of the desired part using computer-aided design (CAD) software. The CAD model is then sliced into thin cross-sectional layers, and the data is transferred to the EBM machine. Inside the EBM machine, a thin layer of metal powder is evenly spread across the build platform. An electron beam, generated by a highly focused electron gun, is directed onto the surface of the powder bed, selectively melting and sintering the metal particles according to the cross-sectional shape of the desired part. As each layer is completed, the build platform descends, and a new layer of powder is spread over the previous layer. This process is repeated layer by layer until the entire part is fabricated. EBM is compatible with a variety of metal materials, including titanium, titanium alloys, cobalt-chromium, and stainless steel. These materials offer excellent mechanical properties, corrosion resistance, and biocompatibility, making them suitable for a wide range of applications in industries such as aerospace, medical, and automotive.

EBM enables the fabrication of complex parts with intricate geometries, internal features, and thin walls that are difficult or impossible to achieve using traditional manufacturing methods. EBM is a high-speed additive manufacturing process capable of producing parts at a rapid rate, making it well-suited for serial production and large-scale manufacturing applications. Parts produced using EBM exhibit exceptional mechanical properties, including high strength, ductility, and fatigue resistance, comparable to those of conventionally manufactured metal components. EBM produces near-net shape parts with minimal material waste, reducing the need for additional machining or finishing operations. EBM technology is scalable and can be adapted to produce parts of varying sizes and complexities, from small intricate components to large structural assemblies.

EBM finds applications across a wide range of industries and sectors. In the aerospace industry, EBM is used to produce lightweight, high-performance components such as turbine blades, engine housings, and structural brackets. The ability to manufacture parts with complex geometries and superior mechanical properties is critical for meeting the stringent requirements of aerospace applications. In the medical field, EBM is employed to fabricate patient-specific implants, prosthetics, and surgical instruments. Titanium and its alloys, commonly used in EBM, offer excellent biocompatibility and osseointegration properties, making them ideal for medical implants and orthopedic devices. EBM is utilized in the automotive industry for the production of engine components, exhaust manifolds, and suspension parts. The lightweight and high-strength properties of parts produced using EBM contribute to improved fuel efficiency, performance, and durability in automotive applications. EBM is used for rapid prototyping, tooling, and manufacturing of molds, dies, and jigs in various industries. The ability to produce functional prototypes and tooling components quickly and cost-effectively accelerates product development cycles and reduces time-to-market.

As EBM technology continues to evolve, we can expect further advancements in process efficiency, material selection, and part quality. Continued research and development efforts are focused on expanding the range of compatible materials, enhancing surface finish and post-processing capabilities, and increasing the scalability and affordability of EBM systems. With ongoing advancements, EBM is poised to play an increasingly prominent role in the manufacturing landscape, driving innovation, efficiency, and competitiveness across industries.

iii. *Binder Jetting*: Binder Jetting involves depositing a liquid binding agent onto layers of powdered material, such as metals, ceramics, or sand, to create 3D printed objects. After printing, the green parts are sintered or infiltrated to achieve the desired mechanical properties. Binder Jetting offers a cost-effective and scalable solution for producing complex parts with a wide range of materials, making it suitable for applications in automotive, architecture, and consumer goods. Binder Jetting is an innovative additive manufacturing technology that revolutionizes the production of complex 3D printed objects using a process involving the deposition of a liquid binding agent onto layers of powdered material. This method, also known as powder bed fusion, offers a versatile and cost-effective solution for creating parts with a wide range of materials, including metals, ceramics, and sand.

The Binder Jetting process begins with the preparation of a digital 3D model of the desired object using computer-aided design (CAD) software. The CAD model is then sliced into thin cross-sectional layers, and the data is transferred to the Binder Jetting machine. Inside the machine, a thin layer of powdered material is evenly spread across the build platform. A print head then selectively deposits a liquid binding agent onto the powdered material, binding the particles together to form a solid layer corresponding to the cross-section of the object being printed. Once a layer is complete, the build platform descends, and a new layer of powdered material is spread over the previous layer. The process is repeated layer by layer until the entire object is fabricated. After printing, the green parts, which are the objects in their initial, unbonded state, undergo post-processing steps to achieve the desired mechanical properties. This typically involves sintering or infiltrating the green parts with additional materials to strengthen and densify the final product.

One of the key advantages of Binder Jetting is its ability to produce complex parts with a wide range of materials. Unlike traditional manufacturing methods that may be limited by the availability of certain materials or the complexity of part geometries, Binder Jetting offers a flexible and scalable solution for creating parts from metals, ceramics, or sand. This versatility makes it suitable for a variety of applications across industries such as automotive, architecture, and consumer goods. In the automotive industry, Binder Jetting is used for the production of lightweight components, such as engine parts, brackets, and exhaust systems. The ability to create complex geometries and intricate designs with Binder Jetting enables automotive manufacturers to reduce weight and improve fuel efficiency without compromising on performance or safety. In architecture, Binder Jetting is utilized for rapid prototyping and the fabrication of scale models, architectural components, and decorative elements. The cost-effectiveness and scalability of Binder Jetting make it an attractive option for architectural firms seeking to streamline the design and manufacturing process while maintaining high levels of precision and detail. In the consumer goods industry, Binder Jetting is employed for the production of customized products, personalized accessories, and limited-edition collectibles. The ability to produce unique, one-of-a-kind items with intricate details and complex shapes appeals to consumers seeking personalized and exclusive products. Overall, Binder Jetting offers a cost-effective and scalable solution for producing complex parts with a wide range of materials. Its versatility and flexibility make it well-suited for a variety of applications in automotive, architecture, consumer goods, and beyond, driving innovation and efficiency in the additive manufacturing landscape.

The adoption of industrial-scale additive manufacturing technologies has had a profound impact on various industries. Additive manufacturing has revolutionized aerospace manufacturing by enabling the production of lightweight, complex components with reduced lead times and material waste. Aerospace companies leverage 3D printing for prototyping, tooling, and production of critical parts such as engine components, brackets, and ducting systems. The automotive industry has embraced additive manufacturing for rapid prototyping, tooling, and production of customized parts. 3D printing enables automotive manufacturers to reduce time-to-market, optimize supply chains, and manufacture lightweight components with enhanced performance characteristics, such as improved fuel efficiency and crash resistance. Additive manufacturing has transformed the healthcare industry by facilitating the production of patient-specific implants, surgical instruments, and prosthetics. Technologies like DMLS and EBM are used to fabricate orthopedic implants, dental crowns, and patient-specific surgical guides, offering improved fit, functionality, and biocompatibility compared to traditional manufacturing methods. Industrial-scale 3D printing has opened up new possibilities in consumer product design and customization. Companies leverage additive manufacturing to create bespoke products, personalized accessories, and limited-edition collectibles, catering to individual preferences and niche markets.

The rise of industrial-scale additive manufacturing has spurred the growth of service bureaus and contract manufacturers specializing in 3D printing. These companies offer on-demand production services, including prototyping, small-batch manufacturing, and customized part production, to a diverse range of clients across industries. Service bureaus leverage advanced additive manufacturing technologies and expertise to deliver high-quality, cost-effective solutions tailored to the specific needs of their clients, further accelerating the adoption of additive manufacturing in industrial settings. As industrial-scale additive manufacturing technologies continue to advance, we can expect further innovation and expansion in applications across industries. Future developments may include improvements in material selection, process automation, and multi-material printing capabilities, as well as increased integration with digital design and simulation tools. The continued evolution of additive manufacturing is poised to drive significant advancements in product development, manufacturing efficiency, and supply chain agility, ultimately shaping the future of production in the digital age.

iv. *Fourth Generation (Current)*:

The current fourth generation of 3D printing is characterized by ongoing innovations in materials, processes, and applications, driving further integration of additive manufacturing into mainstream production workflows. Metal 3D printing technologies like Selective Laser Melting (SLM) and Directed Energy Deposition (DED) have expanded the possibilities of additive manufacturing, enabling the production of functional metal parts with complex geometries and superior mechanical properties. Bioprinting technologies have opened up new frontiers in regenerative medicine, tissue engineering, and drug discovery, offering the potential to revolutionize healthcare and personalized medicine. Hybrid additive-subtractive manufacturing combines the benefits of additive and subtractive processes, allowing for greater design freedom, improved surface finish, and reduced material waste. Digital supply chain solutions leverage 3D printing technology to enable on-demand production, decentralized manufacturing, and mass customization, transforming traditional manufacturing paradigms and enabling more agile and responsive production systems.

The current era of 3D printing, often referred to as the fourth generation, represents the culmination of decades of technological advancements and innovation in additive manufacturing. This generation is characterized by a convergence of cutting-edge technologies, materials, and applications, leading to unprecedented capabilities and widespread adoption across industries. The fourth generation of 3D printing is defined by its emphasis on scalability, sustainability, and customization, ushering in a new era of digital manufacturing and production. One of the defining features of the fourth generation of 3D printing is the scalability of additive manufacturing processes. Industrial-grade 3D printers capable of producing large-scale parts and components have become increasingly prevalent, enabling the adoption of additive manufacturing in mass production and industrial settings. Technologies such as Large-Scale Additive Manufacturing (LSAM) and Continuous Liquid Interface Production (CLIP) have emerged to address the demand for scalable and efficient additive manufacturing solutions. These advancements have revolutionized industries such as aerospace, automotive, and construction, where the production of large, complex parts with high precision and reliability is critical. In the fourth generation of 3D printing, there is a growing emphasis on sustainability and environmental responsibility. Additive manufacturing offers significant advantages over traditional manufacturing methods in terms of material efficiency, waste reduction, and energy savings. By enabling on-demand production and localized manufacturing, 3D printing minimizes the need for inventory stockpiles, transportation, and excess production, leading to a more sustainable and resource-efficient supply chain. Moreover, the use of recycled and biodegradable materials in 3D printing processes further contributes to the advancement of circular economy principles and the reduction of environmental impact. Another hallmark of the fourth generation of 3D printing is the unprecedented level of customization and personalization it enables. Additive manufacturing technologies allow for the creation of highly intricate and unique designs that can be tailored to individual preferences and requirements. From personalized medical implants and prosthetics to customized consumer products and architectural structures, 3D printing empowers designers, engineers, and consumers to create bespoke solutions that meet their specific needs. The ability to produce one-of-a-kind objects with intricate geometries and complex functionalities has revolutionized industries such as healthcare, fashion, and product design, opening up new possibilities for innovation and creativity.

The fourth generation of 3D printing has witnessed significant advancements in materials science, leading to the development of a wide range of advanced materials suitable for additive manufacturing. From high-performance polymers and metal alloys to biocompatible ceramics and composites, 3D printing materials have become increasingly diverse and specialized, catering to a variety of applications and industries. Moreover, multi-material printing technologies have emerged, enabling the simultaneous deposition of multiple materials within a single print job. This capability allows for the creation of complex, multi-functional parts with integrated electronics, sensors, and other components, expanding the scope of additive manufacturing and enabling new applications in fields such as electronics, robotics, and wearable technology. In the fourth generation of 3D printing, there is a growing integration of additive manufacturing with digital technologies such as artificial intelligence (AI), machine learning, and generative design. These digital tools enable designers and engineers to optimize designs for additive manufacturing, minimize material usage, and maximize performance. Additionally, advancements in software algorithms and simulation tools enhance process control, quality assurance, and predictive maintenance in 3D printing workflows. By harnessing the power of digital technologies, manufacturers can unlock new levels of efficiency, productivity, and innovation in additive manufacturing processes. As the fourth generation of 3D printing continues to evolve, we can expect further advancements in scalability, sustainability, customization, materials, and digital integration. Additive manufacturing will increasingly play a central role in the digital transformation of industries, driving innovation, efficiency, and competitiveness in the global marketplace. From revolutionizing traditional manufacturing processes to enabling entirely new forms of production and creativity, 3D printing holds immense promise for shaping the future of manufacturing and ushering in a new era of digital fabrication and production.

Looking ahead, the future of 3D printing holds exciting possibilities for continued innovation and advancement. Emerging technologies such as nanoscale 3D printing, 4D printing, and machine learning-driven design optimization promise to further expand the capabilities and applications of additive manufacturing, unlocking new opportunities for creativity, efficiency, and sustainability. As 3D printing continues to evolve and mature, it is poised to play an increasingly integral role in the fourth industrial revolution and beyond, shaping the future of manufacturing, healthcare, and beyond.

3D printing has revolutionized manufacturing by offering unparalleled design freedom, customization, and efficiency. With advancements in materials, processes, and technologies, the potential applications of 3D printing continue to expand across industries, from aerospace and healthcare to consumer goods and architecture, shaping the future of production in the fourth industrial revolution and beyond.

3.1.2 Rapid Prototyping and Iterative Design: Accelerating Innovation with 3D Printing

Rapid prototyping and iterative design represent essential phases in the product development lifecycle, offering companies a means to refine concepts, test functionalities, and gather feedback before finalizing designs for production. In recent years, 3D printing has emerged as a transformative tool in these processes, fundamentally altering the way products are conceptualized, developed, and brought to market. One of the primary advantages of 3D printing in rapid prototyping lies in its ability to deliver speed and flexibility. Unlike traditional manufacturing methods, which often entail lengthy lead times and significant costs for creating prototypes, 3D printing empowers designers and engineers to swiftly produce physical prototypes directly from digital designs. This accelerated prototyping process enables companies to iterate on designs rapidly, test multiple iterations, and incorporate necessary adjustments in a fraction of the time it would take using conventional methods.

Moreover, 3D printing offers unparalleled freedom of design, permitting the creation of complex geometries and intricate details that would be challenging or impossible to achieve with traditional manufacturing techniques. This flexibility empowers designers to explore innovative concepts and push the boundaries of creativity, resulting in more inventive and user-centric product designs. The iterative design process facilitated by 3D printing enables companies to collect feedback early and frequently, thereby mitigating the risk of costly design flaws or product failures later in the development cycle. By generating multiple prototypes and integrating feedback from stakeholders, designers can refine designs iteratively, ensuring that the final product meets or exceeds customer expectations. Furthermore, 3D printing allows for the prototyping of functional components, enabling designers to evaluate the performance and functionality of parts in real-world conditions. This capability is particularly valuable in industries such as aerospace, automotive, and healthcare, where precise fit and performance are critical considerations. In addition to its speed, flexibility, and functionality, 3D printing offers cost advantages in the prototyping process. By eliminating the need for expensive tooling and setup costs associated with traditional manufacturing methods, 3D printing reduces the overall cost of prototyping, making it more accessible to companies of all sizes.

Overall, rapid prototyping and iterative design facilitated by 3D printing are integral components of the product development process, empowering companies to innovate swiftly, shorten time to market, and ultimately deliver superior products to customers. As 3D printing technology continues to advance and gain broader acceptance, its role in accelerating innovation and driving product development will only continue to expand, reshaping the future of manufacturing in the digital age. The integration of 3D printing technology into rapid prototyping and iterative design processes represents a paradigm shift in product development. By offering unprecedented speed, flexibility, and functionality, 3D printing has revolutionized the way products are conceptualized, designed, and brought to market. As companies continue to leverage the capabilities of 3D printing and explore its full potential, its impact on innovation and product development will undoubtedly be profound, shaping the future of manufacturing in the digital era.

3.2   Advancements in 3D Printing Technology

Advancements in 3D printing technology have revolutionized various industries by offering innovative solutions to complex manufacturing challenges. From prototyping to mass customization, 3D printing has evolved significantly over the years, driven by continuous research and development efforts. One of the most notable advancements lies in the evolution of printing techniques. Traditional methods, such as fused deposition modeling (FDM) and stereolithography (SLA), have paved the way for newer, more efficient processes like selective laser sintering (SLS) and digital light processing (DLP). These advancements have not only improved printing speed but also enhanced the quality and precision of printed objects.

Materials innovation has been another critical aspect of the advancement in 3D printing technology. Initially limited to plastics and resins, the range of materials used in additive manufacturing has expanded significantly. Today, manufacturers can print using metals, ceramics, composites, and even biological materials like living cells. This diversity in materials has unlocked a plethora of applications across industries, from aerospace to healthcare. Researchers are continually exploring new materials and improving existing ones to meet the specific requirements of different applications, such as strength, flexibility, or biocompatibility. High-speed printing has emerged as a game-changer in 3D printing technology. Traditional methods often suffered from slow printing speeds, limiting their scalability for mass production. However, recent advancements have addressed this challenge by introducing techniques like continuous liquid interface production (CLIP) and high-speed sintering (HSS). These methods allow for faster printing without compromising on quality, making additive manufacturing more competitive with traditional manufacturing processes in terms of production rates. Furthermore, advancements in precision and accuracy have significantly enhanced the capabilities of 3D printing technology. Improved resolution and detailing enable the creation of intricate geometries and complex structures with high fidelity. This level of precision is particularly beneficial in industries such as aerospace, where components must meet stringent quality standards and tolerances. Various technologies, including advanced sensors and feedback systems, contribute to ensuring precise printing, minimizing errors, and reducing material waste. Multi-material and multi-color printing have also seen remarkable advancements, enabling more versatile and visually appealing prints. With the ability to print with multiple materials simultaneously, manufacturers can create objects with diverse properties and functionalities in a single print job. Similarly, advancements in multi-color printing technology allow for the creation of vibrant and detailed models with realistic color gradients and textures. These capabilities have expanded the scope of applications for 3D printing, ranging from architectural models to personalized consumer products. Advancements in 3D printing technology have propelled additive manufacturing into the mainstream, offering unprecedented opportunities for innovation and customization across industries. From faster printing speeds to enhanced precision and material versatility, these advancements continue to drive the adoption of 3D printing as a viable manufacturing solution in the fourth industrial revolution and beyond.

3.2.1 Evolution of 3D Printing

The evolution of 3D printing spans over four decades, transforming from a niche technology to a disruptive force across industries. Initially conceptualized in the 1980s by Chuck Hull with the invention of stereolithography (SLA), 3D printing revolutionized traditional manufacturing methods by introducing additive processes, where objects are built layer by layer. Since its inception, 3D printing has undergone remarkable technological advancements, propelling its evolution and expanding its capabilities. Early techniques like SLA laid the foundation for subsequent innovations, including selective laser sintering (SLS) and fused deposition modeling (FDM). These methods offered distinct advantages in terms of materials, precision, and speed, contributing to the diversification of 3D printing applications. Key advancements in materials have been instrumental in broadening the scope of 3D printing. While plastics were the primary materials used in the early stages, the development of metal printing techniques has revolutionized industries such as aerospace, automotive, and healthcare. Metal 3D printing enables the production of complex, high-strength components with superior mechanical properties. Bioprinting has emerged as a frontier in 3D printing, allowing for the fabrication of living tissues, organs, and scaffolds using bioinks containing living cells. This breakthrough technology holds promise for regenerative medicine, drug testing, and personalized healthcare solutions. Furthermore, the advent of composite materials has expanded the range of applications for 3D printing. Composites offer a unique combination of properties, including strength, lightweight, and durability, making them suitable for applications in aerospace, sporting goods, and automotive industries.

Ceramic 3D printing has also gained traction, enabling the creation of intricate designs and functional prototypes with high heat resistance and biocompatibility. Industries such as architecture, art, and dental prosthetics benefit from the versatility and aesthetic appeal of ceramic materials. One of the most significant advantages of 3D printing is its ability to facilitate customization on a mass scale. From personalized consumer products to custom medical implants, 3D printing empowers industries to tailor products to individual preferences and requirements, revolutionizing the concept of mass production. As 3D printing technology continues to advance, fueled by ongoing research and innovation, its impact on manufacturing, healthcare, design, and beyond will only grow. With the ability to create previously unimaginable objects with unparalleled precision and efficiency, 3D printing stands as a testament to human ingenuity, reshaping the way we conceptualize, produce, and interact with the world around us.

3.2.2   Materials Innovation in 3D Printing

Materials innovation has been a cornerstone of the remarkable advancements witnessed in the field of 3D printing. From the early days when plastic filaments dominated the market to today's diverse range of materials spanning metals, ceramics, composites, and even biological substances, the evolution has been nothing short of revolutionary. This section delves into the significance of materials innovation in 3D printing, its impact on various industries, and the future prospects it holds.

At its core, 3D printing, or additive manufacturing, is a process of creating three-dimensional objects by depositing successive layers of material under computer control. The choice of material plays a pivotal role in determining the properties, functionalities, and applications of the printed objects. In the initial stages of 3D printing, thermoplastics such as acrylonitrile butadiene styrene (ABS) and polylactic acid (PLA) were predominantly used due to their ease of use, affordability, and availability. These materials found applications in prototyping, hobbyist projects, and low-cost production. However, as the technology matured and industries began exploring the full potential of 3D printing, the demand for more robust and functional materials grew. This led to a significant expansion of the materials palette available for additive manufacturing. Metals, in particular, emerged as a game-changer in the field, opening up new possibilities in aerospace, automotive, healthcare, and beyond. Metal powders such as titanium, aluminum, stainless steel, and cobalt-chrome alloys are now commonly used in metal additive manufacturing processes like selective laser melting (SLM) and electron beam melting (EBM).

The adoption of metal 3D printing has revolutionized industries that require parts with high strength, durability, and complex geometries. In aerospace, for instance, manufacturers leverage metal additive manufacturing to produce lightweight yet structurally sound components for aircraft and spacecraft. Similarly, in the medical field, implants and prosthetics made from biocompatible metals offer customized solutions for patients, improving comfort and longevity. The ability to print with metals has also spurred innovation in other sectors, such as automotive, where lightweight components contribute to fuel efficiency and performance. Beyond metals, ceramics have also gained traction as a viable material for 3D printing. Ceramic powders such as alumina, zirconia, and silicon carbide are used in processes like binder jetting and stereolithography to create intricate ceramic parts with high temperature resistance, electrical insulation, and corrosion resistance. Ceramic 3D printing finds applications in industries like electronics, aerospace, and biomedical, where precision, reliability, and advanced material properties are paramount. Composites, which combine two or more materials to achieve specific properties, have emerged as another area of focus in materials innovation for 3D printing. Fiber-reinforced composites, for instance, offer enhanced strength, stiffness, and lightweight characteristics compared to traditional materials. Carbon fiber, fiberglass, and Kevlar are commonly used reinforcements in composite 3D printing processes like fused filament fabrication (FFF) and continuous fiber 3D printing. These materials find applications in industries such as automotive, sporting goods, and aerospace, where the demand for high-performance, cost-effective solutions is high.

Moreover, the integration of biological materials into 3D printing has unlocked new frontiers in healthcare, biotechnology, and tissue engineering. Bioinks, which consist of living cells suspended in a biocompatible matrix, enable the printing of complex biological structures such as tissues, organs, and scaffolds for regenerative medicine. Researchers are exploring various bioink formulations, including hydrogels, alginate, and decellularized extracellular matrix (ECM), to mimic the native environment of cells and promote tissue regeneration. Bioprinting holds promise for applications ranging from personalized medicine to drug testing and disease modeling. In addition to the expansion of material options, advancements in material science and engineering have led to the development of hybrid materials with novel properties and functionalities. Smart materials, for example, exhibit responsive behavior to external stimuli such as temperature, light, or electrical signals. Shape memory polymers, shape memory alloys, and conductive polymers are examples of smart materials that can be integrated into 3D printing processes to create self-adaptive structures, sensors, actuators, and other smart devices. These materials find applications in fields like robotics, electronics, and wearable technology, where dynamic functionality and adaptability are desired. Furthermore, the optimization of material properties through additives and post-processing techniques has expanded the scope of applications for 3D-printed parts. Surface treatments such as polishing, coating, and metallization improve the aesthetics, mechanical properties, and surface finish of printed objects, making them suitable for end-use applications. Functional additives like nanoparticles, fibers, and fillers enhance material properties such as strength, conductivity, and thermal resistance, enabling the production of high-performance components for demanding environments. Looking ahead, materials innovation in 3D printing is poised to continue at a rapid pace, driven by advancements in material science, manufacturing technologies, and interdisciplinary collaboration. Researchers are exploring new materials, improving existing ones, and pushing the boundaries of additive manufacturing to address emerging challenges and opportunities. From sustainable materials derived from renewable sources to self-healing materials capable of repairing damage autonomously, the future holds immense potential for materials innovation in 3D printing to reshape industries, foster innovation, and improve lives.

3.3 Industrial Automation and Robotics

Industrial automation and robotics have become integral components of modern manufacturing processes, revolutionizing production efficiency, flexibility, and quality. The integration of 3D printing with robotics represents a significant advancement in automation, offering synergistic benefits that enhance the capabilities of both technologies. Moreover, automation plays a pivotal role in streamlining various manufacturing processes, from design and prototyping to production and post-processing, driving productivity and innovation across industries. The integration of 3D printing with robotics enables automated additive manufacturing workflows that reduce labor costs, increase production throughput, and enhance overall process control. Robotic arms equipped with extrusion, deposition, or powder-bed fusion systems can precisely deposit material layers to build complex geometries with high accuracy and repeatability. This level of automation minimizes human intervention, mitigates errors, and accelerates the manufacturing cycle, making 3D printing more scalable and cost-effective for mass production.

Furthermore, robotic systems can perform auxiliary tasks such as part handling, support removal, and surface finishing, complementing the capabilities of 3D printers and optimizing the entire manufacturing workflow. Automated material handling systems facilitate the loading and unloading of build platforms, powder cartridges, and finished parts, ensuring continuous operation and minimizing downtime. Additionally, robotic arms equipped with tools such as sanders, polishers, and inspection devices can automate post-processing tasks to improve surface quality and dimensional accuracy. Automation in manufacturing processes extends beyond 3D printing to encompass various production stages, including design optimization, material preparation, assembly, and quality control. Computer-aided design (CAD) software equipped with generative design algorithms and artificial intelligence (AI) capabilities enables automated design optimization, allowing manufacturers to create lightweight, structurally efficient components with minimal material usage and production time. Moreover, automated material handling systems facilitate the storage, retrieval, and preparation of raw materials, reducing material waste and inventory costs.

In assembly operations, robotic systems equipped with vision systems, sensors, and adaptive grippers can perform intricate tasks such as part alignment, fastening, and inspection with high precision and reliability. Collaborative robots, or cobots, work alongside human operators to streamline assembly processes, improve ergonomics, and enhance worker safety. Moreover, automated quality control systems leverage machine vision, spectroscopy, and non-destructive testing techniques to inspect parts for defects, dimensional accuracy, and material properties, ensuring compliance with quality standards and specifications. The integration of automation and robotics in manufacturing processes offers numerous benefits, including increased productivity, flexibility, and competitiveness. By automating repetitive tasks and reducing manual labor, manufacturers can optimize resource utilization, minimize production costs, and accelerate time-to-market. Moreover, automation enables greater customization and personalization of products, allowing manufacturers to meet diverse customer demands and preferences effectively. Additionally, automation enhances process consistency and repeatability, resulting in higher product quality, reliability, and consistency.

Industrial automation and robotics play a pivotal role in driving innovation and efficiency in modern manufacturing processes. The integration of 3D printing with robotics enables automated additive manufacturing workflows that enhance production scalability, quality, and control. Moreover, automation streamlines various manufacturing stages, from design and material preparation to assembly and quality control, optimizing resource utilization, reducing production costs, and accelerating time-to-market. As automation technologies continue to evolve, manufacturers must embrace these advancements to remain competitive in an increasingly dynamic and interconnected global marketplace.

3.4   Empowering Customization and Personalization: Unleashing Creativity with 3D Printing

3D printing has emerged as a game-changer in manufacturing, enabling mass customization and personalized solutions across various industries. Its ability to fabricate objects layer by layer with unparalleled precision has revolutionized traditional manufacturing processes, opening up new possibilities for customization and personalization.

1. Facilitating Mass Customization:

3D printing offers unparalleled flexibility in design and production, allowing products to be tailored to individual needs on a mass scale. Unlike traditional manufacturing methods, which often require costly retooling and long lead times for customization, 3D printing enables rapid prototyping and iteration, making it feasible to produce customized products economically. This capability has significant implications for industries ranging from consumer goods to healthcare.

Facilitating mass customization through 3D printing represents a significant advancement in manufacturing and product development. Unlike traditional manufacturing methods that often rely on mass production of standardized products, 3D printing enables the creation of custom-designed items tailored to individual preferences and requirements on a large scale. One of the key advantages of 3D printing in facilitating mass customization is its ability to produce complex geometries and intricate designs with minimal setup and tooling costs. This flexibility allows manufacturers to efficiently produce a wide range of customized products without the need for expensive molds or equipment reconfiguration. As a result, companies can offer personalized variations of their products to meet the diverse needs and preferences of consumers without sacrificing cost-effectiveness. Moreover, 3D printing technology enables on-demand production, eliminating the need for maintaining large inventories of pre-manufactured goods. By leveraging digital design files and additive manufacturing processes, companies can produce items only when they are needed, reducing waste and inventory holding costs. This just-in-time manufacturing approach not only enhances operational efficiency but also enables faster response to changing market demands and trends.

Another benefit of 3D printing in mass customization is the ability to easily iterate and refine designs in response to customer feedback and market preferences. With traditional manufacturing methods, making changes to product designs often involves costly and time-consuming retooling processes. In contrast, 3D printing allows for rapid prototyping and iterative design improvements, enabling companies to quickly adapt their offerings to evolving consumer needs and preferences. Furthermore, 3D printing facilitates localized production and decentralized manufacturing networks, allowing companies to establish production facilities closer to their target markets. This localization strategy reduces shipping distances, transportation costs, and carbon emissions associated with long-distance supply chains. Additionally, it enables faster delivery times and enhances supply chain resilience by reducing reliance on centralized manufacturing facilities vulnerable to disruptions. To fully leverage the potential of 3D printing for mass customization, companies need to invest in advanced digital design capabilities, materials research, and process optimization. By harnessing the power of computer-aided design (CAD) software, artificial intelligence (AI), and generative design algorithms, manufacturers can create highly customized products efficiently and cost-effectively. Furthermore, ongoing advancements in 3D printing materials, such as biodegradable polymers, composite materials, and metal alloys, expand the range of applications and performance capabilities of additive manufacturing technology.

3D printing offers unprecedented opportunities for facilitating mass customization in manufacturing. By harnessing the inherent advantages of additive manufacturing technology, companies can efficiently produce customized products at scale while maintaining cost-effectiveness, agility, and sustainability. As the capabilities of 3D printing continue to evolve, its role in enabling personalized, on-demand manufacturing is poised to revolutionize various industries and transform the way we design, produce, and consume goods.

2. Implications for Healthcare:

In the healthcare sector, 3D printing is revolutionizing patient care by enabling the production of personalized medical devices and implants. From prosthetics to orthopedic implants, each patient's unique anatomical requirements can be addressed with precision through 3D printing. Customized medical solutions not only improve patient outcomes but also reduce recovery times and healthcare costs. The implications of 3D printing for healthcare are profound and far-reaching, offering innovative solutions across various facets of the industry, from patient care to medical research and device manufacturing.

In patient care, 3D printing has revolutionized the field of personalized medicine by enabling the creation of custom-designed medical implants, prosthetics, and anatomical models tailored to individual patient needs. Surgeons can use 3D-printed models of patient-specific anatomy to plan and practice complex surgical procedures with greater precision and accuracy, reducing surgical risks and improving patient outcomes. Additionally, 3D-printed implants, such as dental crowns, orthopedic implants, and cranial plates, can be precisely customized to fit each patient, resulting in better functionality and comfort. Moreover, 3D printing facilitates the development of patient-specific medical devices and assistive technologies, including hearing aids, orthotic braces, and wearable sensors. These devices can be customized to match the unique anatomical features and functional requirements of individual patients, enhancing comfort, usability, and effectiveness. Furthermore, the on-demand nature of 3D printing allows for rapid prototyping and iteration of medical devices, accelerating the pace of innovation and improving accessibility to cutting-edge healthcare solutions. In medical research and education, 3D printing enables the creation of anatomically accurate models and simulations for training healthcare professionals and conducting preclinical studies. Medical schools and training programs utilize 3D-printed anatomical models to teach anatomy, surgical techniques, and medical procedures in a realistic and interactive manner. Additionally, researchers can use 3D printing to create patient-specific disease models, drug delivery systems, and tissue-engineered constructs for studying diseases, testing treatments, and developing personalized therapies. Furthermore, 3D printing has the potential to revolutionize the field of regenerative medicine and tissue engineering by enabling the fabrication of complex biological structures, such as organs, tissues, and scaffolds, using bioinks and living cells. Researchers are exploring the use of 3D printing technology to create patient-specific organ replacements, skin grafts, and bone grafts, offering new hope for patients in need of organ transplants or tissue repair. Additionally, 3D bioprinting techniques allow for the precise deposition of cells and biomaterials to mimic the native architecture and function of biological tissues, paving the way for the development of functional tissue constructs for transplantation and disease modeling. Moreover, 3D printing has implications for improving healthcare access and affordability, particularly in underserved communities and developing countries. The decentralized nature of 3D printing allows for the local production of medical devices, prosthetics, and other healthcare products, reducing dependence on centralized manufacturing facilities and international supply chains. This localization of manufacturing can lower costs, shorten delivery times, and increase accessibility to essential healthcare solutions in remote or resource-limited settings. The implications of 3D printing for healthcare are transformative, offering opportunities to enhance patient care, advance medical research, and improve healthcare accessibility. As 3D printing technology continues to evolve and mature, its potential to revolutionize various aspects of the healthcare industry will undoubtedly lead to improved outcomes, increased efficiency, and greater innovation in the delivery of healthcare services worldwide.

3. Role in Consumer Goods, Fashion, and Architecture:

3D printing has disrupted traditional manufacturing processes in consumer goods, fashion, and architecture, offering designers unprecedented freedom to create intricate and unique designs. In consumer goods, products can be customized to fit individual preferences and lifestyles, enhancing user experience and brand loyalty. In fashion, designers can experiment with avant-garde designs and produce customized apparel and accessories tailored to individual body shapes and style preferences. In architecture, 3D printing enables the fabrication of complex building components and structures with minimal material waste, pushing the boundaries of architectural innovation.

The role of 3D printing in consumer goods, fashion, and architecture has been increasingly significant, revolutionizing traditional manufacturing processes and design practices in these industries.

In consumer goods, 3D printing offers unparalleled opportunities for customization and personalization, allowing consumers to create bespoke products tailored to their unique preferences and requirements. From customized smartphone cases and jewelry to personalized home decor and kitchen gadgets, 3D printing enables individuals to design and manufacture one-of-a-kind items with ease. This shift towards on-demand, customizable production not only enhances consumer satisfaction but also reduces waste and inventory costs for manufacturers.

Furthermore, 3D printing enables rapid prototyping and iterative design in the development of new consumer products, speeding up the product development cycle and facilitating innovation. Designers and engineers can quickly create and test prototypes of new product concepts, iterate on designs based on feedback, and bring products to market faster than ever before. This agile approach to product development fosters creativity, experimentation, and collaboration, driving continuous innovation in consumer goods industries.

In the fashion industry, 3D printing has emerged as a disruptive technology, challenging traditional methods of garment production and design. Fashion designers are leveraging 3D printing to create avant-garde clothing, accessories, and footwear with intricate geometries and novel materials that were previously impossible to achieve with traditional techniques. From 3D-printed dresses and shoes to customizable jewelry and eyewear, 3D printing enables designers to push the boundaries of creativity and expressiveness in fashion design. Moreover, 3D printing offers sustainability benefits in the fashion industry by reducing material waste and energy consumption compared to conventional manufacturing methods. Additive manufacturing allows for precise material deposition and minimal material usage, resulting in less waste and lower environmental impact. Additionally, the ability to produce items on-demand and locally using 3D printing reduces the need for long-distance shipping and excess inventory, further decreasing the carbon footprint of fashion production. In architecture and construction, 3D printing has the potential to revolutionize building design, construction, and urban development. Architects and engineers are exploring the use of large-scale 3D printing technology to create custom-designed building components, prefabricated modules, and even entire structures with unprecedented speed and efficiency. 3D-printed buildings offer advantages such as cost-effectiveness, design flexibility, and sustainability, making them attractive options for housing, infrastructure, and disaster relief projects. Furthermore, 3D printing enables architects and designers to realize complex geometries and organic forms in their designs that would be difficult or impractical to achieve using traditional construction methods. From curvilinear facades and parametric structures to lightweight lattice structures and intricate ornamentation, 3D printing opens up new possibilities for architectural expression and experimentation.

The role of 3D printing in consumer goods, fashion, and architecture is transformative, enabling customization, innovation, and sustainability across these industries. As 3D printing technology continues to advance and become more accessible, its potential to revolutionize design and manufacturing processes will continue to drive creativity, efficiency, and sustainability in consumer products, fashion, and architecture.

Overall, 3D printing's ability to empower customization and personalization is reshaping industries and driving innovation. As the technology continues to advance and become more accessible, the possibilities for creating customized products and personalized solutions are virtually limitless. Whether in healthcare, consumer goods, fashion, or architecture, 3D printing is empowering individuals and businesses to unleash their creativity and meet the diverse needs of an increasingly personalized world.

3.5 Accelerating Innovation and Prototyping with 3D Printing

In today's fast-paced business environment, the ability to innovate quickly and efficiently is essential for staying competitive. 3D printing has emerged as a key enabler of rapid innovation and prototyping, transforming the way companies develop new products and bring them to market. One of the most significant advantages of 3D printing is its ability to expedite the product development cycle through rapid prototyping and iteration. Unlike traditional manufacturing methods, which often involve time-consuming processes and high costs for creating prototypes, 3D printing allows designers and engineers to produce physical prototypes directly from digital designs with remarkable speed and precision.

This accelerated prototyping process facilitates rapid iteration, enabling companies to test multiple design iterations, assess functionalities, and identify potential issues early in the development stage. By quickly iterating on prototypes, companies can reduce the time and resources required to bring a product to market, gaining a competitive edge in the process. Numerous companies across various industries have embraced 3D printing as a tool for accelerating innovation and bringing new ideas to market faster. For example, Nike, a leader in athletic footwear, utilizes 3D printing to prototype and customize shoe designs. By leveraging 3D printing technology, Nike can rapidly iterate on designs, test performance, and gather feedback from athletes, allowing for faster product development and customization options. Similarly, SpaceX, the aerospace manufacturer founded by Elon Musk, relies heavily on 3D printing for manufacturing rocket components. By utilizing 3D printing technology, SpaceX can produce complex parts with intricate geometries more efficiently than traditional manufacturing methods. This approach not only reduces production lead times and costs but also enhances the performance and reliability of SpaceX's rockets. Furthermore, 3D printing has democratized innovation by lowering barriers to entry for entrepreneurs and small businesses. With the accessibility of desktop 3D printers and online platforms for design sharing and collaboration, individuals and startups can turn their ideas into tangible prototypes with minimal upfront investment. This democratization of innovation fosters a culture of creativity and entrepreneurship, empowering innovators to bring their concepts to market without the need for extensive resources or infrastructure. 3D printing has revolutionized the process of innovation and prototyping, offering companies unparalleled speed, flexibility, and cost-effectiveness in product development. By enabling rapid prototyping and iteration, showcasing examples of companies leveraging this technology, and fostering the democratization of innovation, 3D printing is reshaping the landscape of innovation and propelling businesses forward in the digital age.

3.6   Sustainable Manufacturing: Reducing Environmental Impact with 3D Printing

The adoption of 3D printing technology is not only revolutionizing manufacturing processes but also contributing to sustainability efforts across industries. By fundamentally altering the way products are designed, produced, and distributed, 3D printing offers several environmental benefits compared to traditional manufacturing methods.

One of the most significant environmental advantages of 3D printing is the reduction in material waste. Traditional subtractive manufacturing processes often result in significant material wastage, as raw materials are cut away to create the desired shape. In contrast, 3D printing is an additive process, where materials are deposited layer by layer to build the final product. This additive approach minimizes material waste, as only the necessary amount of material is used to create the object, leading to substantial reductions in material consumption and waste generation. Moreover, 3D printing technology has the potential to reduce energy consumption compared to traditional manufacturing methods. In conventional manufacturing, energy-intensive processes such as machining, casting, and molding are commonly used, contributing to high energy consumption and greenhouse gas emissions. 3D printing typically requires less energy, particularly for small-scale production runs, as it eliminates the need for tooling, setup, and transportation associated with traditional manufacturing processes. Another sustainability advantage of 3D printing is the potential for localized production. By decentralizing manufacturing and producing goods closer to the point of consumption, 3D printing can minimize the carbon footprint associated with traditional manufacturing and supply chains. Localized production reduces the need for long-distance transportation of goods, thereby lowering fuel consumption, greenhouse gas emissions, and air pollution. Additionally, localized production enables companies to respond quickly to changes in demand, reduce inventory levels, and eliminate excess inventory obsolescence, further enhancing sustainability. Several initiatives and case studies demonstrate the sustainability advantages of additive manufacturing. For example, some companies have implemented closed-loop recycling systems for 3D printing materials, allowing them to recycle and reuse excess material and failed prints, thereby reducing waste. Others have explored the use of bio-based and recycled materials in 3D printing, further minimizing environmental impact. Furthermore, organizations like the Ellen MacArthur Foundation have recognized the potential of 3D printing to enable a circular economy, where products and materials are designed for reuse, remanufacturing, and recycling. By embracing principles of circular design and leveraging 3D printing technology, companies can reduce resource consumption, extend product lifecycles, and minimize waste generation, contributing to a more sustainable future. 3D printing offers compelling environmental benefits compared to traditional manufacturing methods, including reduced material waste, lower energy consumption, and the potential for localized production. By embracing additive manufacturing and exploring sustainable design practices, companies can minimize their environmental footprint, drive innovation, and create a more sustainable manufacturing ecosystem for future generations.

3.7   Reshaping Supply Chains and Logistics with 3D Printing

The integration of 3D printing technology into manufacturing processes has the potential to revolutionize global supply chains and logistics, fundamentally altering traditional approaches to production, inventory management, and distribution.

One of the most significant impacts of 3D printing on supply chains is the potential for decentralization and on-demand production. Traditional manufacturing often relies on centralized production facilities that mass-produce goods, leading to long lead times, high transportation costs, and excess inventory. 3D printing enables companies to decentralize production by establishing localized manufacturing hubs closer to the point of consumption. This shift towards localized production allows for on-demand manufacturing, where products are produced only when needed, eliminating the need for large warehouses and reducing inventory holding costs. The implications of 3D printing on inventory management, warehousing, and distribution networks are profound. With on-demand production and localized manufacturing, companies can reduce the need for extensive warehousing and inventory stockpiles. Instead of maintaining large inventories of finished goods, companies can leverage digital inventories and produce items as needed, minimizing inventory carrying costs and the risk of obsolescence. Furthermore, 3D printing enables companies to adopt more agile and responsive supply chain strategies. By reducing lead times and enabling rapid prototyping and iteration, 3D printing allows companies to respond quickly to changes in demand and market trends. This agility is particularly valuable in industries with short product lifecycles or unpredictable demand patterns, such as consumer electronics and fashion.

Emerging trends like distributed manufacturing hubs and digital inventories are reshaping the future of supply chains. Distributed manufacturing hubs leverage 3D printing technology to establish localized production facilities in strategic locations, allowing companies to meet regional demand more efficiently while reducing transportation costs and carbon emissions. Digital inventories, enabled by 3D printing and digital design files, provide companies with greater flexibility and agility in managing inventory levels and responding to customer orders in real-time. Overall, 3D printing has the potential to disrupt traditional supply chains and logistics models, offering companies unprecedented flexibility, efficiency, and sustainability. By decentralizing production, enabling on-demand manufacturing, and embracing emerging trends like distributed manufacturing hubs and digital inventories, companies can optimize their supply chains, reduce costs, and enhance their competitive advantage in an increasingly globalized marketplace.

3.8   Applications of 3D Printing: Revolutionizing Industries and Empowering Innovation

The versatility and adaptability of 3D printing technology have revolutionized numerous industries, spanning from manufacturing and healthcare to architecture and aerospace. This transformative tool, also known as additive manufacturing, has unlocked a plethora of applications, pushing the boundaries of what is possible and driving innovation across diverse sectors. Some of the applications of 3D printing are summarized in Fig. 3.8.

3.8.1   Manufacturing and Prototyping

One of the most prominent applications of 3D printing is in manufacturing and rapid prototyping. From automotive and aerospace to consumer electronics, 3D printing enables companies to streamline the product development process by quickly producing prototypes and iterating designs. This agility allows for faster time-to-market and cost-effective testing of concepts before mass production.

In today's fast-paced world, innovation is key to staying ahead in the competitive landscape of manufacturing. Among the myriad of technological advancements shaping the industry, 3D printing stands out as a disruptive force, revolutionizing traditional manufacturing processes and opening new horizons for prototyping and production. Also known as additive manufacturing, 3D printing enables the creation of three-dimensional objects layer by layer from digital models, offering unparalleled flexibility, precision, and customization. This section explores in-depth the transformative impact of 3D printing in manufacturing and prototyping, examining its applications, benefits, challenges, and future prospects.

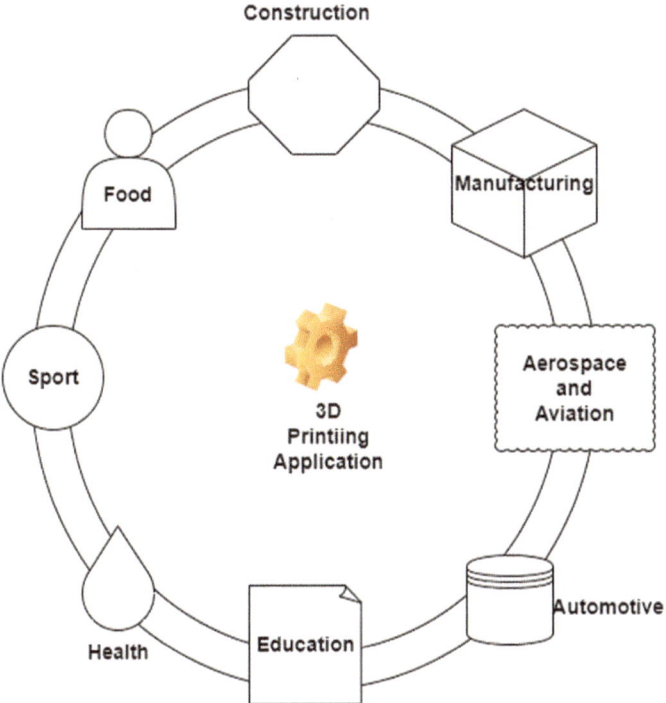

Fig. 3.8   Key applications of 3D printing

Applications of 3D Printing in Manufacturing and Prototyping:

1. *Rapid Prototyping*: Rapid prototyping stands out as one of the most renowned and widely embraced applications of 3D printing technology. Unlike traditional prototyping methods, which typically entail prolonged processes and substantial costs, 3D printing revolutionizes the prototyping landscape by offering designers and engineers the ability to swiftly materialize physical prototypes directly from digital designs. This capability has become instrumental in expediting product development cycles, fostering innovation, and reducing time-to-market across diverse industries. The traditional prototyping process often involves intricate steps such as machining, molding, or casting, which can be time-consuming, labor-intensive, and costly. Additionally, these methods may require specialized equipment, skilled labor, and multiple iterations to refine designs and validate functionalities. In contrast, 3D printing streamlines the prototyping workflow by eliminating many of the logistical challenges associated with traditional methods. With 3D printing, designers can seamlessly translate their digital designs into physical prototypes with unprecedented speed and efficiency. Whether creating intricate mechanical components, architectural models, or consumer

products, 3D printers excel at transforming virtual concepts into tangible objects with remarkable precision and fidelity. This rapid turnaround time empowers designers to iterate on designs swiftly, explore multiple iterations, and refine prototypes based on real-world feedback and testing. The ability to produce physical prototypes quickly and cost-effectively has significant implications for product development and innovation. By leveraging 3D printing for rapid prototyping, companies can accelerate the design iteration process, identify and address design flaws or functional issues early in the development cycle, and ultimately bring high-quality products to market faster. This agility and responsiveness enable companies to stay ahead of competitors, meet evolving customer demands, and capitalize on emerging market opportunities. Automotive and aerospace industries are among the early adopters of 3D printing for rapid prototyping, leveraging the technology to accelerate the development of new vehicle designs, aircraft components, and propulsion systems. In these highly competitive and innovation-driven sectors, speed to market is critical, and 3D printing offers a competitive advantage by enabling faster iteration, validation, and optimization of designs. By producing functional prototypes quickly and cost-effectively, automotive and aerospace companies can streamline their product development processes, reduce development costs, and gain a foothold in the market. Consumer electronics is another industry that extensively utilizes 3D printing for rapid prototyping. From smartphones and wearable devices to home appliances and consumer gadgets, companies in the electronics sector rely on 3D printing to iterate on designs, test form factors, and evaluate user experience. By producing physical prototypes early in the design process, electronics manufacturers can assess product aesthetics, ergonomics, and functionality, ensuring that the final product meets consumer expectations and market demands. Moreover, 3D printing enables designers to create complex geometries, intricate details, and custom features that would be challenging or impossible to achieve using traditional manufacturing methods. This design freedom encourages innovation and creativity, allowing designers to push the boundaries of what is possible and explore new design concepts without constraints. Rapid prototyping represents a cornerstone application of 3D printing technology, offering unparalleled speed, flexibility, and cost-effectiveness in the product development process. By enabling designers and engineers to quickly produce physical prototypes directly from digital designs, 3D printing facilitates faster iteration, validation, and refinement of designs, leading to shorter product development cycles and reduced time-to-market. Across industries such as automotive, aerospace, and consumer electronics, companies leverage 3D printing for rapid prototyping to accelerate innovation, enhance competitiveness, and deliver high-quality products that meet the needs of today's fast-paced market.

2. *Iterative Design:* 3D printing has revolutionized the iterative design process, empowering designers to embark on a journey of creativity and innovation with unparalleled freedom and flexibility. Unlike traditional manufacturing methods that often entail costly tooling and setup for each design iteration, 3D printing offers a rapid and

cost-effective alternative that accelerates the pace of innovation and fosters continuous improvement. At the heart of the iterative design process lies the ability to quickly produce prototypes and evaluate design concepts in a tangible, physical form. With 3D printing, designers can seamlessly translate their digital designs into physical prototypes with remarkable speed and precision. This agility enables designers to explore a myriad of creative concepts, test alternative approaches, and refine designs based on real-world feedback without being encumbered by the constraints imposed by traditional manufacturing processes. The iterative nature of 3D printing facilitates a dynamic and responsive approach to design, where designers can iterate rapidly, evaluate form and fit, and make necessary adjustments on the fly. This iterative cycle of design, prototype, test, and refining enables designers to uncover insights, identify opportunities for improvement, and iterate towards optimal solutions iteratively. By embracing an iterative approach, designers can push the boundaries of what is possible, experiment with new ideas, and evolve designs iteratively to achieve superior outcomes. Moreover, 3D printing empowers designers to overcome the limitations of traditional manufacturing methods and explore innovative design concepts that were previously deemed impractical or unattainable. With the ability to create complex geometries, intricate details, and custom features, designers can unleash their creativity and bring visionary ideas to life without compromise. This design freedom encourages experimentation, fosters innovation, and fuels the development of groundbreaking products that challenge conventions and redefine industry standards. Furthermore, the iterative design process facilitated by 3D printing fosters collaboration and interdisciplinary teamwork, bringing together designers, engineers, and stakeholders to collectively contribute to the design evolution. By embracing a collaborative approach, teams can leverage diverse perspectives, share insights, and co-create solutions that address a wide range of design considerations and user needs. This collaborative ethos promotes cross-pollination of ideas, fosters synergy, and fosters a culture of innovation that drives continuous improvement and excellence. Ultimately, the iterative design process empowered by 3D printing leads to the development of better-designed products that meet or exceed customer expectations. By iterating rapidly, testing iteratively, and refining designs based on real-world feedback, designers can optimize product performance, enhance user experience, and deliver solutions that resonate with customers on a deeper level. This customer-centric approach not only enhances product competitiveness but also builds brand loyalty and trust, driving long-term success and sustainability. 3D printing facilitates an iterative design process that empowers designers to explore creative concepts, iterate rapidly, and refine designs based on real-world feedback. By enabling rapid and cost-effective iterations, 3D printing accelerates the pace of innovation, fosters continuous improvement, and drives the development of better-designed products that meet or exceed customer expectations. Through iterative design, designers can unleash their creativity, push the boundaries of what is possible, and deliver solutions that inspire, delight, and transform the world around us.

3. *Customization and Personalization*: The advent of 3D printing has heralded a new era of customization and personalization in manufacturing, revolutionizing the way products are designed, produced, and consumed. Unlike traditional manufacturing methods, which are often optimized for mass production and standardization, 3D printing offers unparalleled flexibility and versatility, enabling manufacturers to create unique, one-of-a-kind objects tailored to individual preferences and specifications. One of the most significant advantages of 3D printing is its ability to break free from the constraints of mass production and embrace the principles of customization and personalization. Traditional manufacturing processes, such as injection molding or CNC machining, require costly tooling and setup, making customization challenging and economically unfeasible for small-scale production runs. In contrast, 3D printing eliminates the need for tooling and enables on-demand production of highly customized products with minimal setup time and cost. Whether it's customized consumer products, personalized medical devices, or bespoke architectural components, 3D printing empowers manufacturers to cater to the diverse needs and preferences of customers. From custom-fit footwear and tailored fashion accessories to personalized prosthetics and orthotics, 3D printing allows for the creation of products that are uniquely suited to the individual characteristics and requirements of each user. The ability to offer customization and personalization not only enhances the value proposition of products but also fosters stronger customer relationships and brand loyalty. By involving customers in the design process and allowing them to co-create products that reflect their unique tastes and preferences, manufacturers can forge deeper connections with their target audience and build a loyal customer base. Moreover, 3D printing enables manufacturers to differentiate their products in a crowded marketplace by offering unique, customizable solutions that stand out from off-the-shelf alternatives. Whether it's a personalized smartphone case, a custom-designed piece of jewelry, or a bespoke architectural element, 3D-printed products command attention and appeal to consumers seeking individuality and self-expression. In addition to consumer products, 3D printing is also revolutionizing the field of personalized medicine by enabling the production of customized medical devices and implants tailored to the specific anatomy of patients. From patient-specific surgical guides and orthopedic implants to custom-fitted prosthetics and dental restorations, 3D printing is empowering healthcare professionals to deliver personalized care and improve patient outcomes. Furthermore, 3D printing is driving innovation in the field of architecture and construction by enabling the fabrication of custom-designed building components and structures. From intricate facade panels and ornamental details to modular housing units and bespoke furniture, architects and designers are leveraging the capabilities of 3D printing to create buildings and spaces that are uniquely tailored to the needs and preferences of their occupants. 3D printing represents a transformative technology that is revolutionizing the way products are designed, produced, and consumed. By enabling customization and personalization at scale, 3D printing empowers manufacturers to create highly customized products

tailored to individual preferences and specifications. This ability to meet the diverse needs of customers not only enhances customer satisfaction and brand loyalty but also drives innovation, differentiation, and market success.

4. *Small-Batch Production*: The advent of 3D printing has heralded a new era of customization and personalization in manufacturing, revolutionizing the way products are designed, produced, and consumed. Unlike traditional manufacturing methods, which are often optimized for mass production and standardization, 3D printing offers unparalleled flexibility and versatility, enabling manufacturers to create unique, one-of-a-kind objects tailored to individual preferences and specifications. One of the most significant advantages of 3D printing is its ability to break free from the constraints of mass production and embrace the principles of customization and personalization. Traditional manufacturing processes, such as injection molding or CNC machining, require costly tooling and setup, making customization challenging and economically unfeasible for small-scale production runs. In contrast, 3D printing eliminates the need for tooling and enables on-demand production of highly customized products with minimal setup time and cost. Whether it's customized consumer products, personalized medical devices, or bespoke architectural components, 3D printing empowers manufacturers to cater to the diverse needs and preferences of customers. From custom-fit footwear and tailored fashion accessories to personalized prosthetics and orthotics, 3D printing allows for the creation of products that are uniquely suited to the individual characteristics and requirements of each user. The ability to offer customization and personalization not only enhances the value proposition of products but also fosters stronger customer relationships and brand loyalty. By involving customers in the design process and allowing them to co-create products that reflect their unique tastes and preferences, manufacturers can forge deeper connections with their target audience and build a loyal customer base. Moreover, 3D printing enables manufacturers to differentiate their products in a crowded marketplace by offering unique, customizable solutions that stand out from off-the-shelf alternatives. Whether it's a personalized smartphone case, a custom-designed piece of jewelry, or a bespoke architectural element, 3D-printed products command attention and appeal to consumers seeking individuality and self-expression. In addition to consumer products, 3D printing is also revolutionizing the field of personalized medicine by enabling the production of customized medical devices and implants tailored to the specific anatomy of patients. From patient-specific surgical guides and orthopedic implants to custom-fitted prosthetics and dental restorations, 3D printing is empowering healthcare professionals to deliver personalized care and improve patient outcomes. Furthermore, 3D printing is driving innovation in the field of architecture and construction by enabling the fabrication of custom-designed building components and structures. From intricate facade panels and ornamental details to modular housing units and bespoke furniture, architects and designers are leveraging the capabilities of 3D printing to create buildings and spaces that are uniquely tailored to the needs and preferences of their occupants. 3D

printing represents a transformative technology that is revolutionizing the way products are designed, produced, and consumed. By enabling customization and personalization at scale, 3D printing empowers manufacturers to create highly customized products tailored to individual preferences and specifications. This ability to meet the diverse needs of customers not only enhances customer satisfaction and brand loyalty but also drives innovation, differentiation, and market success.

5. *Supply Chain Optimization*: The integration of 3D printing technology into supply chains and logistics holds immense promise for optimizing manufacturing processes and enhancing operational efficiency. Unlike traditional manufacturing methods, which rely on complex and often lengthy supply chains with parts sourced from various suppliers worldwide, 3D printing offers the capability of localized production and on-demand manufacturing, revolutionizing the way products are produced, distributed, and consumed. Traditional manufacturing processes often involve intricate supply chains that span multiple continents, with raw materials and components sourced from disparate suppliers around the globe. This decentralized approach to production results in extended lead times, increased transportation costs, and heightened supply chain vulnerabilities, such as delays due to geopolitical instability, natural disasters, or disruptions in transportation networks. In contrast, 3D printing enables the production of parts on-demand, closer to the point of use, thereby mitigating these challenges and streamlining the supply chain. By leveraging 3D printing technology, manufacturers can produce parts and components precisely when and where they are needed, eliminating the need for long-distance transportation and reducing reliance on global sourcing. This localized production approach not only reduces lead times but also minimizes transportation costs and carbon emissions associated with traditional manufacturing and distribution networks. As a result, companies can achieve greater agility and responsiveness in meeting customer demand while simultaneously reducing their environmental footprint. Moreover, 3D printing enables the on-demand production of spare parts, offering a cost-effective alternative to traditional inventory management and warehousing practices. Instead of maintaining extensive inventories of spare parts to anticipate potential demand, companies can utilize 3D printing to produce spare parts on-demand, as and when needed. This just-in-time manufacturing approach eliminates the need for costly warehousing facilities and reduces the risk of obsolete inventory, thereby optimizing inventory management practices and improving overall supply chain efficiency. Furthermore, 3D printing technology empowers manufacturers to overcome the challenges associated with product obsolescence and supply chain disruptions. By enabling rapid prototyping and iterative design processes, 3D printing allows companies to quickly iterate on product designs, test new concepts, and bring innovative products to market faster. This agility and flexibility in product development enable companies to respond swiftly to changing market dynamics and emerging customer preferences, thereby maintaining a competitive edge in today's fast-paced

business environment. Additionally, 3D printing facilitates the localization of production, enabling companies to establish manufacturing facilities closer to their end markets. By bringing production closer to consumers, companies can reduce shipping distances, lower transportation costs, and enhance supply chain resilience. This localized production model not only improves responsiveness to customer demand but also fosters stronger connections with local communities and reduces dependence on overseas manufacturing facilities. 3D printing has the potential to revolutionize supply chains and logistics by enabling localized production, on-demand manufacturing, and agile inventory management practices. By reducing lead times, transportation costs, and supply chain vulnerabilities, 3D printing offers companies a competitive advantage in today's global marketplace. As the technology continues to evolve and become more accessible, its impact on supply chain optimization and operational efficiency is expected to grow, driving innovation and reshaping the future of manufacturing.

Benefits of 3D Printing in Manufacturing and Prototyping:

1. Speed and Efficiency: One of the most significant benefits of 3D printing is its speed and efficiency. Traditional manufacturing methods often involve lengthy processes and lead times, whereas 3D printing enables rapid production of parts and prototypes. This accelerated turnaround time allows companies to iterate designs quickly, respond to market demands faster, and reduce time-to-market for new products.

2. Cost-Effectiveness: 3D printing offers cost advantages compared to traditional manufacturing methods, particularly for prototyping and small-batch production. By eliminating the need for expensive tooling and setup costs, 3D printing reduces upfront investment and lowers the barrier to entry for new product development. Additionally, 3D printing enables more efficient use of materials, minimizing waste and reducing overall production costs.

3. Design Flexibility: 3D printing provides unparalleled design flexibility, allowing for the creation of complex geometries and intricate details that would be challenging or impossible to achieve with traditional manufacturing methods. Designers can explore innovative concepts, optimize part geometries, and create lightweight structures with minimal material usage. This design freedom enables product innovation and differentiation, leading to the development of more functional, efficient, and aesthetically pleasing products.

4. Customization and Personalization: The ability to customize and personalize products is a significant advantage of 3D printing. Unlike mass-produced goods, which are designed to appeal to a broad audience, 3D printing allows for the creation of bespoke products tailored to individual preferences and specifications. Whether it's custom-fit medical devices, personalized consumer products, or unique architectural components, 3D printing enables manufacturers to cater to the specific needs and desires of customers, enhancing customer satisfaction and brand loyalty.

5. Sustainability: 3D printing offers sustainability benefits compared to traditional manufacturing methods. By enabling on-demand production and localized manufacturing, 3D printing reduces the need for extensive warehousing and transportation, resulting in lower carbon emissions and energy consumption. Additionally, 3D printing allows for the use of recyclable materials and optimized part geometries, further reducing environmental impact and promoting sustainability in manufacturing processes.

Challenges and Considerations:

1. Material Limitations: While the range of printable materials for 3D printing has expanded significantly in recent years, certain materials, such as metals and ceramics, still present challenges in terms of printability, quality, and cost. Developing new materials and improving existing ones to meet the diverse needs of different industries remains a priority for researchers and manufacturers.
2. Scalability: Scaling up 3D printing for mass production presents logistical and technical challenges. Issues such as build volume, throughput, and cost-effectiveness need to be addressed to enable large-scale adoption of 3D printing in manufacturing. Additionally, integrating 3D printing into existing production processes and supply chains may require significant investment in infrastructure and retooling.
3. Quality and Standards: Ensuring consistent quality and adherence to industry standards is essential for the widespread adoption of 3D printing in manufacturing. Variability in print quality, material properties, and dimensional accuracy can pose challenges, particularly in safety–critical applications such as aerospace and healthcare. Establishing standardized processes, testing protocols, and certification requirements is crucial for building trust and confidence in 3D-printed products.

Future Prospects and Outlook: Despite the challenges and considerations, the future prospects of 3D printing in manufacturing and prototyping are promising. As technology continues to advance and become more accessible, the potential for 3D printing to drive innovation, reduce costs, and unlock new possibilities in product development is virtually limitless. With ongoing research and development efforts focused on overcoming material limitations, scaling up production, and ensuring quality and standards compliance, 3D printing is poised to play an increasingly central role in the future of manufacturing. From customized consumer products and personalized medical devices to sustainable production practices and decentralized supply chains, 3D printing has the potential to transform industries, empower creativity, and reshape the way we design, produce, and interact with products in the digital age. 3D printing has emerged as a transformative technology with far-reaching implications for manufacturing and prototyping. Its ability to enable rapid prototyping, iterative design, customization, and small-batch production has revolutionized traditional manufacturing processes, offering unprecedented speed, flexibility, and efficiency. While challenges such as material limitations, scalability, and quality considerations remain, ongoing advancements in technology and research efforts are addressing these issues, paving the way for broader adoption and integration of 3D printing in manufacturing. As we look towards the future, the transformative impact of 3D printing on industries ranging from aerospace and automotive to healthcare and consumer goods is undeniable, signaling a new era of innovation, creativity, and sustainability in manufacturing and prototyping.

3.8.2 Healthcare and Biomedical

In the healthcare sector, 3D printing is revolutionizing patient care by facilitating the production of customized medical devices, prosthetics, and implants. Surgeons can now use patient-specific 3D-printed models to plan complex surgeries with greater accuracy, reducing risks and improving outcomes. Additionally, bioprinting technologies hold promise for regenerative medicine, tissue engineering, and drug discovery, offering new avenues for personalized healthcare solutions.

In recent years, 3D printing has emerged as a revolutionary technology in the field of healthcare and biomedical engineering. The ability to create custom, patient-specific medical devices and anatomical models has transformed the way healthcare professionals approach diagnosis, treatment planning, and surgical interventions. From personalized implants to intricate organ models, 3D printing is making significant strides in improving patient care and advancing medical research. One of the primary applications of 3D printing in healthcare is the production of patient-specific anatomical models. Using medical imaging data such as CT scans and MRIs, healthcare providers can create highly accurate 3D models of patient anatomy. These models enable surgeons to visualize complex anatomical structures, plan surgical procedures, and simulate interventions before entering the operating room. By providing a tangible representation of patient anatomy, 3D-printed models enhance surgical precision, reduce operative time, and improve patient outcomes. In addition to anatomical models, 3D printing is revolutionizing the fabrication of custom medical implants and prosthetics. Traditional manufacturing methods for implants often result in standardized, off-the-shelf devices that may not fully meet the unique anatomical requirements of individual patients. With 3D printing, implants can be designed and fabricated to match the specific anatomy of each patient, leading to better fit, function, and long-term outcomes. Whether it's a hip implant tailored to the patient's bone structure or a cranial implant customized to restore skull integrity, 3D printing offers unparalleled flexibility and customization in implant design and fabrication.

Furthermore, 3D printing is driving innovation in the field of regenerative medicine and tissue engineering. Bioprinting technologies allow researchers to create complex three-dimensional structures using living cells and biomaterials. These bioengineered tissues and organs hold promise for applications such as tissue repair, organ transplantation, and drug testing. By precisely controlling the placement of cells and materials, bioprinting enables the creation of functional tissues and organs that mimic the structure and function of native tissues. While still in the early stages of development, bioprinting has the potential to revolutionize organ transplantation and regenerative medicine, offering new hope for patients with organ failure and tissue damage. Moreover, 3D printing is facilitating innovation in the development of medical devices and surgical instruments. Complex geometries and intricate features that are difficult or impossible to achieve with traditional manufacturing methods can be easily produced using 3D printing. From surgical guides and patient-specific instruments to custom orthotics and prosthetic components, 3D printing enables the rapid prototyping and production of medical devices tailored to individual patient needs. This customization enhances device performance, patient comfort, and overall treatment outcomes, driving advancements in patient care and clinical practice. Beyond clinical applications, 3D printing is also playing a vital role in medical education and training. Anatomical models, surgical simulators, and educational aids produced using 3D printing provide students and healthcare professionals with hands-on learning experiences and realistic representations of human anatomy and pathology. By offering a tangible and interactive learning environment, 3D-printed models enhance understanding, retention, and proficiency in medical procedures, ultimately improving patient safety and quality of care. 3D printing is transforming healthcare and biomedical applications, offering unprecedented opportunities for innovation, customization, and patient care. From personalized anatomical models and custom implants to bioengineered tissues and surgical instruments, 3D printing is revolutionizing the way medical devices are designed, fabricated, and utilized. As technology continues to advance and become more accessible, the potential for 3D printing to address unmet medical needs, improve treatment outcomes, and enhance patient experiences is vast. By harnessing the power of 3D printing, healthcare professionals and researchers are paving the way for a new era of personalized medicine and regenerative therapies, where each patient receives tailored solutions that optimize their health and well-being.

3.8.3 Architecture and Construction

3D printing has emerged as a disruptive force in architecture and construction, allowing for the fabrication of intricate building components and structures with unprecedented speed and precision. Architects and designers can leverage 3D printing to create complex geometries, intricate facades, and sustainable structures, pushing the boundaries of architectural innovation.

In the realm of architecture and construction, 3D printing has emerged as a disruptive technology, revolutionizing traditional building methods and opening up new possibilities for design, construction, and sustainability. Also known as additive manufacturing, 3D printing enables architects, engineers, and designers to create complex structures and components with unprecedented speed, precision, and customization. From intricate architectural models to full-scale building components, 3D printing is reshaping the way buildings are designed, fabricated, and constructed.

One of the primary applications of 3D printing in architecture is the rapid prototyping of architectural models and scale replicas. Traditionally, architects relied on manual model-making techniques or computer-aided design (CAD) software to visualize and communicate their designs. However, 3D printing allows architects to quickly produce physical models directly from digital designs, enabling them to explore design iterations, test structural integrity, and communicate ideas more effectively with clients and stakeholders. These detailed and accurate models provide valuable insights into the spatial qualities, proportions, and aesthetic features of proposed buildings, facilitating informed decision-making throughout the design process. Moreover, 3D printing is revolutionizing the fabrication of architectural components and building elements. Complex geometries, intricate details, and customized features that were once challenging or impossible to achieve with traditional manufacturing methods can now be easily produced using 3D printing. Whether it's custom façade panels, ornamental decorations, or structural components, 3D printing offers architects and designers unparalleled design freedom and flexibility. By eliminating the constraints imposed by traditional manufacturing processes, 3D printing enables the creation of innovative and visually striking architectural elements that enhance the overall aesthetic appeal and functionality of buildings.

In addition to architectural components, 3D printing is also being used to fabricate full-scale building structures and modules. Additive manufacturing technologies such as large-scale concrete printing and robotic fabrication systems enable the construction of entire buildings using 3D printing techniques. By layering successive layers of concrete or other construction materials, 3D printers can build walls, floors, and other structural elements with precision and efficiency. This approach offers several advantages over traditional construction methods, including reduced material waste, shorter construction times, and lower labor costs. Moreover, 3D printing enables architects to create organic shapes and innovative structures that push the boundaries of conventional building design, resulting in more sustainable, efficient, and aesthetically pleasing buildings. Furthermore, 3D printing is driving innovation in sustainable architecture and construction practices. By using locally sourced materials, optimizing material usage, and minimizing construction waste, 3D printing offers a more environmentally friendly alternative to traditional building methods. Additionally, 3D printing enables the integration of sustainable design principles such as passive heating and cooling, natural ventilation, and renewable energy systems into building designs. By leveraging the capabilities of 3D printing, architects and builders can create energy-efficient, resource-efficient, and environmentally sustainable buildings that minimize their carbon footprint and contribute to a more sustainable built environment.

Beyond construction, 3D printing is also playing a role in disaster relief and humanitarian aid efforts. In disaster-prone areas or regions with limited access to traditional construction materials and methods, 3D printing offers a cost-effective and scalable solution for quickly building shelters, emergency housing, and infrastructure. By using locally available materials and portable 3D printing equipment, relief organizations and humanitarian agencies can rapidly deploy housing solutions to displaced populations and communities in need. 3D printing is transforming architecture and construction, offering architects, engineers, and designers unprecedented opportunities for innovation, customization, and sustainability. From rapid prototyping and customized architectural elements to full-scale building construction and disaster relief efforts, 3D printing is reshaping the way buildings are designed, fabricated, and constructed. As technology continues to advance and become more accessible, the potential for 3D printing to revolutionize the architecture and construction industry is limitless. By embracing the capabilities of 3D printing, architects and builders can create buildings that are not only visually stunning and structurally innovative but also sustainable, efficient, and resilient to the challenges of the future.

3.8.4 Consumer Goods and Fashion

In the realm of consumer goods and fashion, 3D printing is enabling customization and personalization on a mass scale. Companies can produce bespoke products tailored to individual preferences, from personalized footwear and accessories to custom-fit eyewear and jewelry. This customization not only enhances the customer experience but also fosters brand loyalty and differentiation in competitive markets. The advent of 3D printing has ushered in a new era of innovation and customization in the consumer goods and fashion industries. This transformative technology, also known as additive manufacturing, enables designers and manufacturers to create intricate, personalized products with unprecedented speed and precision. From customized footwear and accessories to avant-garde fashion pieces and household items, 3D printing is reshaping the way consumer goods are designed, produced, and consumed.

One of the most significant applications of 3D printing in consumer goods and fashion is the customization and personalization of products. Traditional manufacturing methods often rely on mass production techniques that result in standardized, one-size-fits-all products. However, 3D printing allows for the creation of bespoke, tailor-made products that are customized to individual preferences and specifications. Whether it's a pair of custom-fit shoes, a personalized smartphone case, or a unique piece of jewelry, 3D printing enables consumers to express their individuality and style in ways that were previously unimaginable. Moreover, 3D printing is driving innovation in product design and manufacturing processes. Complex geometries, intricate details, and customized features that were once difficult or impossible to achieve with traditional manufacturing methods can now be easily produced using 3D printing. Designers can experiment with unconventional shapes, textures, and materials, pushing the boundaries of creativity and pushing the limits of traditional design aesthetics. By leveraging the capabilities of 3D printing, designers and manufacturers can create products that are not only visually striking but also highly functional, ergonomic, and sustainable. Furthermore, 3D printing is revolutionizing the production of fashion and apparel items. From avant-garde couture to everyday wear, designers are harnessing the power of 3D printing to create innovative and cutting-edge fashion pieces that challenge conventional notions of style and craftsmanship. Additive manufacturing technologies allow designers to experiment with new materials, textures, and construction techniques, resulting in garments and accessories that are both aesthetically captivating and technically advanced. Whether it's a 3D-printed dress made from flexible, biodegradable materials or a pair of intricately designed shoes with customized features, 3D printing is pushing the boundaries of fashion design and manufacturing. Additionally, 3D printing is empowering designers and entrepreneurs to bypass traditional manufacturing and distribution channels, enabling them to bring their products directly to market. With the rise of desktop 3D printers and online marketplaces, designers can create and sell their own 3D-printed products without the need for costly infrastructure or middlemen. This democratization of manufacturing has led to the emergence of a new generation of independent designers and makers who are disrupting the fashion and consumer goods industries with their innovative and unique creations. Moreover, 3D printing is driving sustainability and reducing waste in the consumer goods and fashion industries. Traditional manufacturing methods often result in significant material waste and environmental pollution. However, 3D printing offers a more sustainable alternative by enabling on-demand production, localized manufacturing, and the use of eco-friendly materials. By producing products only when needed and using recyclable or biodegradable materials, 3D printing reduces the environmental impact of manufacturing and contributes to a more sustainable and circular economy. 3D printing is transforming the consumer goods and fashion industries, offering unparalleled opportunities for innovation, customization, and sustainability. From personalized products and avant-garde fashion pieces to sustainable manufacturing practices and independent entrepreneurship, 3D printing is reshaping the way products are designed, produced, and consumed. As technology continues to advance and become more accessible, the potential for 3D printing to revolutionize the consumer goods and fashion industries is limitless. By embracing the capabilities of 3D printing, designers, manufacturers, and consumers can collaborate to create a more diverse, inclusive, and sustainable future for fashion and consumer goods.

3.8.5 Education and Research

3D printing has also found applications in education and research, democratizing access to prototyping and enabling hands-on learning experiences. Educational institutions and research laboratories use 3D printers to create models, prototypes, and experiments, fostering innovation and creativity among students and researchers alike. The integration of

3D printing technology into educational institutions and research laboratories has opened up new horizons for learning, experimentation, and innovation. As additive manufacturing continues to evolve, it has become an indispensable tool for educators, researchers, and students across various disciplines, offering hands-on experiences, practical applications, and opportunities for creativity and discovery.

One of the primary applications of 3D printing in education is in STEM (Science, Technology, Engineering, and Mathematics) fields. By incorporating 3D printing into curriculum, educators can provide students with practical, real-world experiences that reinforce theoretical concepts and foster critical thinking and problem-solving skills. Students can design, prototype, and test their own creations, gaining valuable insights into the design process and engineering principles. Moreover, 3D printing allows for the visualization of complex scientific concepts, making abstract ideas more tangible and accessible to students of all ages. Furthermore, 3D printing is transforming the way students learn about anatomy, biology, and the human body. Medical schools and biology departments use 3D printing to create anatomically accurate models of organs, tissues, and biological structures for educational purposes. These models provide students with a hands-on learning experience, allowing them to explore the intricacies of the human body in a tangible and interactive way. Additionally, 3D printing enables the customization of models to match specific patient demographics or pathological conditions, enhancing the educational value and relevance of the learning experience. Moreover, 3D printing is empowering researchers to explore new frontiers in materials science, engineering, and biomedical research. Additive manufacturing technologies enable researchers to fabricate complex structures and functional prototypes with unprecedented precision and speed. Whether it's developing new materials with specific properties, testing novel medical devices, or simulating biological processes, 3D printing offers researchers a versatile tool for experimentation and innovation. By providing researchers with the ability to rapidly iterate designs, test hypotheses, and visualize data, 3D printing accelerates the pace of scientific discovery and drives advancements in various fields. Additionally, 3D printing is fostering interdisciplinary collaboration and innovation in research and development. Researchers from diverse disciplines, including engineering, medicine, biology, and design, are coming together to explore the potential of 3D printing in addressing complex challenges and solving real-world problems. Collaborative projects involving multiple stakeholders allow for the exchange of ideas, expertise, and resources, leading to the development of innovative solutions that would not be possible through traditional approaches. From developing customized medical implants to designing sustainable architecture, interdisciplinary research powered by 3D printing is pushing the boundaries of innovation and driving positive change in society.

Furthermore, 3D printing is democratizing access to research and education by lowering barriers to entry and promoting inclusivity and diversity. With the availability of affordable desktop 3D printers and open-source design software, students and researchers from all backgrounds can engage in hands-on experimentation and exploration. Moreover, online communities and maker spaces provide platforms for collaboration, knowledge sharing, and skill development, further democratizing access to 3D printing technology and expertise.

3D printing is revolutionizing education and research, offering educators, students, and researchers unparalleled opportunities for learning, experimentation, and innovation. Whether it's enhancing STEM education, advancing medical research, or fostering interdisciplinary collaboration, 3D printing is reshaping the way we teach, learn, and conduct research. As technology continues to advance and become more accessible, the potential for 3D printing to transform education and research is limitless. By embracing the capabilities of 3D printing, we can inspire the next generation of innovators, advance scientific knowledge, and address the challenges of the future with creativity, curiosity, and collaboration.

3.8.6 Food and Culinary

The emerging field of 3D food printing is transforming the way we prepare and consume food. From intricate chocolate sculptures to personalized confections and nutritionally optimized meals, 3D food printing offers novel possibilities for culinary creativity and dietary customization. Moreover, 3D printing can address food sustainability challenges by reducing food waste and optimizing ingredient usage. The intersection of 3D printing and food technology has given rise to innovative approaches in culinary arts, food production, and gastronomy. As additive manufacturing techniques continue to evolve, they are increasingly being applied to the creation of customized food products, intricate culinary designs, and novel dining experiences. From personalized confections to geometrically precise pastries, 3D printing is reshaping the way we think about food preparation, presentation, and consumption.

One of the primary applications of 3D printing in the culinary world is the fabrication of customized food products and edible decorations. Using food-grade materials such as chocolate, sugar, or dough, chefs and food artisans can create intricate designs, shapes, and textures that would be difficult or impossible to achieve with traditional methods. 3D printing allows for the precise deposition of food materials layer by layer, enabling chefs to realize their culinary visions with unparalleled precision and creativity. Whether it's a custom-designed cake topper, a decorative garnish, or a personalized chocolate sculpture, 3D printing offers chefs and pastry artists a versatile tool for culinary expression. Furthermore, 3D printing is revolutionizing the production of food molds, utensils, and kitchen equipment. Additive manufacturing technologies enable the fabrication of custom molds and templates for shaping food items, such as cookies, chocolates, and pastries. Chefs can create unique shapes and designs that reflect their culinary style and creativity, adding a distinctive touch to their creations. Moreover, 3D-printed kitchen utensils and tools, such as spatulas, tongs, and measuring spoons, offer chefs and home cooks ergonomic designs and customized features that enhance efficiency and precision in food preparation. Moreover, 3D printing is driving innovation in the development of functional foods and nutritional supplements. Researchers and food scientists are exploring the use of 3D printing to create personalized dietary products tailored to individual nutritional needs and dietary preferences. By precisely controlling the composition and structure of food materials, 3D printing enables the production of functional foods with specific health benefits, such as enhanced nutrient absorption, controlled release of bioactive compounds, and improved taste and texture. Additionally, 3D printing allows for the incorporation of novel ingredients and formulations into food products, opening up new possibilities for culinary experimentation and innovation. Additionally, 3D printing is playing a role in addressing food security and sustainability challenges. By utilizing alternative food sources and sustainable ingredients, such as algae, insects, and plant-based proteins, researchers and food entrepreneurs are exploring new avenues for food production and consumption. 3D printing offers a flexible and efficient means of processing and transforming these ingredients into nutritious and palatable food products. Moreover, by optimizing the use of resources and minimizing waste, 3D printing contributes to more sustainable and environmentally friendly food systems. Furthermore, 3D printing is enhancing the dining experience through the creation of immersive and interactive culinary presentations. Chefs and restaurateurs are leveraging 3D printing technology to design avant-garde dining experiences that engage all the senses and stimulate the imagination. Whether it's a multi-course tasting menu featuring 3D-printed appetizers, entrees, and desserts or a pop-up dining event showcasing edible sculptures and interactive installations, 3D printing enables chefs to push the boundaries of culinary creativity and innovation. 3D printing is revolutionizing the food and culinary industry, offering chefs, food artisans, and researchers unprecedented opportunities for creativity, customization, and innovation. From personalized confections and functional foods to sustainable ingredients and immersive dining experiences, 3D printing is reshaping the way we produce, prepare, and experience food. As technology continues to advance and become more accessible, the potential for 3D printing to transform the culinary landscape is limitless. By embracing the capabilities of 3D printing, chefs, food entrepreneurs, and researchers can unlock new possibilities for culinary creativity, sustainability, and gastronomic delight.

3.8.7   Defense and Aerospace

3D printing is reshaping the defense and aerospace industries by enabling the production of lightweight, high-performance components and prototypes. Aerospace engineers use 3D printing to fabricate complex aircraft parts, engine components, and satellite components with superior strength-to-weight ratios. Moreover, 3D printing allows for rapid prototyping and customization of defense equipment, enhancing operational readiness and mission effectiveness. The intersection of 3D printing and food technology has given rise to innovative approaches in culinary arts, food production, and gastronomy. As additive manufacturing techniques continue to evolve, they are increasingly being applied to the creation of customized food products, intricate culinary designs, and novel dining experiences. From personalized confections to geometrically precise pastries, 3D printing is reshaping the way we think about food preparation, presentation, and consumption. One of the primary applications of 3D printing in the culinary world is the fabrication of customized food products and edible decorations. Using food-grade materials such as chocolate, sugar, or dough, chefs and food artisans can create intricate designs, shapes, and textures that would be difficult or impossible to achieve with traditional methods. 3D printing allows for the precise deposition of food materials layer by layer, enabling chefs to realize their culinary visions with unparalleled precision and creativity. Whether it's a custom-designed cake topper, a decorative garnish, or a personalized chocolate sculpture, 3D printing offers chefs and pastry artists a versatile tool for culinary expression.

Furthermore, 3D printing is revolutionizing the production of food molds, utensils, and kitchen equipment. Additive manufacturing technologies enable the fabrication of custom molds and templates for shaping food items, such as cookies, chocolates, and pastries. Chefs can create unique shapes and designs that reflect their culinary style and creativity, adding a distinctive touch to their creations. Moreover, 3D-printed kitchen utensils and tools, such as spatulas, tongs, and measuring spoons, offer chefs and home cooks ergonomic designs and customized features that enhance efficiency and precision in food preparation. Moreover, 3D printing is driving innovation in the development of functional foods and nutritional supplements. Researchers and food scientists are exploring the use of 3D printing to create personalized dietary products tailored to individual nutritional needs and dietary preferences. By precisely controlling the composition and structure of food materials, 3D printing enables the production of functional foods with specific health benefits, such as enhanced nutrient absorption, controlled release of bioactive compounds, and improved taste and texture. Additionally, 3D printing allows for the incorporation of novel ingredients and formulations into food products, opening up new possibilities for culinary experimentation and innovation. Additionally, 3D printing is playing a role in addressing food security and sustainability challenges. By utilizing alternative food sources and sustainable ingredients, such as algae, insects, and plant-based proteins, researchers and food entrepreneurs are exploring new avenues for food production and consumption. 3D printing offers a flexible and efficient means of processing and transforming these ingredients into nutritious and palatable food products. Moreover, by optimizing the use of resources and minimizing waste, 3D printing contributes to more sustainable and environmentally friendly food systems. Furthermore, 3D printing is enhancing the dining experience through the creation of immersive and interactive culinary presentations. Chefs and restaurateurs are leveraging 3D printing technology to design avant-garde dining experiences that engage all the senses and stimulate the imagination. Whether it's a multi-course tasting menu featuring 3D-printed appetizers, entrees, and desserts or a pop-up dining event showcasing edible sculptures and interactive installations, 3D printing enables chefs to push the boundaries of culinary creativity and innovation. 3D printing is revolutionizing the food and culinary industry, offering chefs, food artisans, and researchers unprecedented opportunities for creativity, customization, and innovation. From personalized confections and functional foods to sustainable ingredients and immersive dining experiences, 3D printing is reshaping the way we produce, prepare, and experience food. As technology continues to advance and become more accessible, the potential for 3D printing to transform the culinary landscape is limitless. By embracing the capabilities of 3D printing, chefs, food entrepreneurs, and researchers can unlock new possibilities for culinary creativity, sustainability, and gastronomic delight.

3.8.8 Environmental and Sustainable Applications

Beyond traditional industries, 3D printing is being leveraged for environmental and sustainable applications. Companies are exploring the use of biodegradable and recycled materials in 3D printing to reduce environmental impact and promote circular economy principles. Additionally, 3D printing enables localized production, minimizing transportation emissions and energy consumption associated with traditional manufacturing and supply chains. Environmental and sustainable applications have become increasingly crucial in addressing contemporary challenges facing our planet. With the rise in environmental degradation, resource depletion, and climate change, there is a growing need for innovative solutions that promote sustainability and mitigate environmental impacts across various sectors.

One significant area where environmental and sustainable applications play a vital role is in energy production and consumption. The shift towards renewable energy sources such as solar, wind, hydroelectric, and geothermal power has gained momentum in recent years. These sources offer clean alternatives to fossil fuels, significantly reducing greenhouse gas emissions and air pollution. Additionally, advancements in energy-efficient technologies and practices contribute to sustainable energy consumption, promoting conservation and reducing environmental harm. Another critical aspect is sustainable agriculture and food systems. Traditional agricultural practices often result in soil degradation, water pollution, and biodiversity loss. However, sustainable farming methods, including organic farming, agroforestry, and permaculture, prioritize environmental stewardship while ensuring food security and livelihoods for communities. By implementing sustainable agricultural practices, we can minimize chemical inputs, conserve water resources, and preserve ecosystems, thereby fostering long-term resilience in food production systems. Furthermore, the concept of sustainable urban development is gaining traction as the global population increasingly resides in cities. Sustainable urban planning emphasizes creating livable, resource-efficient cities that prioritize public transportation, green spaces, energy-efficient buildings, and waste management systems. These initiatives aim to reduce carbon emissions, enhance air quality, and improve the overall quality of life for urban residents while preserving natural habitats and biodiversity. In manufacturing and production processes, the adoption of eco-friendly practices and technologies is essential for reducing environmental footprints. This includes initiatives such as eco-design, which focuses on designing products that minimize resource consumption, waste generation, and pollution throughout their lifecycle. Additionally, the implementation of circular economy principles promotes the reuse, recycling, and repurposing of materials, thereby minimizing waste and conserving resources.

In the realm of water management, sustainable approaches are critical for ensuring equitable access to clean water resources while protecting aquatic ecosystems. Water conservation measures, such as rainwater harvesting, greywater recycling, and efficient irrigation techniques, help mitigate water scarcity and reduce the strain on freshwater sources. Moreover, the restoration and preservation of wetlands and watersheds play a vital role in maintaining water quality and biodiversity. The adoption of environmental and sustainable applications is not only imperative for addressing current environmental challenges but also for securing a sustainable future for generations to come. By integrating these principles into various sectors and everyday practices, we can work towards building a more resilient, equitable, and environmentally sustainable world.

In summary, the applications of 3D printing are vast and varied, spanning industries and disciplines. From revolutionizing manufacturing and healthcare to empowering creativity and sustainability, 3D printing continues to push the boundaries of innovation and reshape the way we design, create, and interact with the world around us. As technology continues to advance and become more accessible, the potential for 3D printing to transform industries and drive positive change remains boundless.

3.8.9 3D Printing Application in Energy Sector Innovations

The integration of 3D printing technology in the energy sector has sparked significant innovation and transformation. This section explores the diverse applications of 3D printing in renewable energy solutions and enhancing efficiency in energy production.

Renewable Energy Solutions with 3D Printing
Renewable energy solutions are essential components of global efforts to mitigate climate change and transition towards a sustainable energy future. Among the various technologies driving this transition, 3D printing, also known as additive manufacturing, has emerged as a promising tool for innovation and advancement in renewable energy systems. One of the key applications of 3D printing in renewable energy is the fabrication of components for solar energy systems. Solar photovoltaic (PV) panels, which convert sunlight into electricity, rely on various components such as frames, mounts, and junction boxes. With 3D printing, these components can be produced using a variety of materials, including plastics, metals, and composites. Additive manufacturing allows for the creation of complex

geometries and customized designs, optimizing the performance and durability of solar panels. In addition to traditional solar PV panels, 3D printing enables the production of innovative solar energy solutions, such as solar concentrators and receivers. Solar concentrators focus sunlight onto a small area, increasing the intensity of solar radiation and improving energy capture efficiency. By leveraging 3D printing, manufacturers can create intricately designed concentrator structures with precision and accuracy, enhancing the performance of concentrated solar power (CSP) systems. Furthermore, 3D printing facilitates the rapid prototyping and iterative design of solar energy components, accelerating innovation in the renewable energy sector. Engineers and researchers can quickly develop and test new concepts, optimizing designs for efficiency, reliability, and cost-effectiveness. This agility and flexibility are particularly valuable in the dynamic and rapidly evolving field of renewable energy, where technological advancements drive progress and competitiveness. Moreover, 3D printing offers opportunities for local manufacturing and distributed energy production. Unlike traditional manufacturing processes that rely on centralized production facilities and global supply chains, additive manufacturing can be deployed in virtually any location with access to the technology. This decentralization reduces transportation costs, carbon emissions, and reliance on fossil fuels, contributing to the sustainability of renewable energy systems. Another promising application of 3D printing in renewable energy is the production of wind turbine components. Wind energy is a rapidly growing renewable energy source, with wind turbines playing a pivotal role in electricity generation. Turbine blades, towers, and nacelles are critical components of wind turbines, requiring high-performance materials and intricate designs to withstand harsh environmental conditions. With 3D printing, these components can be manufactured with precision and efficiency, optimizing aerodynamic performance and structural integrity. Additionally, 3D printing enables the customization and optimization of wind turbine designs for specific wind conditions and geographic locations. By tailoring turbine blades and other components to local wind patterns, engineers can maximize energy capture and minimize operational costs. This customization capability enhances the competitiveness of wind energy systems and accelerates the deployment of renewable energy projects worldwide. 3D printing offers significant potential for innovation and advancement in renewable energy solutions. From solar panels and concentrators to wind turbine components, additive manufacturing enables the production of high-performance, customized, and cost-effective renewable energy systems. By leveraging the capabilities of 3D printing, researchers, engineers, and manufacturers can drive progress towards a more sustainable and resilient energy future.

Enhancing Efficiency in Energy Production through 3D Printing
In addition to renewable energy solutions, 3D printing is revolutionizing conventional energy production processes, enhancing efficiency, reliability, and cost-effectiveness. Traditional energy production methods, such as those used in fossil fuel-based power plants and nuclear facilities, often rely on complex machinery and equipment. These components

require regular maintenance, replacement, and optimization to ensure optimal performance and safety.

One significant application of 3D printing in conventional energy production is the manufacturing of gas turbine components. Gas turbines are widely used in power generation facilities, industrial plants, and aviation propulsion systems. These turbines operate under extreme conditions, including high temperatures and pressures, which can cause wear and degradation of components over time. With 3D printing, turbine parts can be produced using advanced alloys and intricate designs, enhancing their thermal and mechanical properties. Furthermore, 3D printing enables the rapid repair and replacement of turbine components, minimizing downtime and maintenance costs. Instead of waiting for replacement parts to be manufactured and delivered, operators can produce them on-site using additive manufacturing techniques. This agility and responsiveness improve the reliability and availability of gas turbine systems, reducing the risk of unplanned outages and maximizing energy production. Moreover, 3D printing offers opportunities for optimizing the design and manufacturing of nuclear power plant components. Nuclear reactors operate under stringent safety and regulatory requirements, requiring robust and reliable equipment to ensure the integrity of the system. Additive manufacturing allows for the creation of complex geometries and materials tailored to the specific needs of nuclear applications. For example, 3D printed reactor cores, fuel assemblies, and shielding structures can enhance safety, efficiency, and longevity.

Additionally, 3D printing has applications in the fabrication of components for oil and gas extraction and refining operations. The oil and gas industry relies on a wide range of equipment, including pumps, valves, and pipelines, which are subject to harsh operating conditions and corrosion. With 3D printing, these components can be produced using corrosion-resistant materials and customized designs, improving their performance and reliability. Moreover, 3D printing facilitates the rapid prototyping and iterative design of oil and gas equipment, accelerating innovation and reducing time-to-market. Engineers can quickly develop and test new concepts, optimizing designs for efficiency, durability, and cost-effectiveness. This agility is particularly valuable in the oil and gas industry, where technological advancements drive competitiveness and operational excellence. 3D printing is revolutionizing conventional energy production processes by enhancing efficiency, reliability, and cost-effectiveness. From gas turbine components to nuclear reactor parts and oil and gas equipment, additive manufacturing offers numerous benefits for the energy sector. By leveraging the capabilities of 3D printing, operators and manufacturers can improve the performance, safety, and sustainability of energy production systems, contributing to a more resilient and efficient energy future.

One notable application of 3D printing in conventional energy is the manufacturing of gas turbine components. Turbines used in power generation facilities often operate under extreme conditions, requiring components that can withstand high temperatures and pressures. With 3D printing, turbine parts can be produced using advanced alloys and intricate cooling channels, resulting in enhanced thermal performance and mechanical strength. Furthermore, additive manufacturing enables the rapid repair and replacement of turbine components, minimizing downtime and maintenance costs. Moreover, 3D printing offers opportunities for optimizing the design and manufacturing of nuclear power plant components. By leveraging additive manufacturing techniques, nuclear engineers can create complex geometries for reactor cores, fuel assemblies, and shielding structures. This enables the development of more efficient and compact reactor designs while ensuring safety and reliability. Furthermore, 3D printing plays a crucial role in the fabrication of components for oil and gas extraction and refining operations. Additive manufacturing allows for the production of customized parts with reduced lead times and costs. For instance, 3D printed impellers and valves can improve the performance and reliability of pumps and pipelines in oil and gas facilities. Additionally, 3D printing enables the creation of lightweight and corrosion-resistant components for offshore platforms and subsea infrastructure, enhancing operational efficiency and longevity.

3D printing technology offers immense potential for innovation and advancement in the energy sector. From renewable energy solutions to enhancing efficiency in conventional energy production, additive manufacturing is driving significant improvements in performance, reliability, and sustainability. By leveraging 3D printing, energy companies can accelerate the transition towards a cleaner, more resilient, and more efficient energy future.

3.9   3D Printing: Impacts on Global Economy and Trade

The rise of 3D printing technology has brought about profound changes in global economy and trade dynamics. This section explores the multifaceted impacts of 3D printing, including the disruption of traditional manufacturing models and the economic opportunities and challenges it presents.

Disruption of Traditional Manufacturing Models

Traditional manufacturing models have long relied on centralized production facilities, mass production techniques, and global supply chains. However, 3D printing, also known as additive manufacturing, is challenging these conventional paradigms by offering decentralized, on-demand production capabilities. One significant impact of 3D printing on the global economy is the decentralization of manufacturing. Unlike traditional manufacturing processes that require large-scale production facilities, 3D printing enables localized manufacturing in virtually any location with access to the technology. This decentralization reduces reliance on centralized production hubs and global supply chains, potentially reshaping global trade patterns. Furthermore, 3D printing disrupts traditional inventory management practices by enabling on-demand production of goods. Instead of maintaining large inventories of finished products, businesses can produce items as needed, reducing storage costs and minimizing the risk of overstocking or obsolescence. This just-in-time manufacturing approach enhances supply chain efficiency and flexibility, leading to cost savings and improved responsiveness to market demand.

Moreover, 3D printing facilitates the customization and personalization of products, catering to individual preferences and niche markets. This customization capability disrupts traditional mass production models, where economies of scale often favor standardized, one-size-fits-all products. By offering personalized solutions, businesses can differentiate themselves in the market and create new revenue streams.

Economic Opportunities and Challenges of 3D Printing

While 3D printing presents numerous economic opportunities, it also poses several challenges and considerations for businesses, policymakers, and trade relations. One of the most significant economic opportunities of 3D printing is cost savings through reduced material waste and production time. Traditional subtractive manufacturing processes often result in significant material wastage, whereas additive manufacturing techniques used in 3D printing build objects layer by layer, minimizing material usage. Additionally, 3D printing allows for the production of complex geometries and assemblies as a single, integrated component, eliminating the need for assembly and reducing production time and labor costs. Furthermore, 3D printing opens up new avenues for innovation and entrepreneurship. The technology enables small and medium-sized enterprises (SMEs) and individual entrepreneurs to compete in markets traditionally dominated by large corporations. With relatively low barriers to entry and access to affordable desktop 3D printers, innovators can bring their ideas to life and disrupt established industries.

However, despite its potential benefits, 3D printing also presents challenges and considerations for the global economy and trade relations. One major concern is the impact on traditional manufacturing industries and associated job displacement. As 3D printing becomes more widespread, certain manufacturing sectors may experience disruption, leading to layoffs and restructuring. Moreover, the proliferation of 3D printing could pose intellectual property challenges, as digital design files can be easily shared and replicated, raising questions about copyright infringement and counterfeiting. Additionally, 3D printing has implications for global trade patterns and supply chain dynamics. The decentralization of manufacturing enabled by 3D printing may lead to a reconfiguration of global supply chains, with production shifting closer to end markets. This could have implications for countries that rely heavily on manufacturing exports and may necessitate adjustments in trade policies and agreements.

3D printing is reshaping the global economy and trade landscape in profound ways. By disrupting traditional manufacturing models and offering new economic opportunities, additive manufacturing is driving innovation, efficiency, and customization across industries. However, the widespread adoption of 3D printing also presents challenges and considerations, including job displacement, intellectual property protection, and shifts in global trade dynamics. As the technology continues to evolve and mature, businesses, pol-icymakers, and stakeholders must navigate these complexities to harness the full potential of 3D printing for economic growth and prosperity.

3.10 Addressing Challenges and Future Outlook

Despite its numerous advantages, 3D printing technology still faces several challenges that limit its widespread adoption and potential impact on the manufacturing landscape. Some of the current limitations include speed, scalability, and material constraints. Speed remains a significant challenge in 3D printing, particularly for large-scale production. While the technology has advanced considerably in recent years, printing complex objects can still be time-consuming compared to traditional manufacturing methods. Improving printing speed without compromising print quality is a key area of research and devel-opment. Scalability is another challenge facing 3D printing. While the technology is well-suited for prototyping and small-batch production, scaling up to mass production levels presents logistical and technical hurdles. Issues such as build volume, through-put, and cost-effectiveness need to be addressed to enable large-scale adoption of 3D printing in manufacturing. Material constraints also pose challenges to the widespread adoption of 3D printing. While the range of printable materials has expanded significantly in recent years, certain materials, such as metals and ceramics, still present challenges in terms of printability, quality, and cost. Developing new materials and improving existing ones to meet the diverse needs of different industries is a priority for researchers and manufacturers.

Despite these challenges, ongoing research and development efforts are underway to overcome limitations and unlock the full potential of 3D printing technology. Innova-tions in printing techniques, such as high-speed continuous printing and multi-material printing, aim to improve printing efficiency and expand the range of printable materials. Advances in software algorithms and machine learning are also enhancing the design and optimization process, enabling more complex and efficient printing. Furthermore, collabo-rations between industry, academia, and government organizations are driving innovation in 3D printing technology. Initiatives such as public–private partnerships, research grants, and technology incubators support the development of new technologies and applications, accelerating the pace of innovation in the field.

Looking ahead, the future trajectory of 3D printing is promising. As technology con-tinues to advance and become more accessible, 3D printing is poised to play a central role in shaping the manufacturing landscape in the Fourth Industrial Revolution and beyond. From personalized healthcare solutions to sustainable manufacturing practices, 3D print-ing has the potential to revolutionize industries, drive economic growth, and empower individuals and businesses to innovate and create in unprecedented ways. As researchers and manufacturers continue to overcome challenges and push the boundaries of what is possible, the transformative impact of 3D printing on the future of manufacturing is bound to be profound and far-reaching.

3.11 Summary

3D printing stands as a cornerstone technology driving innovation across diverse industries. Its ability to enable customization, accelerate innovation, promote sustainability, and reshape supply chains underscores its transformative potential. By harnessing the power of 3D printing, businesses can unlock new opportunities for growth, efficiency, and creativity in the evolving digital age. As this technology continues to advance, its impact on manufacturing and society as a whole is poised to be profound and far-reaching. 3D printing, also known as additive manufacturing, has emerged as a transformative technology with diverse applications across various industries in the Fourth Industrial Revolution (4IR). This revolutionary manufacturing process involves creating three-dimensional objects by adding material layer by layer, offering unprecedented design flexibility, cost-effectiveness, and efficiency. The applications of 3D printing in the 4IR span a wide range of sectors, from healthcare and aerospace to automotive and consumer goods. In the healthcare industry, 3D printing has revolutionized patient care by enabling the production of customized medical devices, prosthetics, and implants tailored to individual patient anatomy. Surgeons utilize 3D-printed models for pre-surgical planning, allowing for more precise and personalized treatment strategies. Additionally, bioprinting techniques are being developed to fabricate living tissues and organs, potentially revolutionizing organ transplantation and regenerative medicine. In aerospace and automotive industries, 3D printing offers significant advantages in lightweighting, part consolidation, and rapid prototyping. Aerospace companies utilize 3D printing to manufacture complex engine components, lightweight aircraft structures, and optimized aerodynamic designs. Similarly, automotive manufacturers leverage 3D printing for rapid prototyping, tooling, and production of lightweight parts, enhancing fuel efficiency, performance, and sustainability. The 4IR has also seen a surge in additive manufacturing applications in consumer goods, including fashion, electronics, and home appliances. Designers and manufacturers utilize 3D printing to create innovative product designs, personalized accessories, and limited-edition collectibles. This enables greater customization, faster time-to-market, and reduced waste compared to traditional manufacturing methods. In the industrial sector, 3D printing is transforming supply chains and production processes by enabling on-demand manufacturing, decentralized production, and just-in-time inventory management. Companies can produce parts locally, reducing lead times and transportation costs while minimizing inventory overhead. This agile manufacturing approach enhances flexibility, scalability, and resilience in the face of supply chain disruptions and market fluctuations. Furthermore, 3D printing is driving innovation in sustainable manufacturing practices, with a focus on material efficiency, waste reduction, and eco-friendly production methods.

Additive manufacturing allows for the use of recycled materials, bio-based polymers, and renewable resources, promoting circular economy principles and reducing environmental impact. Overall, the applications of 3D printing in the Fourth Industrial Revolution are vast and far-reaching, revolutionizing traditional manufacturing paradigms and unlocking new possibilities for innovation, customization, and sustainability across industries. As the technology continues to evolve and mature, its transformative potential is expected to accelerate, ushering in a new era of digital manufacturing and industrial disruption.

Challenges and Opportunities of Thin Films and 3D Printing

4

4.1 Challenges and Opportunities of Thin Films and 3D Printing

4.1.1 Thin Films

The potential for cost and space reduction is also a huge advantage. Fabrication of a device using thin film techniques often requires very little material. This means that the process can be not only more cost-effective, but the device itself can be smaller and lighter. This kind of space reduction is the reason for the transition from thermionic vacuum tubes to semiconductors, and it is still an influential factor today. In comparison to a bulk material, the thin film form of that material will often possess quite different properties. A lot of the time, these properties will show an improvement from those of the bulk material, and it is the ability to alter the surface properties of the substrate which is very advantageous. Coating a medical instrument with a thin film may make it resistant to corrosion or staining, and by doing the same to a magnetic material, it is possible to give the surface an increased coercivity and squareness. By altering the parameters of the thin film deposition process (e.g., temperature, pressure), it is possible to create a range of properties in the film, even using the same material. This degree of control, both from experimental and theoretical viewpoints, is another clear advantage of thin film devices. The term 'thin film' is applied to the layer of material, whether conductive or insulative, on a surface. The thickness of this material will usually be in the nanometer to few micron range. Thin film technology is universally recognized; it has been so for decades, largely due to its flexibility and cost. It is a versatile technique which can be incorporated qualitatively to a variety of materials including ceramics, polymers, and other compounds. There are innumerable methods for applying a thin film to a substrate, most commonly using vacuum techniques such as sputtering and electron beam deposition.

K. Ukoba and T.-C. Jen, *Shaping Tomorrow: Thin Films and 3D Printing in the Fourth Industrial Revolution 1*, Synthesis Lectures on Mechanical Engineering, https://doi.org/10.1007/978-3-031-84124-8_4

Although the thin film is used widely throughout the world, there are many advantages and disadvantages to its use, whether it be in the vast area of microelectronics or in other intricate fields of engineering.

The systematic deposition of a coating produces a better ratio of atoms from the source to the substrate, resulting in a dense structure, fewer defects, and a strong adherence of the thin film to the substrate. Consequently, there is better uniformity in composition and thickness, which is essential in applications such as electronic thin films in integrated circuits, corrosion and wear-resistant coatings, and hard coatings on cutting tools. The better adherence of the coating is due to the high mobility of surface atoms and the high energy of the vapor which enables atom movement even at low substrate temperatures. The binding energy between atoms of the same material is high, so once a layer of atoms is formed on the substrate, they are unlikely to move or form into a different structure. This is essential in preventing diffusion or reaction of the coating with the substrate and subsequent degradation of the desired properties of the coating. High binding energy also gives the coating a high cohesive energy, which results in high density and hardness, and is typical of many hard carbon coatings which are increasingly used in industry today. Thin film can also offer very tailorable properties. This is due to the ability to deposit multi-component materials and graded layers, and the generally strong correlation between deposition parameters and thin film microstructure and properties. Finally, a small quantity of starting materials (often in the form of expensive targets) can be used to produce a thin film, with minimal wastage. This is typical in sputter deposition, and in some cases the by-products of thin film deposition processes can be recycled. Therefore, some thin film processes can be environmentally friendly and offer cost-effective solutions to coating and material production.

Thin films offer a multitude of both challenges and opportunities across various industries due to their unique properties and applications. These challenges and opportunities are pivotal in shaping the advancement and utilization of thin film technology. One of the foremost challenges in thin film technology is achieving uniformity and consistency in deposition processes. Thin films are typically deposited onto substrates through methods such as physical vapor deposition (PVD) or chemical vapor deposition (CVD). However, ensuring uniform thickness and composition across large areas can be challenging, leading to defects and inconsistencies in the film. Addressing this challenge requires precise control over deposition parameters and the development of advanced deposition techniques. Moreover, thin film materials often exhibit limited mechanical durability and adhesion to substrates, particularly when subjected to harsh environmental conditions or mechanical stress. Improving the mechanical properties and adhesion of thin films is essential for enhancing their reliability and longevity in real-world applications. This challenge necessitates the development of novel materials and coatings, as well as optimization of deposition processes to enhance film-substrate interactions. Furthermore, the scalability and cost-effectiveness of thin film manufacturing remain significant challenges for widespread adoption. While thin film deposition techniques offer advantages such as high throughput and material efficiency, scaling up production to commercial levels while maintaining quality and cost-effectiveness can be challenging. Overcoming this challenge requires advancements in manufacturing processes, equipment design, and material synthesis to achieve economies of scale.

Despite these challenges, thin films present numerous opportunities across various industries, including electronics, solar energy, and healthcare. In the electronics industry, thin films are integral to the fabrication of microelectronic devices, such as integrated circuits and displays. Advancements in thin film transistor (TFT) technology have enabled the development of high-performance, low-power electronic devices, driving innovation in areas such as flexible displays and wearable electronics.

In the solar energy sector, thin film photovoltaic (PV) cells offer a lightweight, flexible alternative to traditional silicon-based solar panels. Thin film solar cells can be deposited onto flexible substrates, enabling the production of lightweight and portable solar modules for applications such as building-integrated photovoltaics (BIPV) and off-grid power generation. Furthermore, thin film PV technologies have the potential to reduce manufacturing costs and improve the efficiency of solar energy conversion. Moreover, thin films hold promise in biomedical applications, such as drug delivery systems and medical implants. Thin film coatings can be engineered to provide controlled release of therapeutic agents or enhance the biocompatibility of implantable devices, improving patient outcomes and reducing healthcare costs. Additionally, thin film sensors and diagnostic devices offer opportunities for non-invasive monitoring of physiological parameters and early detection of diseases. Thin films present both challenges and opportunities across various industries, from electronics to solar energy and healthcare. Addressing challenges such as uniformity, durability, and scalability is essential for realizing the full potential of thin film technology. However, advancements in materials science, deposition techniques, and manufacturing processes continue to drive innovation and unlock new applications for thin films in diverse fields.

Challenges of Thin Film:

Thin film technology has emerged as a critical component across various industries, offering advantages in areas such as electronics, optics, coatings, and renewable energy. Thin films, typically ranging from nanometers to micrometers in thickness, are utilized in a myriad of applications, including semiconductor devices, photovoltaic cells, optical coatings, and sensors. Despite their widespread use and potential benefits, thin films also pose numerous challenges that researchers, engineers, and manufacturers must address to realize their full potential. In this comprehensive analysis, we will explore the challenges associated with thin film technology across different industries, covering topics such as deposition techniques, material properties, performance optimization, and scalability.

1. **Deposition Techniques**:

   Thin films are typically deposited onto substrates using various deposition techniques, each with its own set of advantages and limitations. Common deposition techniques include physical vapor deposition (PVD), chemical vapor deposition (CVD), atomic layer deposition (ALD), and sputtering. While these techniques offer precise control over film thickness, composition, and uniformity, they also face challenges such as substrate compatibility, deposition rate, and scalability. For example, PVD techniques such as evaporation and magnetron sputtering require high vacuum conditions, limiting their scalability and throughput. CVD and ALD techniques offer excellent conformal coverage and atomic-level control but may require high temperatures or toxic precursor gases, posing challenges for certain applications.

Physical Vapor Deposition (PVD):

PVD is a versatile thin film deposition technique that involves the physical vaporization of a solid material followed by its deposition onto a substrate surface. Common methods of PVD include evaporation and magnetron sputtering. In the evaporation process, the material to be deposited is heated in a vacuum chamber until it reaches its vaporization temperature, forming a vapor cloud that condenses onto the substrate surface. Magnetron sputtering, on the other hand, involves bombarding a target material with energetic ions in the presence of a magnetic field, causing atoms to be ejected from the target and deposited onto the substrate. One of the main advantages of PVD techniques is their ability to deposit thin films with precise control over thickness and composition. Additionally, PVD offers excellent adhesion and uniformity, making it suitable for a wide range of applications, including semiconductor manufacturing, optical coatings, and decorative finishes. However, PVD techniques such as evaporation and magnetron sputtering require high vacuum conditions, which can limit scalability and throughput. Moreover, the need for target materials and complex vacuum systems can increase equipment and operational costs.

Chemical Vapor Deposition (CVD):

CVD is another widely used thin film deposition technique that involves the chemical reaction of precursor gases on a substrate surface to deposit a thin film. In CVD, precursor gases containing the desired film material are introduced into a reactor chamber, where they undergo chemical reactions at elevated temperatures to form a solid film on the substrate surface. CVD offers excellent conformal coverage and atomic-level control, making it suitable for applications such as semiconductor device fabrication, thin film coatings, and surface modification. One of the main challenges associated with CVD is the requirement for high temperatures and/or toxic precursor gases, which can limit its applicability for certain materials and substrates. Additionally, the deposition rate of CVD processes may be slower compared to other techniques, leading to longer processing times and reduced throughput. Furthermore, the need for precise control over deposition conditions and gas flow rates can increase process complexity and equipment costs.

Atomic Layer Deposition (ALD):

ALD is a specialized thin film deposition technique that offers atomic-level control over film thickness and composition. In ALD, precursor gases are introduced into a reactor chamber sequentially, where they react with the substrate surface in a self-limiting manner to form a monolayer of material. By repeating this cycle multiple times, thin films with precise thickness and composition can be deposited layer by layer. ALD offers several advantages, including excellent conformal coverage, uniformity, and control over film properties. Additionally, ALD is compatible with a wide range of materials and substrates, making it suitable for applications such as semiconductor fabrication, barrier coatings, and nanotechnology. However, ALD processes typically require multiple cycles to achieve the desired film thickness, which can result in longer processing times and reduced throughput compared to other techniques. Moreover, ALD equipment can be expensive and complex, requiring precise control over deposition conditions and gas flow rates.

Sputtering:

Sputtering is a versatile thin film deposition technique that involves the bombardment of a target material with energetic ions to eject atoms from the target surface, which are then deposited onto a substrate surface. Sputtering offers several advantages, including high deposition rates, excellent adhesion, and uniformity. Additionally, sputtering is compatible with a wide range of materials and substrates, making it suitable for applications such as semiconductor device fabrication, thin film coatings, and surface modification. However, sputtering techniques may face challenges such as target material consumption and deposition rate variations. Additionally, the need for high-power radiofrequency (RF) or direct current (DC) sources can increase equipment and operational costs. Moreover, the complexity of sputtering systems and the requirement for precise control over process parameters can pose challenges for scalability and process optimization.

Thin film deposition techniques such as PVD, CVD, ALD, and sputtering offer precise control over film thickness, composition, and uniformity, making them indispensable for a wide range of applications. However, each technique has its own set of advantages and limitations, including challenges related to substrate compatibility, deposition rate, scalability, and equipment costs. By understanding the strengths and weaknesses of each deposition technique, researchers, engineers, and manufacturers can select the most suitable approach for their specific application requirements, ultimately advancing the field of thin film technology and driving innovation across various industries.

2. **Material Properties**:

The properties of thin films, including electrical, optical, mechanical, and thermal properties, play a crucial role in determining their performance and suitability for specific applications. Achieving the desired material properties in thin films can be challenging due to factors such as grain boundary effects, defects, impurities, and interface interactions. For example, thin films deposited at high temperatures may exhibit grain growth and crystalline defects, affecting their electrical conductivity or optical transparency. Moreover, the choice of deposition technique, substrate material, and film thickness can influence material properties, requiring careful optimization to meet performance requirements. Thin films are essential components in various industrial sectors, offering a wide array of properties that can be tailored to specific applications. These properties, including electrical, optical, mechanical, and thermal characteristics, are crucial in determining the performance and suitability of thin films for diverse uses. However, achieving the desired material properties in thin films presents numerous challenges due to factors such as grain boundary effects, defects, impurities, and interface interactions. This section will examine the significance of thin film properties, the challenges in achieving them, and the optimization strategies required to meet performance requirements.

Electrical Properties:

The electrical properties of thin films are critical in applications such as semiconductor devices, sensors, and electronic circuits. These properties include conductivity, resistivity, carrier mobility, and dielectric constant. Achieving the desired electrical properties in thin films requires precise control over factors such as dopant concentration, crystal structure, and defect density. For example, thin films deposited at high temperatures may experience grain growth, resulting in increased resistivity due to grain boundary scattering of charge carriers. Moreover, impurities and defects introduced during deposition or post-processing steps can degrade electrical conductivity and device performance. Optimization strategies may involve doping techniques, annealing processes, and defect engineering to enhance electrical properties and minimize performance degradation.

Optical Properties:

Thin films are also utilized for their optical properties in applications such as optical coatings, displays, and photovoltaic devices. Optical properties include transparency, reflectivity, refractive index, and bandgap energy. Achieving the desired optical properties requires precise control over factors such as film thickness, composition, and microstructure. For example, thin films deposited at high temperatures may exhibit crystalline defects or surface roughness, leading to increased light scattering and reduced optical transparency. Additionally, impurities and contaminants can absorb or scatter light, affecting the optical performance of thin films. Optimization strategies may involve controlling deposition parameters, surface treatments, and post-deposition annealing to improve optical properties and enhance device performance.

Mechanical Properties:

Thin films are subjected to mechanical stresses and strains in applications such as coatings, microelectromechanical systems (MEMS), and flexible electronics. Mechanical properties include hardness, adhesion, tensile strength, and fracture toughness. Achieving the desired mechanical properties in thin films requires careful consideration of factors such as film thickness, substrate material, and deposition conditions. For example, thin films deposited at high temperatures may experience residual stress due to thermal expansion mismatch with the substrate, leading to delamination or cracking. Moreover, the presence of defects such as voids or dislocations can weaken the mechanical integrity of thin films. Optimization strategies may involve stress-relief techniques, substrate engineering, and film reinforcement to enhance mechanical properties and ensure reliability under mechanical loading conditions.

Thermal Properties:

Thin films play a crucial role in thermal management applications such as heat sinks, thermal barriers, and thermoelectric devices. Thermal properties include thermal conductivity, thermal expansion coefficient, and specific heat capacity. Achieving the desired thermal properties in thin films requires precise control over factors such as grain size, interface scattering, and phonon scattering mechanisms. For example, thin films deposited at low temperatures may exhibit amorphous or polycrystalline structures with reduced thermal conductivity compared to single-crystal counterparts. Moreover, the presence of impurities or grain boundaries can act as phonon scattering centers, hindering heat transport in thin films. Optimization strategies may involve grain boundary engineering, interface modification, and alloying techniques to enhance thermal properties and improve heat dissipation efficiency.

In conclusion, the properties of thin films, including electrical, optical, mechanical, and thermal characteristics, play a crucial role in determining their performance and suitability for specific applications. Achieving the desired material properties in thin films presents challenges such as grain boundary effects, defects, impurities, and interface interactions. However, through careful optimization of deposition techniques, substrate materials, and film thickness, it is possible to tailor thin films to meet performance requirements in diverse applications. By addressing these challenges and optimizing thin film properties, researchers, engineers, and manufacturers can unlock the full potential of thin film technology and drive innovation across various industries.

3. **Performance Optimization**:

Optimizing the performance of thin films is a multifaceted task that involves balancing various conflicting requirements to meet specific application needs. Thin films are utilized across diverse industries for their unique properties, such as electrical conductivity, optical transparency, adhesion, and durability. However, achieving optimal performance often requires trade-offs between these properties, necessitating careful optimization strategies. In this section, we will explore the challenges and approaches involved in optimizing thin film performance, highlighting examples from different applications and discussing the role of iterative experimentation, process optimization, and advanced characterization techniques.

Balancing Conflicting Requirements:

Thin films are frequently employed in applications where conflicting requirements must be balanced to achieve optimal performance. For instance, transparent conductive films used in displays or solar cells require a delicate balance between electrical conductivity and optical transparency. While high conductivity is desirable for efficient charge transport in electronic devices, excessive thickness or doping can compromise transparency, leading to reduced light transmission and degraded device performance. Similarly, in thin film coatings for corrosion protection or wear resistance, optimizing the balance between hardness and adhesion is crucial for long-term durability. A coating that is too hard may be prone to cracking or delamination, while insufficient adhesion can lead to premature failure or coating detachment.

Optimization Strategies:

Achieving optimal performance in thin films requires a systematic approach that involves iterative experimentation, process optimization, and characterization techniques. Researchers and engineers employ a variety of strategies to fine-tune thin film properties and meet performance requirements.

   i. *Iterative Experimentation:* Iterative experimentation involves systematically varying deposition parameters, such as temperature, pressure, deposition rate, and precursor concentrations, to explore their effects on thin film properties. By conducting multiple deposition runs and analyzing the resulting films, researchers can identify optimal process conditions that maximize desired properties while minimizing undesirable effects. For example, in the case of transparent conductive films, researchers may vary deposition parameters to optimize conductivity while maintaining high optical transparency.
   ii. *Process Optimization:* Process optimization involves refining deposition techniques and parameters to achieve desired thin film properties efficiently and reproducibly. This may include optimizing deposition temperature and pressure, adjusting precursor

flow rates, or implementing post-deposition treatments such as annealing or surface modification. By fine-tuning process parameters, researchers can enhance thin film performance while minimizing production costs and cycle times.

iii. *Characterization Techniques:* Advanced characterization techniques play a crucial role in optimizing thin film performance by providing insights into film structure, composition, and properties. Techniques such as X-ray diffraction (XRD), spectroscopy (UV-Vis, FTIR), and microscopy (SEM, TEM) enable researchers to analyze thin film morphology, crystal structure, chemical composition, and defects. By correlating these findings with film properties, researchers can identify underlying mechanisms and optimize deposition processes accordingly. For instance, XRD can be used to study crystallographic orientation and grain size in thin films, while spectroscopic techniques can provide information on optical and electronic properties.

Examples from Different Applications:

Transparent Conductive Films: In the case of transparent conductive films used in displays or solar cells, optimizing performance involves balancing electrical conductivity and optical transparency. Researchers may experiment with different deposition techniques (e.g., sputtering, CVD), doping concentrations, and post-deposition treatments to achieve the desired balance. Advanced characterization techniques such as spectroscopy and microscopy can be used to analyze film properties and guide optimization efforts.

i. *Corrosion-Resistant Coatings:* In thin film coatings for corrosion protection, optimizing performance requires balancing hardness, adhesion, and chemical resistance. Researchers may explore deposition parameters such as composition, thickness, and substrate preparation techniques to enhance coating properties. Advanced characterization techniques such as microscopy and spectroscopy can be used to assess coating adhesion, morphology, and corrosion resistance.

Optimizing the performance of thin films is a complex and iterative process that involves balancing conflicting requirements such as conductivity, transparency, adhesion, and durability. By employing a systematic approach that integrates iterative experimentation, process optimization, and advanced characterization techniques, researchers and engineers can finc-tune thin film properties to meet specific application needs. Examples from different applications, such as transparent conductive films and corrosion-resistant coatings, highlight the importance of optimization strategies in achieving optimal thin film performance. As thin film technology continues to advance, the development of novel deposition techniques, materials, and characterization methods will further enhance our ability to optimize thin film properties and unlock new applications across various industries.

4. **Scalability and Manufacturing Challenges**:

Scalability is a pivotal consideration in thin film technology, especially when aiming for applications that necessitate large-area coatings or high-throughput production. While laboratory-scale deposition techniques may showcase promising outcomes, transitioning to industrial-scale manufacturing poses significant challenges, including maintaining deposition uniformity, ensuring production yield, and achieving cost-effectiveness. Furthermore, thin film deposition processes must seamlessly integrate with existing manufacturing infrastructure and production workflows, necessitating meticulous integration and optimization efforts. Addressing scalability challenges often involves the development of novel deposition techniques, the implementation of process automation, and the adoption of advanced manufacturing technologies. This section will examine the complexities of scalability in thin film technology, examining the associated challenges and exploring strategies for overcoming them.

Challenges in Scalability:

i. **Deposition Uniformity:** Achieving uniformity in thin film deposition across large areas is critical for ensuring consistent performance and quality. However, scaling up from laboratory-scale to industrial-scale deposition processes often introduces challenges related to maintaining uniformity. Variations in substrate geometry, temperature gradients, and gas flow dynamics can lead to non-uniform film thickness and properties, impacting product performance and yield.

ii. **Production Yield:** Ensuring high production yield is essential for cost-effective manufacturing of thin film-based products. However, as deposition area and throughput increase, the probability of defects and imperfections also rises, potentially reducing yield. Factors such as particle contamination, substrate defects, and process variability can contribute to yield losses, necessitating robust process control and optimization strategies.

iii. **Cost-Effectiveness:** Achieving cost-effectiveness in large-scale thin film manufacturing requires balancing production costs with product performance and quality. Scaling up deposition processes often involves significant capital investment in equipment, infrastructure, and materials. Additionally, operational costs such as energy consumption, maintenance, and waste disposal must be carefully managed to ensure profitability and competitiveness in the market.

iv. **Compatibility with Existing Infrastructure:** Integrating thin film deposition processes into existing manufacturing infrastructure and workflows presents logistical and technical challenges. Industrial-scale production facilities may have specific requirements and constraints that need to be addressed, such as space limitations, safety regulations, and compatibility with other manufacturing processes. Ensuring seamless integration and minimal disruption to existing operations is essential for successful scalability.

Strategies for Addressing Scalability Challenges:

i. **Development of Novel Deposition Techniques:** Innovations in thin film deposition techniques are key to addressing scalability challenges. Novel deposition methods that offer improved throughput, uniformity, and flexibility can enhance the scalability of thin film manufacturing. For example, roll-to-roll deposition techniques enable continuous, high-speed coating of large-area substrates, making them well-suited for mass production of thin film-based products such as flexible electronics and photovoltaic modules.

ii. **Process Automation:** Automation of thin film deposition processes can improve scalability by reducing human intervention, minimizing variability, and increasing throughput. Automated systems for substrate handling, precursor delivery, and process control enable consistent and efficient production at scale. Furthermore, real-time monitoring and feedback mechanisms help identify and mitigate process deviations, enhancing yield and quality.

iii. **Advanced Manufacturing Technologies:** Leveraging advanced manufacturing technologies such as additive manufacturing (3D printing) and digital fabrication can enhance scalability in thin film production. These technologies offer versatility, customization, and rapid prototyping capabilities, enabling agile and cost-effective manufacturing of complex thin film structures. Additionally, advances in materials science, nanotechnology, and surface engineering contribute to the development of new thin film materials and coatings with enhanced performance and scalability.

iv. **Optimization of Production Workflows:** Streamlining production workflows and optimizing process parameters are essential for achieving scalability in thin film manufacturing. Designing efficient production layouts, minimizing material handling steps, and optimizing equipment utilization can improve throughput and reduce production costs. Furthermore, implementing lean manufacturing principles and continuous improvement initiatives help identify and eliminate inefficiencies in production processes, enhancing scalability and competitiveness.

Scalability is a significant challenge in thin film technology, particularly for applications requiring large-area coatings or high-throughput production. Addressing scalability challenges requires a multifaceted approach that involves the development of novel deposition techniques, the implementation of process automation, and the adoption of advanced manufacturing technologies. By addressing challenges related to deposition uniformity, production yield, cost-effectiveness, and compatibility with existing infrastructure, researchers and manufacturers can enhance the scalability of thin film technology and unlock its full potential across various industries. As technology continues to advance and adoption rates increase, the scalability of thin film manufacturing will play a crucial role in driving innovation and competitiveness in the global market.

5. **Stability and Reliability**:

The stability and reliability of thin films are paramount factors in various applications where long-term performance and durability are essential. Thin films often encounter harsh operating conditions, including temperature fluctuations, mechanical stresses, corrosive environments, and exposure to radiation. Ensuring the stability and reliability of thin films under such conditions necessitates careful materials selection, the application of protective coatings, and the implementation of encapsulation techniques. Furthermore, understanding degradation mechanisms such as corrosion, oxidation, and fatigue is crucial for predicting the lifespan and performance degradation of thin film-based devices. In this section, we will examine the significance of stability and reliability in thin films, explore the challenges associated with ensuring their long-term performance, and discuss strategies for mitigating degradation and enhancing durability.

Significance of Stability and Reliability:

Thin films are utilized in a wide range of applications, including electronic devices, sensors, protective coatings, and photovoltaic cells, where stability and reliability are critical considerations. The ability of thin films to maintain their properties and performance over extended periods is essential for ensuring the functionality and longevity of these applications. For example, in electronic devices such as integrated circuits and thin film transistors, the stability of thin film materials is crucial for maintaining device performance and preventing degradation over time. Similarly, in protective coatings for corrosion resistance or wear protection, the reliability of thin films determines their effectiveness in prolonging the lifespan of substrates and components.

Challenges in Ensuring Stability and Reliability:

 i. **Harsh Operating Conditions:** Thin films are often exposed to harsh operating conditions, including temperature variations, mechanical stresses, corrosive environments, and radiation exposure, which can accelerate degradation and reduce reliability. For example, in aerospace applications, thin film coatings on aircraft components are subjected to extreme temperatures, high velocities, and corrosive chemicals, leading to potential degradation and failure.
 ii. **Degradation Mechanisms:** Understanding the degradation mechanisms affecting thin films is crucial for predicting their lifespan and performance degradation. Common degradation mechanisms include corrosion, oxidation, fatigue, and mechanical wear. For instance, in thin film-based sensors exposed to corrosive environments, such as chemical sensors or environmental monitors, corrosion of the thin film material can compromise sensor performance and accuracy over time.
 iii. **Materials Selection:** Choosing the appropriate materials for thin film deposition is essential for ensuring stability and reliability. Factors such as material compatibility with the substrate, chemical resistance, mechanical properties, and thermal

stability must be carefully considered. Additionally, the deposition technique and process parameters can influence thin film properties and performance. For example, in photovoltaic cells, selecting durable and stable thin film materials with high optical transparency and electrical conductivity is critical for achieving long-term device reliability.

Strategies for Enhancing Stability and Reliability:

i. **Protective Coatings:** Applying protective coatings to thin films can enhance their stability and reliability by providing a barrier against environmental factors such as moisture, chemicals, and abrasion. Protective coatings can be deposited using various techniques, including physical vapor deposition (PVD), chemical vapor deposition (CVD), and spray coating. For example, in automotive applications, thin film coatings are used to protect engine components from corrosion and wear, prolonging their lifespan and reliability.

ii. **Encapsulation Techniques:** Encapsulation involves sealing thin film devices or components within protective enclosures to shield them from environmental factors and mechanical stresses. Encapsulation techniques such as hermetic sealing, conformal coating, and encapsulant materials (e.g., polymers, ceramics) provide an additional layer of protection against moisture, dust, and contaminants. For instance, in electronic devices such as MEMS sensors or microfluidic devices, encapsulation prevents moisture ingress and ensures long-term device reliability.

iii. **Advanced Characterization Techniques:** Employing advanced characterization techniques such as X-ray diffraction (XRD), spectroscopy (FTIR, Raman), and microscopy (SEM, TEM) enables researchers to analyze thin film properties, identify degradation mechanisms, and assess performance degradation over time. These techniques provide valuable insights into thin film structure, composition, defects, and interface interactions, facilitating the optimization of deposition processes and materials selection.

iv. **Accelerated Aging Studies:** Conducting accelerated aging studies allows researchers to simulate and evaluate the long-term performance and reliability of thin film-based devices under accelerated environmental conditions. By subjecting thin films to elevated temperatures, humidity, and mechanical stresses, researchers can assess performance degradation, identify failure mechanisms, and develop mitigation strategies. Accelerated aging studies are particularly useful for assessing thin film reliability in critical applications such as aerospace, automotive, and medical devices.

The stability and reliability of thin films are critical considerations in various applications where long-term performance and durability are essential. Thin films are often subjected to harsh operating conditions, including temperature fluctuations, mechanical stresses, corrosive environments, and radiation exposure, which can accelerate degradation and reduce reliability. Ensuring the stability and reliability of thin films requires robust materials selection, protective coatings, encapsulation techniques, and advanced characterization methods. By addressing degradation mechanisms, enhancing protective measures, and conducting accelerated aging studies, researchers and engineers can improve the stability and reliability of thin film-based devices and components, thereby extending their lifespan and ensuring optimal performance in diverse applications.

6. **Cost and Economics**:

Cost considerations are paramount in thin film technology, particularly for large-scale commercialization and widespread adoption. The cost of thin film deposition processes, materials, equipment, and post-processing steps can significantly impact the overall economics of thin film-based products. Achieving cost-effective thin film solutions may require optimization of deposition processes, materials utilization, and production yield. Additionally, advancements in materials synthesis, recycling, and waste management can help reduce the environmental footprint and lifecycle costs of thin film technologies.

Commercialization and widespread adoption of thin film technologies are pivotal for realizing their full potential across various industries. However, the cost of thin film deposition processes, materials, equipment, and post-processing steps can significantly impact the overall economics of thin film-based products. Achieving cost-effective thin film solutions necessitates the optimization of deposition processes, materials utilization, and production yield. Additionally, advancements in materials synthesis, recycling, and waste management can help reduce the environmental footprint and lifecycle costs of thin film technologies. In this section, we will explore the challenges associated with the commercialization of thin film technologies, examine strategies for achieving cost-effectiveness, and discuss the importance of sustainability in thin film manufacturing.

Challenges in Commercialization:

i. **Cost of Deposition Processes:** Thin film deposition processes, such as physical vapor deposition (PVD), chemical vapor deposition (CVD), and atomic layer deposition (ALD), can be capital-intensive and require specialized equipment and facilities. The high cost of deposition equipment, consumables, and maintenance can pose barriers to entry for small and medium-sized enterprises (SMEs) and startups seeking to commercialize thin film-based products.

ii. **Materials Cost:** Thin film materials, such as semiconductors, metals, and dielectrics, can be expensive, particularly for high-performance applications requiring specialized compositions and purity levels. The cost of materials can significantly impact the overall manufacturing cost of thin film-based products, making it challenging to achieve cost-competitiveness in the market.

iii. **Production Yield**: Ensuring high production yield is essential for cost-effective thin film manufacturing. However, variations in deposition conditions, substrate properties, and process parameters can lead to yield losses and waste generation. Achieving consistent and reproducible deposition results is critical for maximizing production yield and minimizing material wastage.

iv. **Post-Processing Costs:** Post-processing steps such as annealing, etching, and surface modification can add to the overall manufacturing cost of thin film-based products. The complexity and time required for post-processing steps can impact production throughput and labor costs, affecting the overall economics of thin film manufacturing.

Strategies for Achieving Cost-Effectiveness:

i. **Optimization of Deposition Processes:** Optimizing deposition processes is essential for reducing the cost of thin film manufacturing. This includes optimizing process parameters such as deposition rate, temperature, pressure, and precursor flow rates to maximize production yield and minimize material consumption. Additionally, process optimization can improve energy efficiency and reduce operational costs associated with thin film deposition.

ii. **Materials Utilization:** Maximizing materials utilization is critical for cost-effective thin film manufacturing. This involves minimizing material wastage during deposition processes and post-processing steps, as well as recycling and reusing materials wherever possible. Advanced materials characterization techniques can help identify opportunities for materials optimization and waste reduction.

iii. **Production Yield Improvement:** Improving production yield is key to reducing the overall manufacturing cost of thin film-based products. This may involve implementing process monitoring and control systems to detect and correct deviations in real-time, as well as implementing quality assurance protocols to ensure consistency and reproducibility in thin film deposition.

iv. **Scale-Up and Automation:** Scaling up thin film manufacturing processes and implementing automation technologies can help reduce labor costs and increase production throughput, leading to economies of scale. Automated systems for substrate handling, precursor delivery, and process control can improve process efficiency and reduce operational costs, making thin film manufacturing more cost-effective.

Importance of Sustainability:

In addition to cost-effectiveness, sustainability is becoming increasingly important in thin film manufacturing. Advances in materials synthesis, recycling, and waste management are essential for reducing the environmental footprint and lifecycle costs of thin film technologies. By adopting sustainable practices such as green chemistry, materials recycling, and waste reduction, thin film manufacturers can minimize resource consumption, energy usage, and environmental impact, while also enhancing their long-term competitiveness and market positioning.

In conclusion, achieving cost-effective thin film solutions is essential for the commercialization and widespread adoption of thin film technologies across various industries. Addressing challenges such as the cost of deposition processes, materials, equipment, and post-processing steps requires optimization of manufacturing processes, materials utilization, and production yield. Additionally, advancements in sustainability practices, including materials recycling, waste reduction, and energy efficiency, are crucial for reducing the environmental footprint and lifecycle costs of thin film technologies. By embracing cost-effectiveness and sustainability, thin film manufacturers can enhance their competitiveness, profitability, and market acceptance, driving innovation and growth in the thin film industry.

7. **Integration and Compatibility**:

Integrating thin film technology into existing systems and applications is a complex process that requires careful consideration of compatibility, interface interactions, and performance requirements. Thin films must interface seamlessly with other components, substrates, or materials without compromising overall system performance. Compatibility challenges may arise from differences in thermal expansion coefficients, mechanical properties, or chemical compatibility between thin films and substrates. Addressing integration challenges may involve interface engineering, surface modification techniques, or the development of hybrid materials and composites. In this section, we will explore the importance of integration in thin film technology, examine the challenges associated with compatibility and interface interactions, and discuss strategies for addressing integration challenges.

Importance of Integration:

Integration is a critical aspect of thin film technology, as thin films are often incorporated into larger systems or devices to enhance functionality or performance. Whether used in electronic devices, sensors, coatings, or optoelectronic components, thin films must seamlessly integrate with other components or materials to ensure optimal performance and reliability. The success of thin film-based applications depends on their ability to interface effectively with existing systems and substrates while meeting performance requirements.

Challenges in Compatibility and Interface Interactions:

  i. **Thermal Compatibility:** Differences in thermal expansion coefficients between thin films and substrates can lead to mechanical stress, delamination, or cracking during temperature cycling. Thermal mismatches can result in thermal strain at the interface, causing mechanical deformation and reducing the reliability of thin film-based devices. For example, in semiconductor devices, thermal stress can affect the performance and lifespan of thin film interconnects or passivation layers.

ii. **Mechanical Compatibility:** Mechanical properties such as stiffness, elasticity, and ductility play a crucial role in determining the compatibility of thin films with substrates or supporting structures. Mismatched mechanical properties can result in stress concentration, deformation, or failure at the interface, compromising device performance and reliability. For instance, in flexible electronics or MEMS devices, differences in mechanical properties between thin films and substrates can lead to mechanical fatigue or fracture under bending or stretching.

iii. **Chemical Compatibility:** Chemical interactions between thin films and substrates can affect adhesion, stability, and performance. Incompatibility between materials can result in chemical reactions, interfacial diffusion, or degradation, leading to device failure or performance degradation over time. For example, in thin film coatings for corrosion protection or barrier films for moisture resistance, chemical compatibility with the substrate is critical for long-term durability and reliability.

Strategies for Addressing Integration Challenges:

i. **Interface Engineering:** Interface engineering involves optimizing the interface between thin films and substrates to minimize stress, improve adhesion, and enhance compatibility. Techniques such as interlayer deposition, surface treatments, and adhesion promoters can be used to tailor interface properties and mitigate compatibility issues. For example, introducing buffer layers or interfacial modifiers can reduce interfacial energy and improve adhesion between thin films and substrates.

ii. **Surface Modification Techniques:** Surface modification techniques such as plasma treatment, chemical functionalization, or self-assembled monolayers (SAMs) can modify the surface properties of thin films or substrates to enhance compatibility. Surface treatments can alter surface energy, wettability, or chemical reactivity, facilitating adhesion and reducing interface interactions. For instance, plasma treatment can promote surface activation and improve bonding between thin films and substrates.

iii. **Development of Hybrid Materials and Composites:** Hybrid materials and composites combine different materials or components to achieve synergistic properties and enhanced compatibility. By integrating thin films with complementary materials or substrates, it is possible to tailor mechanical, thermal, and chemical properties to meet specific application requirements. For example, in flexible electronics or wearable devices, hybrid materials combining thin films with flexible substrates or polymers can offer improved mechanical flexibility and durability.

Integrating thin film technology into existing systems and applications requires careful consideration of compatibility, interface interactions, and performance requirements.

Challenges such as thermal, mechanical, and chemical compatibility between thin films and substrates can impact device performance, reliability, and lifespan. Addressing integration challenges involves interface engineering, surface modification techniques, and the development of hybrid materials and composites. By optimizing interface properties, enhancing compatibility, and tailoring materials selection, researchers and engineers can overcome integration challenges and unlock the full potential of thin film technology in a wide range of applications. Effective integration is essential for realizing the benefits of thin film-based devices and systems, driving innovation and advancement in various industries.

8. **Environmental and Health Concerns**:

Thin film deposition processes and materials may pose environmental and health concerns due to the use of toxic chemicals, hazardous gases, and energy-intensive processes. For example, certain thin film deposition techniques may require the use of volatile organic compounds (VOCs), heavy metals, or rare earth elements, raising environmental and regulatory concerns. Additionally, the disposal of thin film waste materials and byproducts may pose challenges for waste management and environmental sustainability. Addressing environmental and health concerns may involve the development of green and sustainable thin film technologies, alternative materials, and waste reduction strategies.

Thin film deposition processes and materials, while essential for numerous technological advancements, may also raise environmental and health concerns due to the utilization of toxic chemicals, hazardous gases, and energy-intensive procedures. As the demand for thin film-based technologies continues to grow, it becomes imperative to address these concerns and adopt sustainable practices to minimize adverse impacts on the environment and human health. In this section, we will examine the environmental and health implications of thin film deposition processes and materials, explore the challenges they pose, and discuss strategies for mitigating these concerns.

Environmental and Health Implications:

  i. **Toxic Chemicals:** Many thin film deposition processes involve the use of toxic chemicals and solvents, such as metalorganic precursors, volatile organic compounds (VOCs), and hazardous acids or bases. These chemicals pose risks to both human health and the environment, including acute and chronic health effects, air and water pollution, and ecosystem disruption. For example, the use of heavy metals such as cadmium, lead, and mercury in thin film materials can lead to contamination of air, soil, and waterways, posing risks to wildlife and human populations.
  ii. **Hazardous Gases:** Some thin film deposition techniques, such as chemical vapor deposition (CVD) and plasma-enhanced chemical vapor deposition (PECVD), require the use of hazardous gases such as silane, ammonia, and hydrogen chloride. These gases can pose significant health and safety risks to workers and nearby communities,

including respiratory irritation, chemical burns, and asphyxiation. Additionally, the release of these gases into the atmosphere can contribute to air pollution and climate change.

iii. **Energy Intensity:** Thin film deposition processes often require high temperatures, vacuum conditions, and energy-intensive equipment, leading to significant energy consumption and greenhouse gas emissions. The production of thin film materials, such as semiconductors and photovoltaic cells, may also involve energy-intensive manufacturing processes, further exacerbating environmental impacts. Energy-intensive processes contribute to resource depletion, air pollution, and climate change, posing long-term environmental and societal challenges.

Challenges:

i. **Regulatory Compliance:** Compliance with environmental and occupational health regulations is essential for ensuring the safe operation of thin film deposition facilities and protecting workers and the environment from potential hazards. However, regulatory compliance can be challenging due to the complex nature of thin film processes, the use of hazardous materials, and evolving regulatory requirements. Ensuring regulatory compliance requires ongoing monitoring, training, and implementation of safety protocols.

ii. **Material Substitution:** Finding suitable alternatives to toxic chemicals and hazardous gases used in thin film deposition processes can be challenging. While efforts are underway to develop safer alternatives and greener technologies, transitioning away from traditional materials and processes may require significant investment in research and development. Furthermore, new materials must meet performance, cost, and scalability requirements to be viable replacements for existing technologies.

iii. **Waste Management:** Thin film deposition processes generate waste streams containing hazardous chemicals, byproducts, and contaminated materials that require proper management and disposal. Effective waste management strategies are essential for minimizing environmental contamination and protecting public health. However, disposing of hazardous waste safely and responsibly can be costly and complex, requiring adherence to strict regulations and guidelines.

Mitigation Strategies:

i. **Green Chemistry:** Adopting principles of green chemistry can help reduce the environmental and health impacts of thin film deposition processes by minimizing or eliminating the use of hazardous chemicals and solvents. Green chemistry emphasizes the design of safer chemicals, processes, and materials that are less toxic, energy-intensive, and resource-intensive. By prioritizing safer alternatives and sustainable practices, manufacturers can minimize environmental pollution and human exposure to harmful substances.

ii. **Process Optimization:** Optimizing thin film deposition processes for energy efficiency and resource conservation can help reduce environmental impacts and operational costs. Strategies such as substrate recycling, solvent recovery, and process intensification can improve resource utilization and minimize waste generation. Additionally, implementing closed-loop systems and waste-to-energy technologies can further enhance the sustainability of thin film manufacturing operations.

iii. **Lifecycle Assessment:** Conducting lifecycle assessments (LCAs) can help evaluate the environmental impacts of thin film technologies across their entire lifecycle, from raw material extraction to end-of-life disposal. LCAs provide valuable insights into the environmental hotspots and potential areas for improvement in thin film production processes. By identifying opportunities to reduce energy consumption, emissions, and resource usage, manufacturers can develop more sustainable thin film products and processes.

Thin film deposition processes and materials present environmental and health concerns due to the use of toxic chemicals, hazardous gases, and energy-intensive procedures. Addressing these concerns requires a holistic approach that encompasses regulatory compliance, material substitution, waste management, and sustainability initiatives. By adopting green chemistry principles, optimizing process efficiency, and conducting lifecycle assessments, manufacturers can minimize environmental impacts and protect human health while advancing thin film technology. Collaboration between industry, government, academia, and civil society is essential for promoting responsible innovation and ensuring the sustainable development of thin film-based technologies in the years to come.

Thin film technology offers tremendous opportunities for innovation and advancement across various industries, but it also poses numerous challenges that must be addressed to realize its full potential. Addressing challenges related to deposition techniques, material properties, performance optimization, scalability, stability, cost, integration, and environmental concerns requires interdisciplinary collaboration, innovation, and continuous improvement. By overcoming these challenges, researchers, engineers, and manufacturers can unlock the transformative potential of thin film technology and drive progress in areas such as electronics, optics, coatings, and renewable energy.

4.1.1.1 Opportunities of Thin Films

Thin films represent a versatile and rapidly evolving field with a wide range of opportunities across various industries. From electronics and optoelectronics to renewable energy and healthcare, thin films offer unique advantages such as lightweight, flexibility, and scalability, making them attractive for numerous applications. In this section, we will explore the opportunities presented by thin films, examine their potential across different sectors, and discuss the future prospects of this promising technology.

1. **Electronics and Optoelectronics:** Thin films play a crucial role in the electronics industry, where they are used in a variety of applications ranging from integrated circuits to displays and sensors. With the continuous miniaturization of electronic devices, thin films enable the fabrication of compact and lightweight components with enhanced performance. For example, thin film transistors (TFTs) are key components in flat-panel displays, offering high-resolution images and low power consumption. Additionally, thin film materials such as indium tin oxide (ITO) are widely used as transparent conductive coatings in touchscreens, solar cells, and light-emitting diodes (LEDs). As the demand for wearable electronics, flexible displays, and Internet of Things (IoT) devices continues to grow, thin films are expected to play an increasingly important role in shaping the future of electronics and optoelectronics.

Thin films are indispensable in the electronics industry, where they serve a multitude of functions across various applications, including integrated circuits, displays, and sensors. The continuous trend towards miniaturization in electronic devices has made thin films indispensable, allowing for the fabrication of compact, lightweight components with enhanced performance. Thin films play a pivotal role in enabling advanced functionalities and driving innovation in electronics and optoelectronics. This section will examine the significant contributions of thin films in the electronics industry, exploring their diverse applications and the transformative impact they have on modern technology.

Thin Films in Electronics:

i. **Integrated Circuits**:

Thin films are indispensable to the fabrication of integrated circuits (ICs), which are the cornerstone of modern electronic devices. Semiconductor thin films, particularly those composed of silicon, serve as the bedrock for ICs, enabling the construction of transistors, diodes, and various electronic components on silicon wafers. The miniaturization of these components through thin film technology has ushered in an era of increased functionality, higher processing speeds, and reduced power consumption in electronic devices. The semiconductor industry heavily relies on thin film technology to produce ever-smaller electronic components, driving continuous improvements in performance and efficiency. For instance, the advancement of transistor technology has followed Moore's Law, which predicts that the number of transistors on a microchip doubles approximately every two years. This exponential increase in transistor density is made possible by the continuous refinement of thin film fabrication techniques. Moore's Law has held true for several decades, driving the relentless miniaturization of electronic components and the proliferation of computing power in electronic devices. This trend underscores the pivotal role of thin film technology in sustaining the rapid pace of innovation in the semiconductor industry. Moreover, the miniaturization of electronic components facilitated by thin film technology has led to significant advancements in processing speeds. By shrink-ing the dimensions of transistors and interconnects, the propagation delays in ICs are minimized, allowing for faster data processing and higher clock frequencies. This has profound implications for various applications, including microprocessors, memory chips, and telecommunications devices, where speed is paramount.

As transistor sizes shrink, the density of components on a microchip increases, leading to improved performance and efficiency in electronic devices. Furthermore, the reduction in power consumption enabled by thin film technology has profound implications for energy efficiency and battery life in electronic devices. Miniaturized components exhibit lower capacitance and resistance, resulting in decreased power requirements for operation. This is particularly beneficial for portable devices, such as smartphones and tablets, where energy efficiency is critical for prolonging battery life and enhancing user experience. Thin film technology plays a pivotal role in shaping the evolution of integrated circuits and electronic devices. By enabling the miniaturization of electronic components, thin films drive advancements in functionality, processing speeds, and energy efficiency, driving innovation and propelling the semiconductor industry forward. As the demand for smaller, faster, and more energy-efficient electronic devices continues to grow, thin film technology will remain at the forefront of semiconductor fabrication, powering the next generation of technological breakthroughs.

ii. **Displays**:

Thin films play a pivotal role in revolutionizing display technologies, serving as the backbone for high-resolution, energy-efficient displays that adorn televisions, smart-phones, tablets, and a myriad of other electronic devices. Among the key components that rely on thin films are thin film transistors (TFTs), integral to the operation of flat-panel displays, including liquid crystal displays (LCDs) and organic light-emitting diode (OLED) displays. These TFTs facilitate precise control of individual pixels, resulting in vibrant images, rapid response times, and reduced power consumption, thereby enhancing the user experience. In the realm of display technology, thin film transistors (TFTs) play a pivotal role in enabling the intricate control of pixels within flat-panel displays. Whether it's the liquid crystal displays (LCDs) commonly found in televisions, computer moni-tors, and digital signage, or the organic light-emitting diode (OLED) displays increasingly prevalent in premium smartphones and high-end televisions, TFTs serve as the underly-ing mechanism for pixel control. By utilizing thin film semiconductor materials such as amorphous silicon (a-Si), polysilicon (p-Si), or metal oxides, TFTs enable precise modu-lation of electrical currents to individual pixels, allowing for the accurate representation of colors and images across the display surface. This level of control not only ensures high-resolution imagery but also facilitates fast response times, smooth motion rendering, and reduced power consumption, contributing to an immersive viewing experience.

Moreover, thin film materials such as indium tin oxide (ITO) play a crucial role in enhancing the functionality of displays by serving as transparent conductive coatings. ITO-coated thin films are commonly employed as transparent electrodes in LCDs, OLEDs, and other display technologies, allowing for efficient transmission of light while providing electrical conductivity. This transparency enables displays to remain visually appealing while ensuring seamless touch functionality in touchscreen devices, such as smartphones and tablets. Additionally, the conductivity of ITO-coated thin films ensures uniform distribution of electrical currents across the display surface, enabling consistent brightness and color accuracy throughout the viewing experience. In summary, thin films are indispensable in driving advancements in display technologies, facilitating the creation of high-resolution, energy-efficient displays used in various electronic devices. Through the utilization of thin film transistors (TFTs) and transparent conductive coatings such as indium tin oxide (ITO), displays can achieve vibrant imagery, rapid response times, and seamless touch functionality, thereby enhancing the user experience and pushing the boundaries of visual innovation. As display technologies continue to evolve, thin film technology will remain at the forefront, driving further enhancements in performance, efficiency, and functionality.

iii. **Sensors**:

Thin films represent a cornerstone in sensor technologies, playing a pivotal role in the detection and measurement of an array of physical, chemical, and biological parameters. These sensors leverage thin film materials to offer advantages such as high sensitivity, rapid response times, and miniaturization, making them indispensable across diverse applications ranging from environmental monitoring and biomedical diagnostics to industrial automation. By harnessing thin film materials such as piezoelectric and semiconductor thin films, sensors can detect and measure parameters including pressure, temperature, humidity, gas concentration, and biomolecules with remarkable precision and accuracy. In the realm of environmental monitoring, thin film sensors serve as vital tools for detecting and quantifying various environmental parameters, including air and water quality, pollutant levels, and environmental contaminants. These sensors can be deployed in outdoor monitoring stations, industrial facilities, and research laboratories to provide real-time data on environmental conditions, enabling proactive measures to mitigate pollution, protect ecosystems, and safeguard public health.

Biomedical diagnostics represents another crucial application domain for thin film sensors, where they are utilized for monitoring physiological parameters, detecting biomarkers, and facilitating disease diagnosis and treatment. Thin film sensors can detect subtle changes in biological signals, such as heart rate, blood glucose levels, and DNA sequences, enabling early detection of diseases and personalized medical interventions.

Moreover, the miniaturization afforded by thin film technology allows for the development of wearable sensors and implantable devices that can continuously monitor patients' health status and transmit data to healthcare providers in real-time.

Thin film sensor arrays enable multiplexed detection of biomolecules, enabling high-throughput analysis and screening in biomedical research and clinical diagnostics. These sensors offer high sensitivity and specificity, allowing for the detection of multiple analytes simultaneously with minimal sample volumes and processing times. In industrial automation, thin film sensors play a vital role in monitoring and controlling manufacturing processes, ensuring product quality, and optimizing production efficiency. These sensors can detect parameters such as temperature, pressure, and gas concentration in real-time, enabling predictive maintenance, process optimization, and quality control in various industries, including automotive, aerospace, and semiconductor manufacturing. Furthermore, thin film sensors find applications in consumer electronics, automotive systems, and smart devices, where they are integrated into wearable gadgets, smartphones, and IoT devices to enable personalized experiences and enhance user interactions. These sensors can detect gestures, proximity, and environmental conditions, enabling intuitive user interfaces and smart functionalities in electronic devices. Thin film sensors represent a versatile and indispensable technology for detecting and measuring physical, chemical, and biological parameters across a wide range of applications. By leveraging thin film materials and advanced fabrication techniques, sensors can achieve high sensitivity, fast response times, and miniaturization, enabling transformative advancements in environmental monitoring, biomedical diagnostics, industrial automation, and consumer electronics. As sensor technologies continue to evolve, thin film sensors will play an increasingly pivotal role in shaping the future of smart and connected systems, driving innovation and addressing pressing societal challenges.

Applications of Thin Films:

i. **Flat-Panel Displays**:

Thin film transistor (TFT) technology serves as a foundational element in the manufacturing of flat-panel displays, encompassing liquid crystal displays (LCDs), organic light-emitting diodes (OLEDs), and plasma displays. These displays rely on TFTs to regulate the activation of individual pixels, thereby enabling the generation of high-resolution images, wide viewing angles, and rapid response times. The utilization of thin film materials such as amorphous silicon (a-Si), polysilicon (p-Si), and metal oxides lies at the heart of TFT fabrication, ensuring the attainment of superior electrical properties essential for optimal display performance. TFTs play a pivotal role in controlling the transmission of light and color in flat-panel displays, contributing to their exceptional image quality and visual fidelity. By modulating the electrical signals delivered to each pixel, TFTs facilitate precise adjustments in brightness and color intensity, resulting in vibrant images with lifelike clarity and detail. Moreover, TFTs enable the realization of wide viewing angles, ensuring consistent image quality even when viewed from different perspectives, a crucial feature for displays used in diverse applications such as televisions, monitors, and digital signage. The incorporation of thin film materials into TFT fabrication processes enables the attainment of desirable electrical characteristics necessary for optimal display performance. Amorphous silicon (a-Si), characterized by its non-crystalline structure, offers excellent uniformity and stability, making it a popular choice for TFTs in LCDs. Polysilicon (p-Si), on the other hand, exhibits enhanced carrier mobility and faster switching speeds, making it well-suited for high-performance TFTs in applications requiring rapid response times, such as gaming monitors and virtual reality displays. Additionally, metal oxides such as indium gallium zinc oxide (IGZO) and indium zinc oxide (IZO) offer superior transparency and electrical conductivity, enabling the fabrication of TFTs with enhanced performance and energy efficiency in displays. The integration of TFTs into OLED displays enables precise control of individual pixels, allowing for the emission of light with varying intensity and color, thereby producing high-quality images with exceptional contrast and color accuracy. The thin film transistor technology showcased in OLED displays underscores its critical role in driving advancements in display technology, enabling the realization of immersive viewing experiences across a wide range of applications. Thin film transistor technology serves as the cornerstone of flat-panel displays, enabling the creation of high-resolution, visually stunning images with wide viewing angles and fast response times. By leveraging thin film materials such as amorphous silicon, polysilicon, and metal oxides, TFTs ensure optimal display performance, enhancing user experiences and driving innovation in display technology.

ii. **Touchscreens**:

Thin film materials, notably indium tin oxide (ITO), play a pivotal role in the functionality of touchscreens, serving as transparent conductive coatings that enable accurate and responsive touch input. Capacitive touchscreens, a prevalent technology in modern devices, rely on changes in capacitance to detect touch gestures, with thin film electrodes crafted from ITO facilitating precise sensing capabilities. Furthermore, thin film sensors based on resistive or surface acoustic wave (SAW) technology are employed in touchscreens, particularly for applications necessitating multi-touch capabilities and pressure sensitivity. Capacitive touchscreens represent a dominant technology in the realm of touch-sensitive interfaces, adept at detecting touch inputs by measuring changes in capacitance induced by the proximity or contact of a conductive object, such as a finger or stylus. Thin film electrodes composed of indium tin oxide (ITO) are utilized as the sensing elements in capacitive touchscreens due to their exceptional transparency and conductivity. By integrating ITO-coated thin films into the touchscreen's structure, manufacturers can achieve high-resolution touch sensing capabilities while maintaining optical clarity and responsiveness, resulting in seamless user interactions and intuitive input methods. In addition to capacitive touchscreens, thin film sensors based on resistive or surface acoustic wave (SAW) technology are employed to meet specific requirements for multi-touch capabilities and pressure sensitivity in touch-sensitive devices. Resistive touchscreens comprise multiple layers of thin films, typically composed of indium tin oxide (ITO) or other conductive materials, separated by a thin gap filled with insulating material. When pressure is applied to the screen, the top and bottom layers come into contact, causing changes in electrical resistance that are detected by the touchscreen controller, enabling precise localization of touch inputs. Surface acoustic wave (SAW) touchscreens, on the other hand, utilize thin film sensors consisting of piezoelectric materials such as zinc oxide (ZnO) or aluminum nitride (AlN) deposited on the surface of the touchscreen. When a touch input is applied to the screen, surface acoustic waves are generated and propagate across the screen's surface, with changes in wave propagation detected by transducers positioned along the screen's edges. This enables accurate localization of touch inputs and provides pressure sensitivity capabilities, making SAW touchscreens suitable for applications requiring precise and responsive touch interactions. Capacitive touchscreens utilize ITO-coated thin film electrodes to detect touch inputs, enabling precise and responsive touch interactions across the screen's surface. The integration of thin film technology into touchscreens underscores its critical role in enabling advanced touch sensing capabilities, facilitating intuitive user interactions and driving innovation in touch-sensitive devices. In summary, thin film materials such as indium tin oxide (ITO) are integral to the functionality of touchscreens, enabling accurate and responsive touch input in a wide range of devices. Whether utilized in capacitive touchscreens for high-resolution touch sensing or in resistive and surface acoustic wave (SAW) touchscreens for multi-touch capabilities and pressure sensitivity, thin film sensors play a crucial role in shaping the user experience and expanding the capabilities of touch-sensitive interfaces.

iii. **Photovoltaics**:

   Thin film solar cells represent a transformative advancement in solar energy technology, offering a lightweight, flexible alternative to traditional silicon-based solar cells. These innovative solar cells leverage thin film materials such as cadmium telluride (CdTe), copper indium gallium selenide (CIGS), and perovskites to enable the fabrication of solar cells with high efficiency, low cost, and scalability. Thin film solar cells are particularly well-suited for applications where weight, flexibility, and form factor are critical, and they can be seamlessly integrated into building materials, clothing, and portable devices, providing renewable energy solutions for a wide range of applications. One of the primary advantages of thin film solar cells is their lightweight and flexible nature, which allows for greater versatility in deployment and integration. Unlike traditional silicon-based solar cells, which are rigid and bulky, thin film solar cells can be fabricated on flexible substrates, such as plastic or metal foils, enabling them to conform to curved surfaces and irregular shapes. This flexibility opens up new possibilities for solar energy applications in areas such as wearable technology, portable electronics, and building-integrated photovoltaics (BIPV), where traditional solar panels may be impractical or aesthetically unappealing. The utilization of thin film materials such as cadmium telluride (CdTe), copper indium gallium selenide (CIGS), and perovskites enables the fabrication of thin film solar cells with high efficiency and low cost. These materials offer unique advantages, such as tunable bandgaps, excellent light absorption properties, and compatibility with low-cost deposition techniques, making them attractive candidates for thin film photovoltaic applications. CdTe and CIGS thin film solar cells, in particular, have demonstrated efficiencies comparable to traditional silicon-based solar cells while offering the potential for lower manufacturing costs and faster production processes. Furthermore, thin film solar cells offer scalability and manufacturability advantages, allowing for rapid deployment and widespread adoption of solar energy technologies. Thin film deposition techniques such as sputtering, evaporation, and chemical vapor deposition (CVD) enable the high-throughput production of thin film solar cells on large-area substrates, facilitating economies of scale and mass production. This scalability makes thin film solar cells well-suited for utility-scale solar farms, where large quantities of solar panels are required to generate renewable electricity at competitive prices. Flexible thin film solar cells can be seamlessly integrated into building materials such as roofing shingles, providing a renewable energy solution that blends seamlessly with the built environment. This integration offers architects, developers, and homeowners the opportunity to incorporate solar energy generation into their projects without compromising aesthetics or design flexibility. Thin film solar cells represent a promising and versatile technology for harnessing solar energy, offering lightweight, flexible, and scalable solutions for a wide range of applications. By leveraging thin film materials such as CdTe, CIGS, and perovskites, thin film solar cells enable the cost-effective production of high-efficiency solar panels that can be integrated into various surfaces and form factors. As the demand for renewable energy continues to grow, thin film solar cells are poised to play a significant role in driving the transition to a sustainable energy future.

Future Trends and Opportunities:

i. **Wearable Electronics:** The growing demand for wearable electronics, including smartwatches, fitness trackers, and augmented reality devices, presents significant opportunities for thin film technology. Thin film materials and fabrication techniques enable the development of lightweight, flexible electronic components that can be integrated into clothing, accessories, and personal devices. Wearable sensors, displays, and energy harvesting devices based on thin film technology offer new possibilities for monitoring health, enhancing productivity, and providing immersive experiences.

ii. **Flexible Displays**: Flexible display technologies, enabled by thin film materials such as organic semiconductors and polymers, are poised to revolutionize the consumer electronics market. Flexible displays offer advantages such as lightweight, bendability, and durability, making them suitable for applications such as foldable smartphones, rollable televisions, and wearable displays. Thin film transistor technology plays a crucial role in driving the development of flexible displays, enabling the fabrication of high-performance, energy-efficient devices with innovative form factors.

iii. **Internet of Things (IoT) Devices**: The proliferation of IoT devices, encompassing connected sensors, actuators, and smart appliances, presents new opportunities for thin film technology. Thin film sensors and energy harvesting devices enable the development of low-power, wireless IoT devices capable of sensing, monitoring, and controlling various aspects of the physical environment. Thin film materials such as ferroelectrics, piezoelectrics, and thermoelectrics offer unique functionalities for IoT applications, including energy harvesting, sensing, and actuation.

Thin films play a pivotal role in the electronics industry, facilitating the development of compact, lightweight components with enhanced performance. From integrated circuits to displays and sensors, thin film technology underpins a wide range of applications critical to modern technology. As the demand for wearable electronics, flexible displays, and IoT devices continues to grow, thin films are expected to play an increasingly important role in shaping the future of electronics and optoelectronics. With ongoing advancements in materials science, fabrication techniques, and device integration, thin films offer endless possibilities for innovation and transformation across diverse industries.

2. **Renewable Energy**: Thin films hold immense potential for renewable energy applications, particularly in photovoltaics and solar energy harvesting. Thin film solar cells, such as cadmium telluride (CdTe) and copper indium gallium selenide (CIGS) thin film solar cells, offer advantages such as lightweight, flexibility, and cost-effectiveness compared to traditional silicon-based solar cells. These thin film technologies enable the production of lightweight and flexible solar panels that can be integrated into building facades, rooftops, and portable devices. Furthermore, thin film materials such as perovskites have emerged as promising candidates for next-generation solar cells, offering high efficiency and low-cost manufacturing processes. As the world transitions towards renewable energy sources, thin film technologies are expected to play a significant role in driving the adoption of solar energy and reducing carbon emissions.

Thin films represent a promising avenue for advancing renewable energy applications, particularly in the realm of photovoltaics and solar energy harvesting. With their lightweight, flexible, and cost-effective properties, thin film solar cells offer significant advantages over traditional silicon-based solar cells, paving the way for innovative solutions in solar power generation. This section will examine the immense potential of thin films in renewable energy, focusing on their applications in photovoltaics, the advantages they offer, and their role in driving the transition towards sustainable energy sources.

Thin Films in Renewable Energy:

1. **Photovoltaics**: Thin film solar cells are at the forefront of renewable energy technologies, offering a viable alternative to conventional silicon-based solar cells. These thin film technologies leverage materials such as cadmium telluride (CdTe), copper indium gallium selenide (CIGS), and perovskites to convert sunlight into electricity efficiently. Thin film solar cells have the advantage of being lightweight and flexible, allowing for their integration into a variety of surfaces and structures.
2. **Advantages of Thin Film Solar Cells**:
   - *Lightweight*: Thin film solar cells are significantly lighter than traditional silicon-based solar cells, making them suitable for applications where weight is a concern, such as on rooftops or portable devices.
   - *Flexibility*: Thin film solar cells can be manufactured on flexible substrates, enabling the production of bendable and conformable solar panels. This flexibility opens up new possibilities for integrating solar cells into curved or irregular surfaces, such as building facades or vehicle exteriors.
   - *Cost-effectiveness*: Thin film solar cells offer cost advantages over silicon-based solar cells due to their simpler manufacturing processes and lower material requirements. This cost-effectiveness makes thin film technology an attractive option for large-scale solar energy projects and off-grid applications.

Thin Film Technologies:

i. **Cadmium Telluride (CdTe) Thin Film Solar Cells:** CdTe thin film solar cells are one of the most mature thin film technologies, known for their high efficiency and low manufacturing costs. CdTe solar cells offer excellent light-absorbing properties and can be deposited on inexpensive substrates such as glass or flexible plastic. As a result, CdTe solar panels are widely used in utility-scale solar power plants and rooftop installations.
ii. **Copper Indium Gallium Selenide (CIGS) Thin Film Solar Cells:** CIGS thin film solar cells combine copper, indium, gallium, and selenium to form a highly efficient photovoltaic material. CIGS solar cells offer the advantages of high efficiency, flexibility, and scalability, making them suitable for a wide range of applications. CIGS solar panels can be integrated into building materials, portable electronics, and off-grid power systems, providing renewable energy solutions in diverse environments.
iii. **Perovskite Thin Film Solar Cells:** Perovskite solar cells have emerged as a promising next-generation thin film technology, offering high efficiency and low-cost manufacturing processes. Perovskite materials can be deposited using solution-based methods, allowing for the fabrication of thin film solar cells with simple and scalable manufacturing processes. Perovskite solar cells hold the potential to surpass the efficiency of traditional silicon-based solar cells while offering cost advantages and flexibility in design and integration.

Applications of Thin Film Solar Cells:

i. **Building Integrated Photovoltaics (BIPV):** Thin film solar cells can be seamlessly integrated into building materials such as glass windows, roofing tiles, and facade panels, providing renewable energy solutions for commercial and residential buildings. Building integrated photovoltaics (BIPV) enable the generation of clean energy while enhancing the aesthetic appeal and functionality of building structures.

ii. **Off-grid Power Systems:** Thin film solar panels are well-suited for off-grid power systems, where access to traditional electricity infrastructure is limited or unavailable. Off-grid applications such as rural electrification, remote monitoring systems, and portable power solutions benefit from the lightweight and flexible nature of thin film solar cells, allowing for easy deployment in remote locations.

iii. **Portable Electronics:** Thin film solar panels integrated into portable electronic devices such as smartphones, tablets, and wearables enable self-charging capabilities, extending battery life and enhancing device functionality. Portable solar chargers and power banks equipped with thin film solar cells provide convenient and sustainable energy solutions for outdoor enthusiasts, travelers, and emergency situations.

Role in Driving Renewable Energy Adoption:

Thin film technologies are poised to play a significant role in accelerating the adoption of solar energy and reducing carbon emissions. By offering lightweight, flexible, and cost-effective solutions, thin film solar cells address key challenges in solar power generation, such as installation costs, efficiency, and scalability. As the world transitions towards renewable energy sources to mitigate climate change and meet energy demand, thin film technologies will continue to drive innovation and growth in the solar energy sector.

Thin film solar cells hold immense potential for advancing renewable energy applications, particularly in photovoltaics and solar energy harvesting. With their lightweight, flexible, and cost-effective properties, thin film technologies offer innovative solutions for generating clean energy and reducing carbon emissions. As the demand for sustainable energy sources continues to grow, thin film solar cells will play a pivotal role in driving the transition towards a more sustainable and resilient energy future. With ongoing advancements in materials science, manufacturing processes, and system integration, thin film technologies are poised to revolutionize the renewable energy landscape and contribute to a greener and more sustainable world.

3. **Healthcare and Biomedical Applications:** Thin films offer exciting opportunities in healthcare and biomedical applications, where they are used for drug delivery, medical diagnostics, and tissue engineering. Thin film coatings can be engineered to provide controlled release of drugs or therapeutic agents, enabling targeted treatment of diseases such as cancer and diabetes. Additionally, thin film sensors and biosensors offer sensitive and selective detection of biomolecules and analytes, making them valuable

tools for medical diagnostics and monitoring. In tissue engineering, thin film scaffolds provide a platform for cell growth and tissue regeneration, offering potential solutions for wound healing and organ transplantation. With ongoing advances in biomaterials science and nanotechnology, thin films are poised to revolutionize healthcare and biomedical research, leading to improved patient outcomes and personalized medicine.

Thin films hold tremendous promise in healthcare and biomedical applications, offering innovative solutions for drug delivery, medical diagnostics, tissue engineering, and regenerative medicine. Engineered thin film coatings enable precise control over drug release, while thin film sensors provide sensitive and selective detection of biomolecules, enhancing medical diagnostics and monitoring. Additionally, thin film scaffolds serve as platforms for cell growth and tissue regeneration, addressing critical needs in wound healing and organ transplantation. This section will explore the diverse applications of thin films in healthcare and biomedical research, highlighting their potential to revolutionize patient care and personalized medicine.

Thin Films in Healthcare and Biomedical Applications:

  i. **Drug Delivery:** Thin film coatings offer a versatile platform for controlled drug delivery, enabling the targeted release of therapeutic agents to specific tissues or organs. By engineering thin film matrices with biocompatible polymers or hydrogels, drug molecules can be encapsulated and released in a sustained or controlled manner. This approach allows for precise dosage control, reduced side effects, and improved patient compliance. Thin film drug delivery systems have applications in various medical conditions, including cancer, diabetes, cardiovascular diseases, and infectious diseases.

 ii. **Medical Diagnostics:** Thin film sensors and biosensors play a vital role in medical diagnostics, offering rapid and accurate detection of biomarkers, pathogens, and disease indicators. Thin film sensor platforms, such as surface plasmon resonance (SPR) sensors, quartz crystal microbalance (QCM) sensors, and electrochemical sensors, provide sensitive and selective detection capabilities. These sensors can be integrated into diagnostic devices for point-of-care testing, disease screening, and monitoring of physiological parameters. Thin film-based diagnostic technologies have applications in clinical diagnostics, biomedical research, and public health surveillance.

iii. **Tissue Engineering and Regenerative Medicine:** Thin films serve as scaffolds for tissue engineering and regenerative medicine, providing a supportive environment for cell growth, differentiation, and tissue regeneration. By mimicking the extracellular matrix (ECM) of native tissues, thin film scaffolds promote cell adhesion, proliferation, and tissue integration. These scaffolds can be functionalized with bioactive molecules, growth factors, and signaling cues to guide tissue formation and regeneration. Thin film-based tissue engineering approaches have applications in wound healing, bone regeneration, cartilage repair, and organ transplantation.

Applications of Thin Films in Healthcare:

i. **Cancer Therapy**:

Thin film drug delivery systems represent a promising approach for enhancing the effectiveness of cancer treatment by enabling targeted delivery of chemotherapeutic agents to cancerous tissues while minimizing systemic toxicity. These innovative drug delivery systems leverage thin film coatings that can be engineered to release drugs in response to specific stimuli, such as pH, temperature, or enzymatic activity, thereby providing precise control over drug release kinetics and optimizing therapeutic outcomes. One of the key advantages of thin film drug delivery systems is their ability to target cancerous tissues while sparing healthy cells from exposure to cytotoxic agents. By encapsulating chemotherapeutic drugs within thin film coatings, researchers can design drug delivery systems that selectively release their payload in response to the unique biochemical characteristics of tumor microenvironments. For instance, acidic conditions prevalent in tumor tissues can trigger the release of drugs from pH-responsive thin film coatings, ensuring localized drug delivery to cancer cells while minimizing off-target effects. Moreover, thin film drug delivery systems offer the flexibility to tailor drug release kinetics to suit specific therapeutic requirements. By adjusting the composition and properties of thin film coatings, researchers can modulate the rate and duration of drug release, allowing for sustained drug delivery over extended periods or rapid release of therapeutic agents in response to acute treatment needs. This versatility enables precise control over drug concentrations at the site of action, maximizing therapeutic efficacy while minimizing adverse effects on healthy tissues. Thin film drug delivery systems can be engineered to encapsulate chemotherapeutic agents and selectively release them in response to specific stimuli present in the tumor microenvironment, enabling targeted delivery of drugs to cancerous tissues while minimizing systemic toxicity. Furthermore, thin film sensors integrated into drug delivery systems enable real-time monitoring of cancer biomarkers, tumor microenvironment, and treatment response, providing valuable insights for personalized cancer therapy and disease monitoring. These sensors can detect changes in tumor physiology, such as alterations in pH, oxygenation levels, or metabolic activity, allowing clinicians to assess treatment efficacy, predict treatment outcomes, and adjust therapeutic regimens accordingly. Thin film drug delivery systems offer a versatile and targeted approach for improving the efficacy of cancer treatment while minimizing side effects. By enabling precise control over drug release kinetics and incorporating real-time monitoring capabilities, these innovative drug delivery systems hold promise for personalized cancer therapy and monitoring of disease progression. As research in this field continues to advance, thin film drug delivery systems are poised to play a significant role in shaping the future of cancer treatment and patient care.

ii. **Diabetes Management**:

Thin film biosensors have emerged as invaluable tools for monitoring blood glucose levels in diabetic patients, offering non-invasive or minimally invasive methods that enhance convenience and comfort. Specifically, thin film glucose sensors, integrated into wearable devices or skin patches, enable continuous glucose monitoring (CGM) and feedback, facilitating timely adjustments of insulin dosages and dietary interventions to manage diabetes effectively. The integration of thin film glucose sensors into wearable devices or skin patches revolutionizes the way diabetic patients monitor their blood glucose levels. Unlike traditional finger-prick methods, which provide intermittent measurements and can be cumbersome and painful, thin film biosensors offer continuous monitoring capabilities, providing real-time data on glucose levels throughout the day. This continuous monitoring allows for a more comprehensive understanding of glucose fluctuations and trends, empowering patients to make informed decisions regarding their diabetes management. Thin film glucose sensors integrated into wearable devices enable continuous glucose monitoring, providing diabetic patients with real-time data on their blood glucose levels. The convenience and comfort offered by thin film biosensors contribute to improved patient compliance with glucose monitoring protocols. Wearable devices or skin patches equipped with thin film sensors can be discreetly worn on the body, allowing patients to monitor their glucose levels without interrupting their daily activities. This non-invasive or minimally invasive approach to glucose monitoring reduces the discomfort associated with traditional finger-prick methods, leading to greater patient acceptance and adherence to monitoring regimens. Furthermore, continuous glucose monitoring provided by thin film biosensors enables proactive diabetes management, helping to prevent hyperglycemic or hypoglycemic episodes and reduce the risk of long-term complications. By providing real-time feedback on glucose levels, these sensors allow patients and healthcare providers to identify trends and patterns in glucose fluctuations, facilitating timely interventions such as insulin adjustments or dietary modifications. This proactive approach to diabetes management helps to optimize glycemic control and minimize the risk of adverse outcomes associated with poorly managed diabetes. Thin film biosensors represent a significant advancement in diabetes management, offering non-invasive or minimally invasive methods for continuous glucose monitoring. Integrated into wearable devices or skin patches, thin film glucose sensors provide diabetic patients with convenient and comfortable options for monitoring their blood glucose levels. By enabling continuous monitoring and real-time feedback, these sensors enhance diabetes management, improve patient compliance, and reduce the risk of complications associated with poorly controlled diabetes. As research in this field continues to advance, thin film biosensors hold promise for further improving the quality of life for individuals living with diabetes.

iii. **Wound Healing**:

Thin film scaffolds represent a promising approach for promoting wound healing by providing a supportive matrix for cell migration, proliferation, and tissue regeneration. These scaffolds, typically composed of biocompatible polymers or hydrogels, offer a conducive environment for the natural wound healing process while protecting the wound from external contaminants and maintaining optimal moisture levels. Engineered to release bioactive molecules such as growth factors or antimicrobial agents, thin film dressings further enhance the wound healing process and tissue regeneration. Thin film dressings offer several advantages over traditional wound care methods. Their flexibility and conformability enable them to adapt to the contours of the wound site, ensuring proper coverage and contact with the underlying tissue. Additionally, thin film dressings are breathable, allowing for the exchange of gases and moisture vapor between the wound and the surrounding environment. This feature helps to maintain an optimal moisture balance, which is essential for promoting cell migration, proliferation, and tissue regeneration. Moreover, thin film dressings can be engineered to incorporate bioactive molecules that enhance the wound healing process. For example, growth factors such as platelet-derived growth factor (PDGF) or vascular endothelial growth factor (VEGF) can stimulate cell proliferation and angiogenesis, accelerating wound closure and tissue repair. Similarly, antimicrobial agents such as silver nanoparticles or antibiotics can help prevent infection and promote a sterile environment conducive to healing. By providing a supportive and biocompatible matrix for wound healing, thin film scaffolds offer a versatile and effective solution for managing a wide range of wounds, including acute injuries, chronic ulcers, and surgical incisions. Their flexibility, conformability, and ability to deliver bioactive molecules make them valuable tools in wound care management, improving patient comfort and facilitating the healing process. Thin film scaffolds represent a promising approach for promoting wound healing and tissue regeneration. Composed of biocompatible polymers or hydrogels, these scaffolds provide a supportive matrix for cell migration, proliferation, and tissue repair while protecting the wound from infection and maintaining moisture balance. Engineered to release bioactive molecules, thin film dressings further enhance the wound healing process, offering advantages such as flexibility, conformability, and breathability. As research in this field continues to advance, thin film scaffolds hold promise for improving the outcomes of wound care and enhancing patient quality of life.

Future Perspectives:

i. **Personalized Medicine:** Thin film technologies have the potential to revolutionize personalized medicine by enabling tailored therapies and diagnostics based on individual patient characteristics. Thin film drug delivery systems can be customized to deliver precise dosages of therapeutics to specific tissues or cells, maximizing treatment efficacy while minimizing side effects. Similarly, thin film sensors and diagnostic platforms can be personalized to detect disease biomarkers, monitor treatment response, and guide clinical decision-making, leading to improved patient outcomes and healthcare outcomes.

ii. **Point-of-Care Testing:** Thin film sensors and diagnostic devices offer rapid and decentralized testing capabilities, enabling point-of-care testing in clinical settings, remote areas, and resource-limited settings. By integrating thin film sensors into portable and handheld devices, healthcare providers can perform rapid diagnostics, disease screening, and monitoring at the patient's bedside or in the community. Point-of-care testing using thin film technologies enhances healthcare accessibility, reduces turnaround times, and improves patient outcomes, particularly in underserved populations.

iii. **Biomimetic Materials:** Future advancements in biomaterials science and nanotechnology are expected to drive the development of biomimetic thin films that closely mimic the structure and function of native tissues and organs. These biomimetic materials offer enhanced biocompatibility, bioactivity, and regenerative properties, making them ideal for tissue engineering, regenerative medicine, and organ-on-chip platforms. Biomimetic thin films hold promise for applications such as organ regeneration, disease modeling, and drug screening, offering new avenues for understanding complex biological systems and developing innovative therapeutic strategies.

Thin films offer exciting opportunities in healthcare and biomedical applications, spanning drug delivery, medical diagnostics, tissue engineering, and regenerative medicine. Engineered thin film coatings enable targeted drug delivery, while thin film sensors provide sensitive and selective detection capabilities for disease diagnosis and monitoring. Thin film scaffolds serve as platforms for tissue engineering and regenerative medicine, offering potential solutions for wound healing, organ transplantation, and personalized medicine. With ongoing advances in biomaterials science and nanotechnology, thin films are poised to revolutionize healthcare and biomedical research, leading to improved patient outcomes, enhanced quality of life, and personalized healthcare solutions.

4. **Environmental Monitoring and Sensing:** Thin films are increasingly being utilized for environmental monitoring and sensing applications, where they offer advantages such as high sensitivity, rapid response, and miniaturization. Thin film sensors can detect a wide range of environmental parameters, including temperature, humidity, gas concentration, and pollutant levels. These sensors find applications in air quality monitoring, water quality assessment, and industrial safety, enabling real-time detection and mitigation of environmental hazards. Furthermore, thin film-based sensor arrays and microfluidic devices offer multiplexed detection capabilities, allowing simultaneous monitoring of multiple analytes with high precision. As concerns about environmental

pollution and climate change continue to escalate, thin film sensors are expected to play a crucial role in addressing these challenges and promoting sustainability.

Thin films are increasingly recognized for their significant contributions to environmental monitoring and sensing applications, offering unique advantages such as high sensitivity, rapid response, and miniaturization. In a world facing escalating concerns about environmental pollution and climate change, thin film sensors play a crucial role in detecting and mitigating environmental hazards. This section will explore the diverse applications of thin films in environmental monitoring and sensing, highlighting their capabilities, advantages, and contributions to promoting sustainability.

Applications of Thin Films in Environmental Monitoring and Sensing:

i. **Air Quality Monitoring**:

Thin film sensors have emerged as essential tools for monitoring air quality, playing a critical role in detecting various pollutants and contaminants present in the atmosphere. These sensors offer advantages such as high sensitivity and rapid response, enabling real-time monitoring of air quality in diverse settings, including urban areas, industrial facilities, and indoor environments. By detecting pollutants such as carbon monoxide (CO), nitrogen dioxide (NO_2), sulfur dioxide (SO_2), particulate matter (PM), and volatile organic compounds (VOCs), thin film sensors provide accurate and timely data to support efforts aimed at assessing air quality, identifying pollution sources, and implementing measures to improve public health. Thin film sensors are highly sensitive to changes in pollutant concentrations, allowing for the detection of trace levels of contaminants in the air. This high sensitivity enables early detection of pollution events and provides valuable insights into pollutant dynamics and distribution patterns. By continuously monitoring pollutant levels, thin film sensors help to identify pollution hotspots and track changes in air quality over time, facilitating targeted interventions to mitigate pollution and protect public health. Moreover, thin film sensors offer rapid response times, providing real-time data on pollutant levels to support timely decision-making and emergency response efforts. This rapid response capability is particularly critical in urban areas and industrial facilities, where sudden increases in pollutant concentrations can pose immediate risks to public health and safety. By detecting changes in pollutant levels in real-time, thin film sensors enable authorities to issue timely alerts, implement pollution control measures, and mitigate the impact of pollution-related incidents. Thin film sensors are versatile and can be deployed in various environments to monitor air quality effectively. Whether integrated into stationary monitoring stations, portable devices, or wearable sensors, thin film sensors provide flexibility and scalability for air quality monitoring applications. This versatility allows for comprehensive coverage of different geographical areas and enables targeted monitoring of specific pollution sources or emission sources. Furthermore, thin film sensors can detect a wide range of pollutants commonly found in the atmosphere, including carbon monoxide (CO), nitrogen dioxide (NO_2), sulfur dioxide (SO_2), particulate matter (PM), and volatile organic compounds (VOCs). By monitoring multiple pollutants simultaneously, thin film sensor arrays provide a comprehensive assessment of air quality and pollutant levels, allowing for a more accurate understanding of environmental conditions and pollution trends. Thin film sensors play a vital role in monitoring air quality and detecting pollutants in the atmosphere. With their high sensitivity, rapid response, and versatility, thin film sensors enable real-time monitoring of air quality in diverse settings, supporting efforts to assess environmental conditions, identify pollution sources, and implement measures to improve public health and safety. As air quality concerns continue to grow, thin film sensors will remain essential tools for monitoring and managing pollution levels, contributing to efforts to create cleaner, healthier environments for communities worldwide.

ii. **Water Quality Assessment**:

Thin film sensors are indispensable tools for monitoring water quality parameters across various aquatic environments, including freshwater bodies, marine ecosystems, and wastewater treatment facilities. These sensors play a crucial role in assessing key water quality parameters such as pH, temperature, dissolved oxygen (DO), conductivity, and the presence of various contaminants. Thin film-based sensors offer several advantages, including high sensitivity and selectivity, which enable the detection of trace levels of pollutants, heavy metals, and organic compounds. By providing accurate and reliable data on water quality, thin film sensors support efforts to protect aquatic ecosystems, safeguard public health, and ensure compliance with regulatory standards. Thin film sensor arrays enable the monitoring of various water quality parameters, facilitating the assessment of aquatic ecosystem health. Thin film sensors are widely employed in freshwater, marine, and wastewater monitoring applications to assess water quality and identify potential sources of pollution. In freshwater bodies such as rivers, lakes, and reservoirs, thin film sensors provide real-time data on parameters such as pH, temperature, and DO levels, allowing for the monitoring of water quality and the detection of changes caused by pollution events or environmental stressors. Similarly, in marine ecosystems, thin film sensors help researchers and environmental agencies monitor key parameters such as salinity, nutrient levels, and contaminant concentrations to assess the health of coastal waters and marine habitats. Moreover, thin film sensors play a crucial role in monitoring wastewater effluents to ensure compliance with environmental regulations and prevent pollution of receiving water bodies. These sensors can detect contaminants such as heavy metals, organic compounds, and pathogens, providing valuable insights into the quality of wastewater discharge and the effectiveness of treatment processes. By monitoring key water quality parameters in real-time, thin film sensors enable wastewater treatment facilities to optimize treatment processes, minimize environmental impact, and protect public health. Thin film sensors offer several advantages over traditional water quality monitoring methods, including high sensitivity and selectivity, rapid response times, and the ability to detect trace levels of contaminants. These sensors can be customized to target specific pollutants or parameters of interest, allowing for targeted monitoring and early detection of water quality issues. Additionally, thin film sensors are robust and durable, making them suitable for long-term monitoring applications in harsh environmental conditions. Thin film sensors play a vital role in monitoring water quality parameters and detecting contaminants in freshwater, marine, and wastewater environments. With their high sensitivity, selectivity, and real-time monitoring capabilities, thin film sensors provide valuable data to support efforts aimed at protecting aquatic ecosystems, safeguarding public health, and ensuring compliance with regulatory standards. As water quality concerns continue to grow, thin film sensors will remain essential tools for monitoring and managing water resources, contributing to the sustainable management of aquatic ecosystems and the protection of human health and the environment.

iii. **Industrial Safety**:

Thin film sensors are indispensable tools for monitoring environmental conditions in industrial settings, playing a crucial role in ensuring workplace safety and regulatory compliance. These sensors are capable of detecting a wide range of parameters, including temperature, humidity, gas concentration, and airborne contaminants, making them essential for monitoring environmental conditions in manufacturing facilities, chemical plants, refineries, and other industrial environments. Thin film-based sensors offer several advantages, such as miniaturization and portability, allowing for distributed monitoring of environmental conditions in hazardous or remote locations. By providing early warning of potential hazards such as leaks, spills, and emissions, thin film sensors help prevent accidents, minimize environmental impact, and protect workers' health and safety. Thin film sensor arrays enable the monitoring of environmental conditions in industrial settings, facilitating early detection of potential hazards and ensuring workplace safety. Thin film sensors play a vital role in monitoring temperature and humidity levels in industrial environments to ensure optimal working conditions and prevent equipment malfunction or damage. These sensors can detect fluctuations in temperature and humidity that may indicate equipment failures, leaks, or other potential hazards. By providing real-time data on environmental conditions, thin film sensors enable proactive maintenance and troubleshooting, helping to prevent costly downtime and ensure uninterrupted operations. Moreover, thin film sensors are used to monitor gas concentrations and detect airborne contaminants in industrial settings, including hazardous gases, volatile organic compounds (VOCs), and particulate matter. These sensors provide early warning of potential exposure to harmful substances, allowing for prompt evacuation or mitigation measures to protect workers' health and safety. Additionally, thin film sensors can be integrated into air quality monitoring systems to track emissions and ensure compliance with environmental regulations. One of the key advantages of thin film sensors is their miniaturization and portability, which allows for distributed monitoring of environmental conditions in hazardous or remote locations. These sensors can be deployed in hard-to-reach areas or integrated into wearable devices for personal monitoring, providing real-time data to workers and safety personnel. By enabling continuous monitoring of environmental conditions, thin film sensors help identify potential hazards before they escalate into emergencies, reducing the risk of accidents and injuries in industrial settings. In summary, thin film sensors are essential tools for monitoring environmental conditions in industrial settings to ensure workplace safety and regulatory compliance. With their ability to detect a wide range of parameters and their miniaturization and portability, thin film sensors enable distributed monitoring of environmental conditions in hazardous or remote locations. By providing early warning of potential hazards such as leaks, spills, and emissions, thin film sensors help prevent accidents, minimize environmental impact, and protect workers' health and safety in industrial environments. As industrial operations continue to evolve, thin film sensors will remain critical components of environmental monitoring systems, contributing to the safe and sustainable operation of industrial facilities.

iv. **Multiplexed Detection**:

Thin film-based sensor arrays and microfluidic devices represent a significant advancement in environmental monitoring and sensing, enabling multiplexed detection of multiple analytes simultaneously. These sensor platforms integrate multiple thin film sensors with microfluidic channels or arrays, allowing for parallel analysis of different environmental parameters. Multiplexed detection offers enhanced capabilities for monitoring complex environmental systems, such as aquatic ecosystems, air quality networks, and industrial processes, by providing a holistic view of environmental conditions and facilitating data-driven decision-making and proactive management of environmental resources. Thin film sensor arrays integrated with microfluidic devices enable multiplexed detection of multiple analytes, enhancing environmental monitoring capabilities. Thin film-based sensor arrays and microfluidic devices offer several advantages for environmental monitoring and sensing. Firstly, these sensor platforms enable multiplexed detection, allowing for simultaneous analysis of multiple analytes in a single measurement. This capability is particularly valuable for monitoring complex environmental systems where multiple parameters may influence ecosystem health or industrial processes. By detecting multiple analytes simultaneously, thin film sensor arrays and microfluidic devices provide a comprehensive understanding of environmental conditions, facilitating more accurate assessments and informed decision-making. Moreover, thin film-based sensor arrays and microfluidic devices offer high sensitivity and selectivity, enabling detection of trace levels of contaminants and pollutants in environmental samples. These sensors can detect a wide range of analytes, including gases, liquids, and particulate matter, with excellent sensitivity and specificity. This sensitivity allows for early detection of environmental changes or pollution events, enabling timely interventions to mitigate risks and protect environmental resources. Furthermore, the integration of thin film sensor arrays with microfluidic devices allows for efficient sample processing and analysis. Microfluidic channels or arrays enable precise control of sample flow and manipulation, facilitating rapid and accurate measurements of environmental parameters. This integration enhances the efficiency and throughput of environmental monitoring systems, enabling real-time monitoring of dynamic environmental processes and events. Thin film sensor arrays and microfluidic devices have diverse applications in environmental monitoring and sensing. In aquatic ecosystems, these sensor platforms can be deployed to monitor water quality parameters such as pH, dissolved oxygen, nutrient levels, and contaminant concentrations, enabling comprehensive assessment of ecosystem health and water quality. In air quality networks, thin film sensor arrays integrated with microfluidic devices can be used to monitor gas concentrations, particulate matter levels, and volatile organic compounds in urban and industrial environments, providing valuable data for air quality management and pollution control. In industrial processes, these sensor platforms enable real-time monitoring of process parameters such as temperature, pressure, and chemical composition, facilitating process optimization and environmental compliance. Thin film-based sensor arrays and microfluidic devices offer enhanced capabilities for multiplexed detection of multiple analytes in environmental monitoring and sensing applications. By providing a holistic view of environmental conditions and facilitating data-driven decision-making, these sensor platforms contribute to proactive management of environmental resources and protection of ecosystem health. As technology continues to advance, thin film sensor arrays and microfluidic devices will play an increasingly important role in addressing environmental challenges and promoting sustainable development.

Advantages of Thin Film Sensors:

Thin film sensors offer numerous advantages that make them highly attractive for a wide range of applications. These advantages stem from their unique properties and capabilities, which include:

i. **High Sensitivity**: Thin film sensors exhibit exceptional sensitivity, allowing them to detect even trace amounts of target analytes or environmental parameters. This high sensitivity enables accurate measurements and early detection of potential hazards, making thin film sensors invaluable for environmental monitoring, medical diagnostics, and industrial safety applications.

ii. **Rapid Response**: Thin film sensors provide rapid response times, offering real-time monitoring capabilities for dynamic environments. Whether detecting changes in temperature, humidity, gas concentrations, or pollutant levels, thin film sensors offer quick feedback, enabling timely interventions and preventive measures.

iii. **Miniaturization**: Thin film sensors can be miniaturized to extremely small sizes, allowing for their integration into compact, portable devices. This miniaturization enables on-site monitoring in remote or confined spaces where traditional bulky sensors may be impractical. Additionally, miniaturized thin film sensors facilitate wearable and implantable sensor technologies, opening up new possibilities for continuous health monitoring and personalized medicine.

iv. **Selective Detection**: Thin film sensors can be engineered to selectively detect specific analytes or environmental parameters, minimizing interference from background noise or contaminants. By tailoring the composition, structure, or surface chemistry of thin film sensor materials, researchers can achieve high selectivity, ensuring accurate measurement of target compounds in complex matrices.

v. **Multiplexed Detection**: Thin film sensor arrays and microfluidic devices enable multiplexed detection of multiple analytes simultaneously. This multiplexed detection capability allows for comprehensive monitoring of complex systems, such as environmental samples or biological fluids, where multiple parameters need to be analyzed concurrently. Multiplexed thin film sensors enhance data richness and provide a more holistic understanding of the system under study.

vi. **Cost-Effectiveness**: Thin film sensors offer cost-effective solutions for various applications due to their simplified fabrication processes, reduced material consumption, and scalability. Compared to traditional bulky sensors or analytical techniques, thin film sensors often require fewer resources and offer higher throughput, making them economically viable for large-scale deployment in environmental monitoring networks, medical diagnostics, and industrial process control.

vii. **Flexibility and Adaptability**: Thin film sensors can be fabricated on flexible substrates, allowing for their integration into conformal or irregular surfaces. This flexibility enables the development of bendable, stretchable, and wearable sensor devices that can conform to the contours of the human body or complex structures. Additionally, thin film sensors can be adapted to different environmental conditions or specific applications by modifying their design parameters or sensor materials, enhancing their versatility and utility.

In summary, thin film sensors offer a host of advantages, including high sensitivity, rapid response, miniaturization, selective detection, multiplexed capabilities, cost-effectiveness, and flexibility. These advantages make thin film sensors indispensable tools for a wide range of applications, from environmental monitoring and medical diagnostics to industrial process control and consumer electronics. As research and development in thin film sensor technology continue to advance, these sensors are poised to play an increasingly critical role in addressing pressing societal challenges and advancing scientific knowledge.

Future Perspectives:

   i. **Advancements in Sensor Technology:** Ongoing advancements in materials science, nanotechnology, and sensor fabrication techniques are expected to drive innovation in thin film sensor technology. Future developments may include enhanced sensor performance, improved selectivity, and increased integration with data analytics and wireless communication technologies.
   ii. **Environmental Monitoring Networks:** Thin film sensors have the potential to be deployed in networks of interconnected monitoring stations to create comprehensive environmental monitoring networks. These networks can provide real-time data on environmental conditions, support predictive modeling of environmental processes, and facilitate decision-making for environmental management and policy development.
   iii. **Integration with IoT and AI:** Thin film sensors can be integrated with Internet of Things (IoT) platforms and artificial intelligence (AI) algorithms to create smart environmental monitoring systems. These systems can autonomously collect, analyze, and act on environmental data, providing insights into environmental trends, patterns, and anomalies.

In conclusion, thin film sensors offer significant opportunities for environmental monitoring and sensing applications, providing high sensitivity, rapid response, and miniaturization advantages. These sensors play a crucial role in monitoring air quality, water quality, and industrial safety, enabling real-time detection and mitigation of environmental hazards. With ongoing advancements in sensor technology and integration with IoT and AI, thin film sensors are poised to revolutionize environmental monitoring, promote sustainability, and safeguard environmental resources for future generations.

5. **Emerging Technologies:** In addition to the sectors, thin films hold promise for a wide range of emerging technologies and applications. For example, in quantum computing and nanotechnology, thin film materials such as superconductors and magnetic materials are being explored for their unique quantum properties and functionalities. In flexible and stretchable electronics, thin film materials enable the development of conformable and wearable devices for healthcare monitoring, human–machine interfaces, and augmented reality. Furthermore, in additive manufacturing and 3D printing, thin film deposition techniques offer precise control over material deposition and patterning, enabling the fabrication of complex and customized structures with high resolution and accuracy. Thin films offer tremendous potential across a myriad of emerging technologies and applications, extending beyond the sectors previously discussed. From quantum computing and nanotechnology to flexible electronics and additive manufacturing, thin film materials play a crucial role in driving innovation and advancing the

frontiers of science and engineering. This section will explore the diverse applications of thin films in emerging technologies, highlighting their unique properties and functionalities that enable groundbreaking advancements.

i. **Quantum Computing and Nanotechnology**:

Thin film materials, particularly superconductors and magnetic materials, are of significant interest in the fields of quantum computing and nanotechnology. Superconducting thin films exhibit quantum phenomena such as zero electrical resistance and magnetic flux quantization, making them ideal candidates for qubits—the fundamental units of quantum information in quantum computing. These superconducting thin films enable the development of superconducting quantum bits (qubits) and other components essential for building quantum computers, which have the potential to revolutionize computing by performing complex calculations exponentially faster than classical computers. In nanotechnology, thin film materials are explored for their unique quantum properties and functionalities. Thin film deposition techniques, such as molecular beam epitaxy (MBE) and atomic layer deposition (ALD), allow precise control over film thickness and composition at the atomic scale, enabling the fabrication of nanostructures and nanodevices with tailored properties. Thin film nanomaterials find applications in nanoelectronics, nanophotonics, quantum dots, and spintronics, paving the way for novel technologies with enhanced performance and functionality at the nanoscale.

ii. **Flexible and Stretchable Electronics**:

Thin film materials play a pivotal role in the development of flexible and stretchable electronics, enabling the fabrication of conformable and wearable devices for diverse applications. Flexible electronics incorporate thin film materials on flexible substrates, such as polymers or textiles, allowing for bending, folding, and stretching without compromising functionality. These devices find applications in healthcare monitoring, human–machine interfaces, augmented reality, and electronic skins. For instance, thin film sensors integrated into wearable devices enable continuous monitoring of vital signs, such as heart rate, blood pressure, and body temperature, facilitating remote healthcare monitoring and early detection of health conditions. Flexible displays and touchscreens based on thin film transistor (TFT) technology offer seamless integration into clothing, accessories, and portable devices, enhancing user experience and interaction in augmented reality environments. Furthermore, thin film batteries and energy harvesting devices power these wearable electronics, providing on-the-go energy solutions for mobile and wearable technology.

iii. **Additive Manufacturing and 3D Printing**:

Thin film deposition techniques play a crucial role in additive manufacturing and 3D printing, offering precise control over material deposition and patterning. Thin film-based additive manufacturing processes, such as inkjet printing, aerosol jet printing, and laser-induced forward transfer (LIFT), enable the fabrication of complex and customized structures with high resolution and accuracy. These processes allow for the deposition of functional thin films onto substrates or directly onto 3D-printed objects, adding additional functionalities or enhancing performance. For example, thin film coatings applied through additive manufacturing processes can impart properties such as conductivity, corrosion resistance, or biocompatibility to 3D-printed components, expanding their applications in electronics, aerospace, biomedical devices, and automotive industries. Additionally, thin film-based 3D printing techniques, such as direct-write printing and electrohydrodynamic printing, offer versatility in printing a wide range of materials, including polymers, metals, ceramics, and biomaterials, opening up new possibilities for creating intricate and multifunctional 3D structures. Thin films hold promise for a wide range of emerging technologies and applications, including quantum computing, nanotechnology, flexible electronics, and additive manufacturing. These materials offer unique properties and functionalities that enable groundbreaking advancements in diverse fields, from enabling quantum computing to revolutionizing wearable electronics and enhancing the capabilities of 3D printing technologies. As research and development in thin film materials and deposition techniques continue to advance, the potential for innovation and disruption across these emerging sectors remains vast, driving progress towards a more technologically advanced and interconnected world.

Future Prospects: The future of thin films looks promising, with ongoing advancements in materials science, nanotechnology, and manufacturing processes driving innovation and growth. As researchers continue to explore novel materials, deposition techniques, and applications, thin films are expected to become even more ubiquitous in our daily lives, powering next-generation electronics, renewable energy systems, healthcare devices, and environmental sensors. Moreover, as the demand for sustainable and eco-friendly technologies grows, thin films are poised to play a key role in addressing global challenges such as climate change, healthcare disparities, and resource scarcity. With interdisciplinary collaboration and investment in research and development, thin films have the potential to revolutionize multiple industries and create a brighter and more sustainable future for generations to come.

Challenges of 3D printing:

While 3D printing has revolutionized manufacturing and design processes, it also presents several challenges that need to be addressed for its widespread adoption and continued advancement.

One significant challenge in 3D printing is the limited range of materials available for printing. While there has been progress in developing new materials suitable for additive manufacturing, the selection remains relatively limited compared to traditional manufacturing methods. Many high-performance materials used in industries such as aerospace and automotive are difficult to process using 3D printing techniques, limiting the technology's applicability in certain sectors. Moreover, achieving consistent quality and reliability in 3D printed parts can be challenging. Factors such as layer adhesion, porosity, and dimensional accuracy can vary depending on printing parameters, machine calibration, and material properties. Variations in part quality may lead to defects, surface roughness, or mechanical weaknesses, compromising the performance and reliability of printed components. Addressing these challenges requires rigorous process optimization, quality control measures, and standards development to ensure consistent and repeatable results. Another challenge in 3D printing is scalability and production speed. While additive manufacturing offers advantages such as design flexibility and customization, it typically lags behind traditional manufacturing methods in terms of production throughput and efficiency. The layer-by-layer deposition process used in 3D printing can be time-consuming, particularly for large or complex parts. Scaling up production to meet demand while maintaining cost-effectiveness and turnaround times remains a significant challenge for many industries.

Furthermore, 3D printing technologies face challenges related to post-processing and finishing of printed parts. Depending on the printing method and material used, printed parts may require additional processing steps such as support removal, surface smoothing, or heat treatment to achieve the desired properties and surface finish. Manual post-processing tasks can be labor-intensive and time-consuming, adding to production costs and lead times. Developing automated post-processing solutions and integrating them into the additive manufacturing workflow is essential for streamlining production and improving efficiency. Additionally, intellectual property (IP) concerns pose challenges for the widespread adoption of 3D printing technology. The ease of digital file sharing and replication in additive manufacturing raises concerns about copyright infringement, counterfeiting, and unauthorized reproduction of patented designs. Protecting intellectual property in the digital age requires robust legal frameworks, digital rights management systems, and enforcement mechanisms to safeguard innovation and creativity in 3D printing. While 3D printing offers numerous benefits and opportunities for innovation, it also presents several challenges that need to be addressed for its successful implementation. Overcoming limitations in materials availability, quality control, scalability, post-processing, and IP protection is essential for realizing the full potential of 3D printing technology across industries. Continued research, development, and collaboration are key to addressing these challenges and advancing the capabilities of additive manufacturing.

1. **Material Selection and Development**:

Expanding the range of materials suitable for 3D printing remains a significant challenge. While there has been progress in printing with metals, polymers, ceramics, and composites, developing new materials with tailored properties for specific applications is ongoing.

Material selection and development in 3D printing, also known as additive manufacturing, play a crucial role in the production of high-quality, functional parts and components. As 3D printing technology continues to advance, researchers and engineers are constantly exploring new materials and improving existing ones to meet the diverse needs of various industries. Material selection involves considerations such as mechanical properties, thermal stability, chemical resistance, and cost-effectiveness, among others. Additionally, the development of new materials opens up possibilities for novel applications and enhanced performance in 3D printing. One of the key considerations in material selection for 3D printing is mechanical properties. Different applications require materials with specific mechanical characteristics, such as strength, flexibility, and durability. For example, in aerospace and automotive industries, where lightweight yet strong materials are essential, engineers often opt for polymers reinforced with carbon fibers or metal alloys with high strength-to-weight ratios. By carefully selecting materials based on their mechanical properties, manufacturers can ensure that 3D-printed parts meet performance requirements and withstand operational stresses. Thermal stability is another critical factor in material selection for 3D printing. Some applications involve exposure to high temperatures or thermal cycling, requiring materials that can maintain their structural integrity under such conditions. Heat-resistant polymers like polyether ether ketone (PEEK) and polyetherimide (PEI) are commonly used in industries such as aerospace, automotive, and electronics for their ability to withstand elevated temperatures without deformation or degradation. By choosing thermally stable materials, manufacturers can produce parts suitable for demanding operating environments.

Chemical resistance is essential in applications where 3D-printed parts come into contact with corrosive substances or harsh chemicals. Materials such as acrylonitrile butadiene styrene (ABS), polycarbonate (PC), and polypropylene (PP) offer excellent chemical resistance and are commonly used in industries like chemical processing, pharmaceuticals, and oil and gas. By selecting chemically resistant materials, manufacturers can ensure the longevity and reliability of 3D-printed components in corrosive environments, minimizing maintenance and replacement costs. Cost-effectiveness is a significant consideration in material selection for 3D printing, particularly for large-scale production or consumer-oriented applications. While some high-performance materials may offer superior properties, they can also come with higher costs, making them less economical for certain applications. Engineers often balance performance requirements with cost considerations to optimize material selection for specific use cases. Additionally, advancements in material development and production processes continue to drive down costs, making 3D printing more accessible to a broader range of industries and applications.

The development of new materials is a continuous process driven by innovation and research. Researchers are constantly exploring novel materials and formulations tailored for 3D printing, unlocking new possibilities for enhanced performance and functionality. For example, bio-based polymers derived from renewable sources are gaining popularity for their sustainability and biodegradability, making them attractive options for environmentally conscious applications. Similarly, metal powders with customized compositions and microstructures enable the production of high-strength, lightweight parts with superior mechanical properties. In addition to new materials, advancements in material processing and post-processing techniques contribute to the evolution of 3D printing technology. Additive manufacturing processes such as selective laser sintering (SLS), electron beam melting (EBM), and binder jetting offer unique capabilities for processing a wide range of materials, including metals, polymers, ceramics, and composites. Post-processing methods such as heat treatment, surface finishing, and coating further enhance the properties and performance of 3D-printed parts, expanding their applicability across industries.

In conclusion, material selection and development are critical aspects of 3D printing, influencing the performance, functionality, and cost-effectiveness of printed parts and components. By carefully evaluating mechanical properties, thermal stability, chemical resistance, and cost considerations, manufacturers can choose materials that meet specific requirements for their applications. Furthermore, ongoing research and innovation in material development continue to drive advancements in additive manufacturing, unlocking new possibilities for enhanced performance, sustainability, and functionality in various industries.

2. **Printing Speed and Efficiency**:

Increasing printing speed and efficiency without compromising quality is a challenge. Traditional layer-by-layer printing methods can be time-consuming, especially for complex geometries. Improving printing techniques, hardware design, and software algorithms is necessary to enhance throughput and reduce production times. Printing speed and efficiency are crucial factors in the realm of 3D printing, influencing production timelines, cost-effectiveness, and overall workflow optimization. As the demand for rapid prototyping, custom manufacturing, and on-demand production continues to grow, enhancing printing speed and efficiency has become a primary focus for manufacturers and researchers alike. Achieving higher printing speeds and efficiency involves optimizing various aspects of the 3D printing process, including hardware, software, materials, and operational strategies.

One of the key determinants of printing speed is the printing technology itself. Different 3D printing technologies, such as fused deposition modeling (FDM), stereolithography (SLA), selective laser sintering (SLS), and digital light processing (DLP), offer varying levels of speed and efficiency. For instance, FDM printers typically have faster printing speeds compared to SLA or SLS printers due to their simpler layer-by-layer deposition process. However, advancements in SLA and DLP technology, such as faster curing times and improved resin formulations, have led to significant improvements in printing speeds and throughput, making them more competitive with traditional FDM methods. Hardware plays a critical role in determining printing speed and efficiency. Factors such as printer design, build volume, nozzle size, and motion control systems influence the rate at which parts can be produced. High-speed motion systems, precision linear guides, and optimized heating and cooling mechanisms contribute to faster print speeds and more consistent print quality. Additionally, advancements in extrusion systems, such as dual-extrusion and multi-material printing, enable simultaneous deposition of multiple materials, reducing print times and increasing efficiency.

Software optimization is another key aspect of improving printing speed and efficiency. 3D printing software, including slicing software and printer firmware, plays a crucial role in generating optimized toolpaths, minimizing print time, and maximizing printer utilization. Features such as variable layer height, adaptive infill patterns, and optimized support structures help streamline the printing process and reduce material usage without compromising part quality. Furthermore, real-time monitoring and control systems enable operators to adjust print parameters on the fly, optimizing print settings for maximum speed and efficiency. Materials selection and optimization are essential for improving printing speed and efficiency. Choosing materials with suitable flow characteristics, melting points, and cooling rates can significantly impact printability and processability. Low-viscosity thermoplastics and resins facilitate faster extrusion and curing times, enabling higher printing speeds and throughput. Additionally, advancements in material formulations, such as reinforced filaments and high-performance resins, offer improved mechanical properties and surface finish, reducing post-processing requirements and enhancing overall print efficiency. Operational strategies also play a crucial role in maximizing printing speed and efficiency. Batch printing, job scheduling, and print queue management help optimize printer utilization and minimize downtime. Implementing automated print removal and bed leveling systems reduces manual intervention and streamlines the printing workflow. Furthermore, optimizing post-processing techniques, such as support removal, surface finishing, and curing, reduces overall cycle times and increases throughput. By adopting lean manufacturing principles and continuous improvement methodologies, manufacturers can identify bottlenecks, streamline workflows, and optimize resource utilization for maximum efficiency. Printing speed and efficiency are critical considerations in 3D printing, impacting production timelines, cost-effectiveness, and overall productivity. Achieving higher printing speeds and efficiency involves optimizing various aspects of the printing process, including hardware, software, materials, and operational strategies. By leveraging advancements in technology, materials, and workflow optimization, manufacturers can enhance printing speed and efficiency, enabling faster prototyping, production, and delivery of high-quality 3D-printed parts and components.

3. Resolution and Surface Finish:

Achieving high resolution and smooth surface finishes in 3D printed objects presents a significant challenge, especially when dealing with intricate geometries and fine details. Several factors come into play, including layer thickness, nozzle size, and printing parameters, all of which influence print quality and surface finish. To attain optimal results, careful optimization tailored to each specific application is necessary.

Layer thickness is a critical parameter affecting print resolution and surface finish. Typically, thinner layers result in finer details and smoother surfaces, but they also increase printing time. Conversely, thicker layers lead to faster printing but may sacrifice detail and surface quality. Finding the right balance between layer thickness and printing speed is essential for achieving the desired resolution and surface finish. For intricate geometries and fine details, a smaller layer thickness is often preferred to ensure accuracy and precision in the final printed object. Nozzle size plays a crucial role in determining print resolution and surface finish. Smaller nozzles allow for finer extrusion and greater detail, resulting in higher resolution prints with smoother surfaces. However, smaller nozzles may also increase printing time due to the finer deposition of material. On the other hand, larger nozzles enable faster printing but may compromise detail and surface quality, particularly in intricate geometries. Selecting the appropriate nozzle size based on the desired resolution and surface finish is essential for optimizing print quality.

Printing parameters such as print speed, temperature, and cooling settings also impact print quality and surface finish. Adjusting these parameters allows for fine-tuning of the printing process to achieve optimal results. Slower print speeds generally lead to higher resolution and smoother surfaces, as they allow more time for each layer to solidify and bond properly. Similarly, controlling the temperature of the printing environment and the extruded material can minimize warping and improve surface finish. Additionally, optimizing cooling settings, such as fan speed and part cooling, helps prevent overheating and ensure uniform cooling, resulting in smoother surfaces and reduced defects. Furthermore, selecting the appropriate printing technology and material is crucial for achieving high-resolution prints with smooth surface finishes. Different 3D printing technologies, such as Fused Deposition Modeling (FDM), Stereolithography (SLA), and Selective Laser Sintering (SLS), offer varying levels of resolution and surface quality. For example, SLA and SLS technologies typically produce higher resolution prints with smoother surfaces compared to FDM due to their layer-less printing process. Additionally, using high-quality materials with consistent properties and minimal impurities helps ensure uniform extrusion and optimal surface finish.

Post-processing techniques such as sanding, polishing, and surface treatments can further enhance the resolution and surface finish of 3D printed objects. Sanding and polishing remove layer lines and imperfections, resulting in smoother surfaces and improved aesthetics. Surface treatments such as chemical smoothing or vapor polishing can also be employed to achieve a glossy finish and eliminate surface roughness. Additionally, applying primer or paint can further enhance the appearance and durability of 3D printed objects, making them suitable for various applications. Achieving high resolution and smooth surface finishes in 3D printed objects requires careful consideration of several factors, including layer thickness, nozzle size, printing parameters, printing technology, and material selection. Optimization of these parameters tailored to each specific application is essential for maximizing print quality and surface finish. Additionally, post-processing techniques can be employed to further enhance the resolution and surface finish of 3D printed objects, resulting in high-quality, visually appealing final products.

4. **Process Control and Reliability**:

Ensuring consistent print quality and reliability across different printing jobs and machines is paramount for industrial applications. Variations in material properties, printing parameters, and environmental conditions can significantly impact print outcomes, underscoring the importance of robust process control and monitoring systems. By implementing stringent quality control measures and leveraging advanced monitoring technologies, manufacturers can maintain consistency, reliability, and repeatability in their additive manufacturing processes.

Material properties play a fundamental role in determining print quality and reliability. Variations in material composition, moisture content, and filament diameter can affect extrusion behavior, layer adhesion, and final part characteristics. To mitigate these challenges, manufacturers must carefully select and qualify materials for specific applications, ensuring consistency and reliability in their performance. Additionally, establishing strict material handling protocols, such as proper storage conditions and filament drying procedures, helps minimize material degradation and maintain optimal printing properties. Printing parameters, including layer height, print speed, temperature, and infill density, significantly influence print quality and reliability. Variations in these parameters can lead to inconsistencies in layer adhesion, dimensional accuracy, and surface finish. To achieve consistent results, manufacturers must establish standardized printing parameters based on material properties, part geometry, and application requirements. Furthermore, implementing advanced slicing software with predictive modeling capabilities allows for real-time adjustments and optimization of printing parameters, ensuring uniformity and reliability across different printing jobs and machines.

Environmental conditions, such as temperature, humidity, and airflow, also impact print quality and reliability. Fluctuations in ambient conditions can affect material properties, thermal stability, and print adhesion, leading to variations in print outcomes. To mitigate these effects, manufacturers must control and monitor environmental conditions within the printing environment. Implementing climate-controlled chambers, enclosure systems, and humidity sensors helps maintain stable printing conditions and minimize environmental influences on print quality. Additionally, integrating real-time monitoring systems enables proactive identification and mitigation of environmental factors that may impact print reliability. Robust process control and monitoring systems are essential for maintaining consistency and reliability in additive manufacturing processes. Automated quality control measures, such as in-line inspection systems and closed-loop feedback mechanisms, enable real-time monitoring and correction of print defects. By integrating quality control checks at key stages of the printing process, manufacturers can identify deviations from the desired specifications and take corrective actions to ensure consistent print quality. Additionally, implementing comprehensive data analytics and machine learning algorithms allows for predictive maintenance and optimization of printing parameters, further enhancing reliability and efficiency.

Standardization and certification play a crucial role in ensuring consistent print quality and reliability across different printing jobs and machines. Establishing industry-wide standards and best practices for material characterization, process validation, and quality assurance facilitates interoperability and consistency in additive manufacturing. Certification programs, such as ISO standards and ASTM specifications, provide guidelines and benchmarks for evaluating print quality, reliability, and performance. By adhering to recognized standards and obtaining certifications, manufacturers demonstrate their commitment to quality excellence and customer satisfaction. Ensuring consistent print quality and reliability in additive manufacturing processes is essential for industrial applications. Variations in material properties, printing parameters, and environmental conditions can impact print outcomes, highlighting the need for robust process control and monitoring systems. By implementing stringent quality control measures, standardizing printing parameters, and leveraging advanced monitoring technologies, manufacturers can maintain consistency, reliability, and repeatability in their additive manufacturing processes, ultimately delivering high-quality, reliable parts and components for a wide range of applications.

5. Post-processing:

Post-processing steps are integral to the production of high-quality 3D printed parts. While additive manufacturing offers unparalleled design freedom and rapid prototyping capabilities, the raw output often requires additional refinement to meet aesthetic and functional requirements. Post-processing steps such as support removal, surface finishing, and painting are commonly employed to enhance the appearance, mechanical properties, and functionality of 3D printed parts. However, streamlining post-processing workflows and developing automated finishing techniques remain ongoing challenges in the field of 3D printing.

Support removal is a critical post-processing step for parts printed using additive manufacturing technologies like Fused Deposition Modeling (FDM) or Stereolithography (SLA). During the printing process, supports are added to overhanging features and intricate geometries to prevent sagging and ensure print accuracy. However, these supports often leave behind unwanted marks or rough surfaces on the finished part. Manual removal of supports can be time-consuming and labor-intensive, particularly for complex parts with intricate support structures. Developing automated support removal techniques, such as water jetting, ultrasonic vibration, or chemical dissolution, can significantly improve efficiency and consistency in post-processing workflows. Surface finishing is another essential post-processing step aimed at improving the visual appearance and tactile feel of 3D printed parts. Layer lines, surface roughness, and imperfections inherent in the printing process can detract from the overall quality of the part. Various surface finishing techniques, including sanding, polishing, and vapor smoothing, are employed to achieve smoother surfaces and finer details. Manual surface finishing methods can be labor-intensive and require skilled labor, leading to increased production costs and variability in quality. Automating surface finishing processes using robotic systems, abrasive media tumbling, or electrochemical polishing can help streamline post-processing workflows and ensure consistent results across different parts and batches.

Painting is often used as a final post-processing step to enhance the aesthetics and functionality of 3D printed parts. Painting allows for customization of color, texture, and surface properties, as well as adding protective coatings or markings. Traditional painting methods involve manual spraying, brushing, or dipping, which can be time-consuming and require skilled labor to achieve desired results. Developing automated painting systems with programmable robots, spray guns, and paint booths enables faster and more precise application of coatings, reducing production time and improving consistency. Additionally, advancements in digital printing technologies, such as inkjet printing or direct coloration, offer alternative approaches for adding color and graphics directly onto 3D printed parts, eliminating the need for separate painting processes. Despite the importance of post-processing in achieving high-quality 3D printed parts, streamlining post-processing workflows and developing automated finishing techniques remain ongoing challenges in the field of additive manufacturing. The complexity and variability of parts produced through additive manufacturing present unique challenges for automation and standardization of post-processing operations. Additionally, the diverse range of materials, geometries, and applications further complicates the development of universal post-processing solutions. However, advancements in robotics, machine vision, and digitalization are driving innovation in automated finishing technologies, offering promising opportunities for improving efficiency, consistency, and quality in post-processing workflows. Furthermore, integrating post-processing considerations into the design phase of additive manufacturing can help minimize the need for extensive post-processing and reduce overall production costs. Designing parts with self-supporting features, optimized surface finishes, and integrated textures or patterns can mitigate post-processing requirements and improve printability. Additionally, leveraging advanced design software and simulation tools enables designers to anticipate and optimize for post-processing considerations, such as support structures, orientation, and surface roughness, during the design iteration process.

Post-processing steps such as support removal, surface finishing, and painting are essential for achieving high-quality and functional 3D printed parts. Streamlining post-processing workflows and developing automated finishing techniques are ongoing challenges in the field of additive manufacturing. However, advancements in automation, robotics, and digitalization offer promising opportunities for improving efficiency, consistency, and quality in post-processing operations. By integrating post-processing considerations into the design phase and leveraging advanced technologies, manufacturers can optimize post-processing workflows, reduce production costs, and enhance the overall quality of 3D printed parts.

Opportunities of 3D Printing:

The advent of 3D printing, also known as additive manufacturing, has sparked a revolution in various industries, offering unprecedented opportunities for innovation, customization, and efficiency. Unlike traditional subtractive manufacturing methods, which involve cutting away material from a solid block, 3D printing builds objects layer by layer from digital designs. This technology has transformed the way products are designed, prototyped, and manufactured, opening up new possibilities across a wide range of sectors. In this section, we will explore the opportunities presented by 3D printing in manufacturing and beyond, examining its impact on product development, supply chain management, healthcare, aerospace, automotive, and education.

i. **Product Development and Rapid Prototyping**:

One of the most significant opportunities of 3D printing lies in product development and rapid prototyping. Traditional manufacturing methods often involve lengthy and costly prototyping processes, which can slow down innovation and time-to-market. 3D printing enables designers and engineers to quickly iterate designs, produce prototypes on-demand, and test concepts with minimal lead time. This accelerated product development cycle allows companies to bring new products to market faster and more cost-effectively, giving them a competitive edge in today's fast-paced business environment.

Utilizing 3D printing in any phase of product development or rapid prototyping process gives advanced solutions and parts within a very short period of time. It is always hard to determine, with all competitiveness, to choose ways of creating mind-blowing innovative solutions and quality parts. With utilizing 3D printing, it is much easier to develop new ideas, iterate concepts, and produce products in comparison with conventional methods. The importance of 3D printing lies in the realm of rapid prototyping, which encourages iterative design processes in less time with more complex details and designs, easy simulated ergonomic tests, less costly mold developments, visualization of internal components and assembling, and to validate the product itself. With a series of evolutions and types of 3D printing, which will be discussed further, product developers and engineers may choose a suitable 3D printing method to produce prototypes or even up to end products.

In the current global market scenario, it has been a common need between enterprises or even small-medium industries to get their product to the market in the fastest time frame. With traditional or conventional methods, it is really time-consuming to develop a prototype and much more on tooling for end products. 3D printing will expedite time to market by producing rapid and low-cost prototypes in the early phase of product development. With no additional tooling preparation, parts can be built directly from design to function, and assembly tests can be performed simultaneously.

3D printing makes it easier to create custom products where mass production would not be cost-effective. It has been used in many manufacturing processes to reduce costs. There have been recent medical breakthroughs with 3D printing as well. It has provided doctors the ability to create custom devices for specific individuals with a new, more complex, and less costly method. Rapid prototyping is an umbrella term used for a host of technologies that can create parts and tools in an automated manner, with the use of 3D data directly from a design. The use of 3D data is done using computer-aided design (CAD) or animation modeling software and usually results in a 3D printed output. This technology helps in product visualization, design conceptualization, and product simulation. A designer can create an idea and the computer can produce a model, thus making it easy to change and manipulate the product. This is a relatively inexpensive and time-saving process. The next step in the process is taking the conceptualized idea to an actual 3D model. This is usually done using CNC or many layers of 2D slices. The recent idea is implementing the 2D image onto 3D using 3D printing. This can significantly cut costs and time. Numerous factors are responsible for the upsurge of development in the fields of product development and rapid prototyping. During the last decade, the war of competition has focused on the time factor. Customers are demanding high-quality products in less time. This is the era of time-based competition. This has led to the development of newer technologies in design and manufacturing. A very popular and widely used technique today is 3D printing. 3D printing has revolutionized the way designs are being developed and executed for new product development. Many companies are using this technology for new product development as well as for improvement in the existing product design.

ii. **Customization and Personalization**:

Another key opportunity offered by 3D printing is customization and personalization. Traditional manufacturing processes are typically geared towards mass production, resulting in standardized products with limited variation. In contrast, 3D printing allows for the creation of highly customized and personalized products tailored to individual preferences and requirements. From customized medical implants to personalized consumer goods, 3D printing empowers consumers to participate in the design process and create products that meet their unique needs, preferences, and specifications.

The advent of 3D printing technology has opened up unprecedented opportunities for customization and personalization across various industries. Unlike traditional manufacturing methods, which often rely on mass production of standardized products, 3D printing, also known as additive manufacturing, allows for the creation of highly customized and personalized items tailored to individual preferences and requirements. This has led to a paradigm shift in product design, manufacturing, and consumer engagement, offering businesses new avenues for innovation and differentiation. In this section, we will explore the opportunities presented by 3D printing in customization and personalization, examining its impact on product design, consumer experiences, and market trends. 3D printing, also known as additive manufacturing, has revolutionized the manufacturing industry by offering unparalleled opportunities for customization and personalization. Unlike traditional manufacturing processes, which are often geared towards mass production of standardized products, 3D printing enables the creation of highly customized and personalized items tailored to individual preferences and requirements. This transformative capability has far-reaching implications across various sectors, from healthcare and consumer goods to aerospace and automotive industries. This section will explore the key opportunities presented by 3D printing in customization and personalization, highlighting its impact on product design, manufacturing, and consumer engagement.

One of the most significant applications of 3D printing in healthcare is the production of customized medical implants. Traditional implants, such as hip or knee replacements, are typically manufactured in standard sizes and shapes, which may not perfectly match the patient's anatomy. With 3D printing, medical implants can be precisely customized based on patient-specific data obtained from medical imaging, such as CT scans or MRIs. This personalized approach ensures a perfect fit and alignment, reducing the risk of complications and improving patient outcomes. Furthermore, 3D printing enables the fabrication of complex geometries and porous structures that promote osseointegration and tissue ingrowth, enhancing the long-term performance of medical implants. In the consumer goods industry, 3D printing has enabled the creation of personalized products tailored to individual tastes and preferences. Consumers can now customize various items, including jewelry, accessories, footwear, and home decor, by choosing specific designs, colors, and materials. Online platforms and retail stores offer customization tools that allow customers to personalize their products in real-time, creating a unique shopping experience. Additionally, 3D printing enables on-demand manufacturing, eliminating the need for large inventories and reducing waste associated with mass production. This direct-to-consumer approach enables brands to build stronger relationships with their customers and cater to niche markets with specialized products. 3D printing has also made significant inroads in the fashion and apparel industry, enabling designers to create bespoke garments and accessories with intricate details and complex geometries. Designers leverage 3D printing technology to produce custom-fit clothing, footwear, and accessories that conform to the wearer's body shape and size. This personalized approach to fashion not only enhances comfort and fit but also opens up new avenues for creative expression and artistic innovation. Furthermore, 3D printing allows for the integration of functional elements, such as ventilation channels or cushioning structures, into garments, enhancing performance and functionality.

4. **Customized Automotive Parts**:

In the automotive industry, 3D printing is revolutionizing the production of customized and specialty parts for vehicles. From interior components and exterior trim to engine components and prototypes, 3D printing enables manufacturers to create bespoke parts with intricate designs and optimized performance characteristics. Automotive enthusiasts and aftermarket suppliers use 3D printing to produce custom modifications, upgrades, and accessories for vehicles, catering to individual preferences and aesthetic preferences. Moreover, 3D printing facilitates rapid prototyping and iterative design processes, allowing engineers to test and refine designs quickly before mass production. In the field of education, 3D printing has emerged as a powerful tool for creating personalized learning aids and educational resources. Teachers and educators use 3D printers to produce visual aids, tactile models, and interactive simulations that enhance student engagement and comprehension. For example, 3D-printed anatomical models enable students to explore complex anatomical structures in a hands-on manner, facilitating learning and retention. Additionally, 3D printing allows for the customization of educational materials to meet the diverse needs of students with different learning styles and abilities, promoting inclusive and accessible education. In product development and prototyping, 3D printing offers unparalleled flexibility and speed, enabling designers and engineers to create customized prototypes and iterate designs rapidly. Unlike traditional prototyping methods, which are often time-consuming and costly, 3D printing allows for the quick production of functional prototypes with complex geometries and intricate details. This accelerated prototyping process enables companies to bring products to market faster and more cost-effectively, reducing lead times and enhancing innovation. Additionally, 3D printing facilitates customization of prototypes to meet specific project requirements and client feedback, ensuring a more iterative and collaborative design process.

3D printing presents a myriad of opportunities for customization and personalization across various industries and applications. From customized medical implants and personalized consumer goods to bespoke fashion and automotive parts, 3D printing enables the creation of tailored products that meet individual preferences and requirements. This transformative technology empowers consumers to participate in the design process and co-create products that reflect their unique tastes and lifestyles. As 3D printing continues to evolve and become more accessible, the possibilities for customization and personalization are virtually limitless, paving the way for a more personalized and consumer-centric future.

1. **Customized Product Design**:

The advent of 3D printing has ushered in a new era of customization and personalization, offering designers and consumers unprecedented freedom to create highly tailored and unique products. Unlike traditional manufacturing methods, which often produce standardized items with limited variation, 3D printing enables the fabrication of custom products with intricate designs, complex geometries, and personalized features. This transformative capability has revolutionized various industries, from jewelry and furniture to consumer electronics and beyond, empowering designers to unleash their creativity and cater to the diverse preferences and requirements of today's consumers.

One of the most notable applications of 3D printing in customization and personalization is in the jewelry industry. Traditionally, jewelry production involved labor-intensive processes such as casting, molding, and handcrafting, limiting design possibilities and customization options. With 3D printing, designers can create intricate and bespoke jewelry pieces that reflect individual tastes and styles. Using computer-aided design (CAD) software, designers can digitally sculpt jewelry designs with precise detail and complexity. These digital designs are then translated into physical objects using 3D printing technology, allowing for the production of one-of-a-kind pieces that cannot be replicated through traditional methods. Whether it's personalized wedding bands, custom-engraved pendants, or unique statement earrings, 3D printing enables designers to offer consumers truly unique and meaningful jewelry that resonates with their personal stories and preferences. In the realm of furniture design, 3D printing has opened up new possibilities for creating bespoke and customizable pieces that cater to individual lifestyles and spaces. Traditional furniture manufacturing typically involves mass production of standardized designs, which may not always align with the specific needs and preferences of consumers. 3D printing allows designers to experiment with unconventional shapes, materials, and textures, resulting in furniture pieces that are both aesthetically striking and functionally innovative. Whether it's custom-designed chairs, personalized lighting fixtures, or modular shelving systems, 3D printing enables designers to push the boundaries of traditional furniture design and offer consumers highly customized solutions that enhance their living environments and reflect their unique personalities. The consumer electronics industry has also embraced 3D printing as a means of delivering highly customized and personalized products to consumers. From smartphone cases and laptop accessories to wearable devices and smart home gadgets, 3D printing allows for the creation of customized accessories and add-ons that complement and enhance electronic devices. Designers can tailor the form, function, and aesthetics of these accessories to meet individual preferences and lifestyle needs. For example, 3D-printed smartphone cases can be customized with personalized designs, colors, and textures, providing consumers with a unique and stylish way to protect their devices. Similarly, wearable devices such as fitness trackers and smartwatches can be personalized with custom-fit bands and personalized engraving, offering users a personalized and comfortable wearing experience. 3D printing has also revolutionized the home decor industry by enabling the creation of unique and customizable accessories and decor items. From decorative vases and wall art to functional kitchen gadgets and tableware, 3D printing allows designers to explore innovative designs and materials that were previously inaccessible. Designers can experiment with novel shapes, patterns, and textures to create custom decor items that add personality and character to living spaces. Whether it's a custom-designed lampshade, a personalized wall clock, or a one-of-a-kind sculpture, 3D printing enables designers to bring their creative visions to life and offer consumers highly personalized home decor solutions that reflect their individual tastes and lifestyles. In the fashion industry, 3D printing is revolutionizing the way clothing and accessories are designed, produced, and customized. Traditionally, fashion design and production have been constrained by standardized sizing and limited customization options. With 3D printing, designers can create custom-fit garments, footwear, and accessories that conform to the unique measurements and proportions of individual consumers. This personalized approach to fashion not only enhances comfort and fit but also allows for greater creative expression and individuality. Designers can experiment with innovative materials, textures, and patterns to create bespoke fashion pieces that reflect the wearer's personality and style. Whether it's a custom-made dress, a personalized pair of shoes, or a unique handbag, 3D printing empowers consumers to express themselves through fashion and embrace their individuality.

In conclusion, 3D printing offers unparalleled opportunities for customization and personalization across a wide range of industries and applications. From customized jewelry and bespoke furniture to personalized consumer electronics and unique home decor, 3D printing enables designers to push the boundaries of creativity and offer consumers truly unique and meaningful products that resonate with their individual preferences and lifestyles. As 3D printing technology continues to advance and become more accessible, the possibilities for customization and personalization are virtually limitless, ushering in a new era of consumer-centric design and manufacturing.

2. **Tailored Consumer Experiences**:

The integration of 3D printing technology into business operations has transformed the way companies engage with consumers, offering unprecedented opportunities for tailored consumer experiences and mass customization. By leveraging 3D printing, businesses can empower consumers to participate in the design process and customize products to their liking, fostering interactive and engaging interactions that enhance brand loyalty and customer satisfaction. This section will examine how 3D printing enables businesses to offer personalized consumer experiences, facilitates mass customization, and strengthens brand-consumer relationships.

One of the key ways in which 3D printing enables tailored consumer experiences is through online customization platforms and design tools. These platforms allow consumers to personalize products by selecting colors, materials, and features according to their preferences. Whether it's customizing the design of a smartphone case, selecting the color of a pair of shoes, or adding personalized engravings to a piece of jewelry, online customization platforms offer consumers unprecedented control over the final product. Through intuitive user interfaces and interactive design tools, consumers can experiment with different options and visualize the customized product in real-time, ensuring that it meets their expectations before placing an order. This level of customization not only enhances the consumer experience but also creates a sense of ownership and emotional attachment to the product. By allowing consumers to participate in the design process, 3D printing technology empowers individuals to express their creativity and create products that reflect their personality and style. Unlike traditional manufacturing methods, which often limit customization options, 3D printing enables consumers to bring their unique visions to life and create one-of-a-kind products that are truly tailored to their preferences. Whether it's designing custom jewelry, personalized home decor, or bespoke fashion accessories, consumers can unleash their creativity and design products that are as unique as they are. This level of creative freedom fosters a deeper connection between consumers and the products they purchase, leading to greater satisfaction and loyalty.

In addition to enabling individualized consumer experiences, 3D printing facilitates mass customization, allowing businesses to produce customized products at scale. Unlike traditional manufacturing processes, which rely on mass production of standardized items, 3D printing enables on-demand manufacturing of personalized products in response to customer orders. This eliminates the need for large inventories of pre-made products, reducing storage costs and minimizing the risk of overstocking. Furthermore, 3D printing allows businesses to offer a wider range of customization options without significantly increasing production complexity or costs. By leveraging digital design files and additive manufacturing technology, companies can efficiently produce customized products on-demand, ensuring that each item meets the unique preferences and requirements of individual consumers. The ability to customize products to their liking enhances customer satisfaction and strengthens brand loyalty. When consumers are able to personalize products according to their preferences, they feel a greater sense of ownership and connection to the brand. This emotional attachment translates into higher levels of satisfaction and loyalty, as consumers are more likely to return to a brand that offers personalized experiences and products that resonate with their individual tastes and preferences. Additionally, the interactive and engaging nature of online customization platforms creates a positive shopping experience, further enhancing customer satisfaction and encouraging repeat purchases. By offering tailored consumer experiences and personalized products, businesses can strengthen their relationships with consumers and build brand loyalty. When consumers feel that a brand understands their unique needs and preferences, they are more likely to develop a sense of trust and loyalty towards that brand. 3D printing allows businesses to engage with consumers on a deeper level, inviting them to co-create products and become active participants in the design process. This collaborative approach fosters a sense of community and belonging among consumers, as they feel that their input is valued and their voices are heard. As a result, consumers are more likely to become brand advocates and ambassadors, spreading positive word-of-mouth and driving further growth and success for the brand.

In conclusion, 3D printing enables businesses to offer tailored consumer experiences, facilitate mass customization, and strengthen brand-consumer relationships. By leveraging online customization platforms and design tools, businesses can empower consumers to participate in the design process and create products that reflect their personality and style. This level of customization enhances customer satisfaction, fosters brand loyalty, and drives growth and success for businesses in an increasingly competitive marketplace. As 3D printing technology continues to advance and become more accessible, the opportunities for tailored consumer experiences and mass customization are virtually limitless, offering businesses new ways to engage with consumers and differentiate themselves in the market.

3. **Personalized Healthcare Solutions**:

The integration of 3D printing technology in the healthcare industry has revolutionized patient care, offering unparalleled opportunities for personalized treatment solutions tailored to individual patient needs. By leveraging 3D printing, healthcare professionals can create custom-fitted orthopedic implants, patient-specific medical devices, and personalized drug formulations, leading to improved patient outcomes and enhanced quality of life. This section will examine how 3D printing is transforming healthcare by enabling personalized treatment solutions and enhancing clinical outcomes.

One of the most significant applications of 3D printing in healthcare is in the fabrication of custom-fitted orthopedic implants. Traditional orthopedic implants are often manufactured in standard sizes and shapes, which may not always fit the unique anatomy of individual patients. With 3D printing, medical imaging techniques such as CT scans and MRI scans can be used to generate detailed 3D models of patient anatomy. These models serve as the basis for designing and fabricating custom implants that precisely match the patient's anatomy. Whether it's a hip replacement, knee replacement, or spinal fusion surgery, custom-fitted orthopedic implants reduce the risk of complications such as implant loosening, implant wear, and soft tissue damage. By improving the fit and alignment of implants, 3D printing enables surgeons to achieve better surgical outcomes and enhance patient mobility and function. In addition to orthopaedic implants, 3D printing enables the production of patient-specific medical devices tailored to individual anatomical variations and clinical requirements. This includes devices such as prosthetics, dental implants, and craniofacial implants, which can be customized to fit the unique contours of a patient's body. By using 3D printing technology, healthcare providers can design and fabricate medical devices that are optimized for patient comfort, functionality, and aesthetics. For example, 3D-printed prosthetic limbs can be precisely tailored to match the residual limb shape and size of an amputee, improving fit and comfort compared to off-the-shelf prosthetics. Similarly, 3D-printed dental implants can be customized to match the shape and color of natural teeth, enhancing the aesthetic outcome of dental restorations. By offering patient-specific medical devices, 3D printing enhances patient satisfaction and improves overall treatment outcomes.

Another promising application of 3D printing in healthcare is in the production of personalized drug formulations tailored to individual patient requirements and dosage needs. Traditional pharmaceutical manufacturing methods typically involve mass production of standardized drug formulations, which may not always meet the specific needs of individual patients. With 3D printing, pharmaceutical companies can create custom drug formulations that take into account factors such as patient age, weight, medical history, and drug tolerance. This personalized approach to drug manufacturing allows clinicians to optimize drug dosages and delivery methods based on individual patient characteristics, improving treatment efficacy and minimizing side effects. Additionally, 3D printing enables the production of complex drug formulations, such as multi-layered tablets and sustained-release formulations, which are difficult to achieve using traditional manufacturing techniques. By offering personalized drug formulations, 3D printing enhances medication adherence, reduces medication errors, and improves patient compliance with treatment regimens.

Beyond the fabrication of implants and medical devices, 3D printing plays a crucial role in precision medicine and treatment planning. By using patient-specific anatomical models generated from medical imaging data, clinicians can visualize complex anatomical structures and plan surgical procedures with greater precision and accuracy. Whether it's a complex cardiac surgery, a neurosurgical procedure, or a reconstructive surgery, 3D-printed anatomical models enable surgeons to simulate surgical interventions, optimize surgical approaches, and anticipate potential challenges before entering the operating room. This proactive approach to treatment planning reduces surgical risks, shortens operating times, and improves clinical outcomes for patients. Additionally, 3D-printed anatomical models can be used for patient education and informed consent, allowing patients to better understand their condition and treatment options. In addition to improving patient outcomes, 3D printing in healthcare has the potential to reduce healthcare costs and optimize resource utilization. By customizing treatment solutions and streamlining clinical workflows, 3D printing minimizes the need for repeat surgeries, revisions, and post-operative complications, leading to cost savings for healthcare providers and payers. Furthermore, the on-demand nature of 3D printing allows for efficient inventory management and reduces the need for stockpiling of medical implants and devices. With 3D printing, healthcare facilities can produce implants and devices as needed, minimizing waste and reducing the financial burden associated with maintaining large inventories. Additionally, the scalability of 3D printing technology enables cost-effective production of personalized medical devices, making advanced treatments more accessible and affordable for patients.

In conclusion, 3D printing offers unparalleled opportunities for personalized treatment solutions in the healthcare industry. By enabling the creation of custom-fitted orthopedic implants, patient-specific medical devices, personalized drug formulations, and anatomical models for treatment planning, 3D printing enhances clinical outcomes, improves patient satisfaction, and reduces healthcare costs. As 3D printing technology continues to advance and become more widespread, its impact on healthcare delivery and patient care will continue to grow, ushering in a new era of personalized medicine and precision healthcare.

4. **Customized Fashion and Apparel**:

In the fashion and apparel industry, 3D printing offers opportunities for customization and personalization, allowing designers to create custom-fit garments and accessories that cater to individual body shapes and sizes. Traditional manufacturing methods often involve producing clothing in standard sizes, which may not fit all body types equally well. 3D printing enables designers to create custom-fit garments using body scanning technology and digital design tools, ensuring a perfect fit for each individual customer. Additionally, 3D printing allows for the creation of intricate and complex textile patterns and structures that cannot be achieved using traditional textile manufacturing techniques. This opens up new possibilities for creative expression and design innovation in the fashion industry.

The fashion and apparel industry is experiencing a transformative shift with the integration of 3D printing technology, offering unprecedented opportunities for customization, personalization, and design innovation. Traditionally, clothing manufacturing has relied on mass production of standardized garments in limited sizes, which may not always accommodate the diverse range of body shapes and sizes. However, with 3D printing, designers can overcome these limitations and create custom-fit garments and accessories that cater to individual preferences and anatomies. This section will explore how 3D printing is revolutionizing the fashion industry by enabling personalized clothing solutions, fostering creative expression, and driving design innovation.

One of the most significant advantages of 3D printing in the fashion industry is its ability to create custom-fit garments that perfectly match the unique contours of an individual's body. Traditional clothing manufacturing methods often rely on standardized sizing charts, resulting in garments that may not fit all body types equally well. With 3D printing, designers can utilize body scanning technology to capture precise measurements of a customer's body, which can then be used to generate digital models for custom garment production. These digital models are translated into physical garments using 3D printing technology, ensuring a perfect fit and superior comfort for the wearer. Whether it's a custom-tailored suit, a made-to-measure dress, or personalized activewear, 3D printing allows designers to offer bespoke clothing solutions that cater to the unique needs and preferences of each individual customer. In addition to custom-fit garments, 3D printing enables the creation of personalized accessories and embellishments that add a unique touch to fashion ensembles. From custom jewelry and handbags to bespoke footwear and eyewear, 3D printing offers endless possibilities for creative expression and individualization. Designers can leverage 3D printing technology to experiment with novel shapes, textures, and materials, resulting in accessories that stand out and make a statement. Whether it's a pair of 3D-printed earrings with intricate geometric patterns or a personalized smartphone case with custom engraving, 3D printing allows consumers to express their personality and style through fashion accessories that are as unique as they are.

3D printing technology opens up new avenues for design innovation and creative freedom in the fashion industry. Unlike traditional textile manufacturing techniques, which are often limited by the constraints of weaving, knitting, and sewing, 3D printing allows designers to create intricate and complex textile patterns and structures that were previously impossible to achieve. With 3D printing, designers can experiment with unconventional materials, textures, and forms, pushing the boundaries of traditional fashion design and creating garments that are truly innovative and avant-garde. Whether it's a 3D-printed dress adorned with geometric shapes and sculptural elements or a pair of shoes with intricate lattice patterns, 3D printing enables designers to explore new possibilities and redefine the aesthetics of fashion. In addition to fostering design innovation and creative expression, 3D printing offers opportunities for sustainable fashion practices. Traditional clothing manufacturing processes are often resource-intensive and generate significant waste, particularly in the form of fabric scraps and offcuts. With 3D printing, designers can minimize material waste by precisely controlling the amount of material used in the production process. Unlike traditional manufacturing methods, which require cutting and sewing fabric pieces together, 3D printing builds garments layer by layer, using only the necessary amount of material. This reduces material waste and contributes to a more sustainable fashion ecosystem. Additionally, 3D printing enables designers to produce garments on-demand, eliminating the need for large inventories and reducing the environmental impact associated with overproduction and excess inventory. Furthermore, 3D printing democratizes fashion design by making the design process more accessible and inclusive. Traditionally, fashion design has been a highly specialized and exclusive field, limited to professional designers and industry insiders. However, with the advent of 3D printing technology and digital design tools, anyone with access to a computer and a 3D printer can create their own fashion designs from the comfort of their own home. This democratization of fashion design empowers individuals to express their creativity and design their own clothing and accessories, bypassing the traditional barriers to entry in the fashion industry. Whether it's a hobbyist experimenting with DIY fashion projects or an aspiring designer launching their own fashion label, 3D printing enables greater participation and diversity in the fashion design process.

In conclusion, 3D printing is revolutionizing the fashion industry by enabling personalized clothing solutions, fostering design innovation, and promoting sustainable fashion practices. From custom-fit garments and personalized accessories to creative design possibilities and democratized fashion design, 3D printing offers endless opportunities for creative expression and individualization. As 3D printing technology continues to advance and become more accessible, its impact on the fashion industry will continue to grow, reshaping the way we design, produce, and consume fashion in the twenty-first century.

5. **Personalized Consumer Electronics**:

In the consumer electronics industry, 3D printing offers opportunities for customization and personalization, allowing consumers to create custom-designed electronic devices and accessories tailored to their specific needs and preferences. From customized smartphone cases to personalized wearable devices, 3D printing enables consumers to design and manufacture unique electronic products that reflect their personality and style. Additionally, 3D printing enables the production of custom-fit ergonomic designs that improve user comfort and usability. This personalized approach to product design and manufacturing enhances consumer satisfaction and loyalty, driving demand for customized electronic devices and accessories. 3D printing presents vast opportunities for customization and personalization across various industries, from product design and manufacturing to healthcare and consumer electronics. By enabling designers to create highly customized and personalized products, 3D printing allows for greater creative expression, consumer engagement, and brand differentiation. Additionally, 3D printing empowers consumers to participate in the design process and create products that meet their individual preferences and requirements. As technology continues to advance and adoption rates increase, the opportunities presented by 3D printing in customization and personalization are boundless, offering endless possibilities for innovation, creativity, and consumer empowerment.

The consumer electronics industry is witnessing a paradigm shift with the integration of 3D printing technology, offering unprecedented opportunities for customization, personalization, and innovation. Traditionally, consumer electronics have been mass-produced in standardized designs, limiting consumer choice and customization options. However, with 3D printing, consumers can now design and create custom-designed electronic devices and accessories that cater to their specific needs and preferences. This section will explore how 3D printing is revolutionizing the consumer electronics industry by enabling personalized products, fostering creative expression, and driving design innovation.

1. **Custom-Designed Electronic Devices**:

One of the most significant advantages of 3D printing in the consumer electronics industry is its ability to create custom-designed electronic devices tailored to individual preferences and requirements. Traditional electronics manufacturing methods often involve producing devices in standardized designs, which may not always meet the diverse needs of consumers. With 3D printing, consumers can now design and fabricate electronic devices that are optimized for their specific use cases and preferences. Whether it's a custom-designed smartphone case, a personalized laptop stand, or a bespoke gaming controller, 3D printing allows consumers to create electronic devices that reflect their unique style and functionality requirements.

In addition to custom-designed electronic devices, 3D printing enables the creation of personalized accessories and embellishments that complement consumer electronics products. From custom-designed smartphone cases and tablet stands to personalized headphone holders and charging docks, 3D printing offers endless possibilities for creative expression and individualization. Consumers can leverage 3D printing technology to design and fabricate accessories that match their personal style, preferences, and lifestyle. Whether it's adding custom logos, textures, or patterns to electronic accessories, 3D printing allows consumers to personalize their devices and make them uniquely their own. 3D printing technology opens up new avenues for design innovation and creative freedom in the consumer electronics industry. Unlike traditional manufacturing methods, which are often limited by the constraints of mass production and tooling, 3D printing enables designers to create intricate and complex electronic components and enclosures with unparalleled precision and detail. With 3D printing, designers can experiment with novel shapes, textures, and materials, pushing the boundaries of traditional electronics design and creating products that are truly innovative and unique. Whether it's a modular smart home device, a customizable wearable gadget, or a futuristic drone prototype, 3D printing allows designers to explore new possibilities and redefine the aesthetics of consumer electronics.

Furthermore, 3D printing offers opportunities for sustainable electronics manufacturing practices. Traditional electronics manufacturing processes are often resource-intensive and generate significant waste, particularly in the form of plastic waste from injection molding and CNC machining. With 3D printing, manufacturers can minimize material waste by precisely controlling the amount of material used in the production process. Additionally, 3D printing enables on-demand manufacturing of electronic components and accessories, eliminating the need for large inventories and reducing the environmental impact associated with overproduction and excess inventory. By adopting 3D printing technology, consumer electronics companies can reduce their carbon footprint, minimize waste, and contribute to a more sustainable future. Moreover, 3D printing democratizes electronics design by making the design process more accessible and inclusive. Traditionally, electronics design has been a highly specialized and exclusive field, limited to professional engineers and designers. However, with the advent of 3D printing technology and digital design tools, anyone with access to a computer and a 3D printer can create their own electronic devices and accessories. This democratization of electronics design empowers individuals to express their creativity and design custom electronics solutions for their specific needs and preferences. Whether it's a hobbyist experimenting with DIY electronics projects or a startup launching their own line of custom-designed gadgets, 3D printing enables greater participation and diversity in the electronics design process.

In conclusion, 3D printing is revolutionizing the consumer electronics industry by enabling personalized products, fostering design innovation, and promoting sustainability. From custom-designed electronic devices and personalized accessories to creative design possibilities and democratized electronics design, 3D printing offers endless opportunities for customization, personalization, and creativity. As 3D printing technology continues to advance and become more accessible, its impact on the consumer electronics industry will continue to grow, reshaping the way we design, produce, and interact with electronic devices in the digital age.

iii. Supply Chain Management and Distributed Manufacturing:

3D printing has the potential to revolutionize supply chain management and distribution networks by enabling distributed manufacturing. Traditional manufacturing relies on centralized production facilities and extensive logistics networks to transport goods to various locations. In contrast, 3D printing decentralizes production, allowing products to be manufactured on-site or near the point of consumption. This reduces lead times, transportation costs, and inventory holding costs, while also minimizing the environmental impact associated with long-distance shipping. Furthermore, distributed manufacturing can enhance supply chain resilience by reducing reliance on centralized production facilities and mitigating risks associated with disruptions in global supply chains.

The rise of 3D printing technology has brought about transformative opportunities in supply chain management and distributed manufacturing. Traditionally, manufacturing has been centralized, with production facilities located in specific regions and products transported to various locations through extensive supply chains. However, 3D printing, also known as additive manufacturing, enables decentralized production by allowing products to be manufactured on-site or near the point of consumption. This shift has the potential to revolutionize supply chain management, reduce lead times, lower transportation costs, and enhance supply chain resilience. In this section, we will explore the opportunities presented by 3D printing in supply chain management and distributed manufacturing, examining its impact on inventory management, customization, and sustainability.

1. **On-Demand Manufacturing and Inventory Management**:

One of the most significant opportunities of 3D printing in supply chain management is the ability to produce goods on-demand, eliminating the need for large inventories and reducing inventory holding costs. Traditional manufacturing methods often require companies to produce goods in bulk and maintain extensive inventories to meet fluctuating demand. This can lead to overstocking, obsolescence, and inventory write-offs. 3D printing enables companies to produce goods as needed, in response to customer orders or demand signals, reducing the need for large inventories and minimizing the risk of excess inventory. By shifting from mass production to on-demand manufacturing, companies can achieve significant cost savings, improve inventory turnover, and increase responsiveness to changing market conditions.

One of the most significant opportunities of 3D printing in supply chain management lies in its ability to revolutionize the traditional manufacturing model by enabling the production of goods on-demand. This transformative capability eliminates the need for maintaining large inventories of finished goods, thereby reducing inventory holding costs and streamlining the supply chain process. Traditional manufacturing processes typically involve producing goods in large batches, which requires manufacturers to maintain significant inventory levels to meet fluctuating demand. However, this approach often results in excess inventory, storage costs, and the risk of obsolescence. Moreover, long lead times and supply chain disruptions can further exacerbate inefficiencies and increase operational costs. In contrast, 3D printing offers a decentralized and agile manufacturing solution that allows companies to produce goods precisely when and where they are needed. By leveraging additive manufacturing technology, companies can quickly and cost-effectively produce customized and complex products on-demand, eliminating the need for large-scale production runs and warehousing facilities. One of the key advantages of on-demand manufacturing through 3D printing is the ability to reduce inventory holding costs. With traditional manufacturing methods, companies must maintain large inventories of finished goods to meet anticipated demand, tying up capital and incurring storage costs. However, with 3D printing, companies can produce goods on-demand in response to actual customer orders, minimizing the need for excess inventory and reducing carrying costs. Furthermore, 3D printing enables companies to implement a just-in-time (JIT) manufacturing approach, where products are manufactured only when there is demand for them. This lean manufacturing strategy helps to optimize inventory levels, reduce lead times, and improve supply chain efficiency. By aligning production with actual demand, companies can minimize inventory holding costs while ensuring timely delivery of products to customers. Additionally, 3D printing facilitates localization and customization of production, allowing companies to manufacture goods closer to the point of consumption and tailor products to meet specific customer requirements. This enables companies to respond quickly to changes in customer preferences, market trends, and demand fluctuations, enhancing agility and competitiveness in the marketplace. Moreover, 3D printing offers opportunities for supply chain resilience and risk mitigation. By decentralizing production and reducing reliance on centralized manufacturing facilities, companies can mitigate the impact of disruptions such as natural disasters, geopolitical events, and supply chain disruptions. Furthermore, the digital nature of 3D printing enables companies to store and transfer design files electronically, reducing the risk of intellectual property theft and counterfeit products. Despite the numerous benefits of on-demand manufacturing through 3D printing, several challenges remain to be addressed. These include limitations in materials, process scalability, and production speed, as well as regulatory and quality assurance considerations. Additionally, the upfront investment in 3D printing technology and expertise may pose barriers to adoption for some companies. 3D printing offers significant opportunities for transforming supply chain management by enabling on-demand manufacturing of goods. By eliminating the need for large inventories and reducing inventory holding costs, 3D printing helps companies optimize their supply chain operations, improve agility, and enhance competitiveness in the marketplace. As 3D printing technology continues to advance and mature, its impact on supply chain management is expected to grow, driving innovation and reshaping the future of manufacturing.

2. **Customization and Personalization**:

Another key opportunity presented by 3D printing in supply chain management is the ability to customize and personalize products to meet individual customer preferences and requirements. Traditional manufacturing methods often involve producing standardized products with limited variation, which may not fully meet the diverse needs of today's consumers. 3D printing allows for the creation of highly customized and personalized products tailored to individual specifications, whether it's customized medical devices, personalized consumer goods, or bespoke industrial components. By leveraging the design flexibility afforded by 3D printing, companies can offer a wider range of product options, differentiate themselves in the market, and meet the growing demand for personalized products.

Another key opportunity presented by 3D printing in supply chain management is the ability to customize and personalize products to meet individual customer preferences and requirements. Traditional manufacturing processes often rely on mass production of standardized products, which may not fully meet the diverse and evolving needs of consumers. However, with 3D printing technology, companies can leverage additive manufacturing to create highly customized and personalized products that cater to the unique preferences and specifications of each customer. One of the primary advantages of 3D printing in customization is its ability to produce complex geometries and intricate designs that are difficult or impossible to achieve with traditional manufacturing methods. By layering materials in a precise and controlled manner, 3D printers can create intricate shapes, textures, and patterns with unparalleled precision and detail. This allows companies to offer a wide range of customizable options, from personalized jewelry and fashion accessories to bespoke medical devices and architectural models. Moreover, 3D printing enables companies to implement a mass customization strategy, where products are tailored to individual customer preferences while still benefiting from economies of scale. By leveraging digital design files and automated manufacturing processes, companies can efficiently produce customized products in small batch sizes, reducing setup costs and lead times associated with traditional manufacturing methods. This flexibility allows companies to offer personalized products at competitive prices, enhancing customer satisfaction and loyalty. Furthermore, 3D printing facilitates rapid prototyping and iteration of product designs, enabling companies to quickly develop and test new concepts before full-scale production. This iterative design process allows companies to gather feedback from customers and stakeholders early in the product development cycle, reducing the risk of costly design errors and rework. By accelerating the product development process, 3D printing helps companies bring innovative and customized products to market faster, gaining a competitive edge in the marketplace. Additionally, 3D printing technology enables on-demand production of customized products, eliminating the need for companies to maintain large inventories of finished goods. Instead of stocking pre-made products in anticipation of customer orders, companies can produce items as needed, reducing inventory holding costs and minimizing the risk of overstocking or obsolescence. This lean manufacturing approach helps companies optimize their supply chain operations, improve cash flow, and respond quickly to changes in customer demand. Moreover, 3D printing offers opportunities for co-creation and collaboration between companies and customers, empowering consumers to participate in the design and customization process. Through online platforms and design tools, customers can customize products according to their preferences, selecting colors, materials, and features to create personalized items that reflect their individual style and personality. This interactive and participatory approach to product customization enhances customer engagement and brand loyalty, driving repeat business and word-of-mouth referrals. Despite the numerous benefits of product customization through 3D printing, several challenges remain to be addressed. These include limitations in material selection, production speed, and quality assurance, as well as the need for investment in 3D printing technology and expertise. Additionally, companies must carefully manage intellectual property rights and data privacy concerns when engaging customers in the customization process. 3D printing technology offers significant opportunities for customization and personalization in supply chain management. By leveraging additive manufacturing, companies can create highly customized and personalized products that meet individual customer preferences and requirements. From customized consumer goods to bespoke industrial components, 3D printing enables companies to differentiate their offerings, enhance customer satisfaction, and drive innovation in the marketplace. As 3D printing technology continues to advance and become more accessible, its role in product customization and supply chain management is expected to grow, reshaping the future of manufacturing and commerce.

3. **Reducing Transportation Costs and Carbon Footprint**:

3D printing, also known as additive manufacturing, has emerged as a transformative technology with the potential to revolutionize traditional manufacturing and distribution networks. One significant opportunity presented by 3D printing is the ability to reduce transportation costs and carbon emissions associated with the production and distribution of goods.

Traditional manufacturing processes typically involve centralized production facilities that mass-produce goods in large quantities before shipping them to distribution centers and retail outlets. This centralized manufacturing model often requires extensive transportation networks to move raw materials, components, and finished products between various locations, resulting in significant transportation costs and carbon emissions. In contrast, 3D printing enables decentralized and localized manufacturing, where products are produced on-site or near the point of consumption. By leveraging additive manufacturing technology, companies can fabricate goods directly from digital design files, eliminating the need for long-distance transportation of raw materials and finished products. This localized production model reduces transportation distances, transportation-related costs, and greenhouse gas emissions associated with traditional supply chains. Moreover, 3D printing offers opportunities for supply chain optimization and inventory reduction, further reducing the need for transportation and warehousing. With traditional manufacturing methods, companies often maintain large inventories of finished goods to meet fluctuating demand and ensure timely delivery to customers. However, this approach can lead to excess inventory, inventory holding costs, and the risk of obsolescence. With 3D printing, companies can implement a just-in-time (JIT) manufacturing approach, where products are manufactured on-demand in response to actual customer orders. By producing goods only when there is demand for them, companies can minimize inventory levels, reduce the need for warehousing facilities, and eliminate the need for long-distance transportation of finished products. This lean manufacturing strategy optimizes supply chain efficiency, reduces carbon footprint, and enhances sustainability. Furthermore, 3D printing enables companies to implement sustainable materials and design practices, further reducing the environmental impact of manufacturing and distribution. Additive manufacturing technologies allow for the use of recycled materials, bio-based polymers, and environmentally friendly additives, minimizing resource consumption and waste generation. Additionally, 3D printing enables designers to optimize product designs for material efficiency, lightweighting, and durability, reducing material usage and energy consumption throughout the product lifecycle. Additionally, 3D printing offers opportunities for distributed manufacturing and on-demand production of spare parts and replacement components, reducing the need for centralized manufacturing facilities and global supply chains. Instead of sourcing spare parts from distant suppliers or warehouses, companies can produce parts locally using 3D printing technology, reducing lead times, transportation costs, and carbon emissions associated with traditional supply chains. Despite the numerous benefits of 3D printing in reducing transportation costs and carbon emissions, several challenges remain to be addressed. These include limitations in material selection, process scalability, and production speed, as well as the need for investment in 3D printing technology and expertise. Additionally, companies must consider regulatory requirements, intellectual property rights, and quality assurance considerations when implementing 3D printing in their manufacturing operations. 3D printing offers significant opportunities to reduce transportation costs and carbon emissions associated with traditional manufacturing and distribution networks. By enabling localized production, supply chain optimization, and sustainable manufacturing practices, 3D printing helps companies minimize their environmental footprint while enhancing operational efficiency and competitiveness in the marketplace. As 3D printing technology continues to advance and become more widespread, its role in reducing transportation-related costs and emissions is expected to grow, driving innovation and sustainability in the manufacturing industry.

4. **Enhancing Supply Chain Resilience**:

3D printing, also referred to as additive manufacturing, serves as a powerful tool for enhancing supply chain resilience by decentralizing production and reducing reliance on centralized manufacturing facilities. This innovative technology mitigates risks associated with disruptions in global supply chains, such as natural disasters, geopolitical instability, and transportation bottlenecks, which can severely impact traditional manufacturing methods.

Traditional manufacturing processes often rely on centralized production facilities that mass-produce goods in large quantities before distributing them to various locations. However, this centralized manufacturing model is susceptible to disruptions, as a single point of failure can lead to widespread supply chain disruptions and shortages. Natural disasters, such as earthquakes, hurricanes, or floods, can damage or destroy production facilities, resulting in production delays and supply shortages. Moreover, geopolitical instability and trade disputes can disrupt global supply chains, leading to tariffs, trade restrictions, and export/import bans that hinder the movement of goods and raw materials. Additionally, transportation bottlenecks, such as port closures, shipping delays, or fuel shortages, can disrupt the flow of goods and increase lead times, impacting production schedules and supply chain efficiency. In contrast, 3D printing enables distributed manufacturing and localized production, reducing the reliance on centralized production facilities and mitigating the risks associated with global supply chain disruptions. With additive manufacturing technology, companies can produce goods on-site or near the point of consumption, eliminating the need for long-distance transportation of raw materials and finished products. By decentralizing production, 3D printing enhances supply chain resilience by reducing the vulnerability to disruptions at specific locations. Even in the event of a disaster or disruption at one manufacturing site, production can be quickly shifted to alternative locations or distributed across multiple facilities, minimizing the impact on supply chain operations and ensuring continuity of production. Furthermore, 3D printing offers opportunities for rapid response and agile production in the face of supply chain disruptions. With traditional manufacturing methods, companies may face challenges in quickly adjusting production schedules or reconfiguring production lines to meet changing demand or unforeseen disruptions. However, with 3D printing, companies can rapidly prototype, iterate, and produce parts and products on-demand, allowing for flexible and responsive manufacturing in dynamic and uncertain environments. Additionally, 3D printing enables companies to implement inventory-free manufacturing, where products are produced on-demand in response to actual customer orders. This eliminates the need for maintaining large inventories of finished goods, reducing the risk of excess inventory, inventory holding costs, and inventory write-offs in the event of supply chain disruptions. Despite the numerous benefits of 3D printing in enhancing supply chain resilience, several challenges remain to be addressed. These include limitations in material selection, process scalability, and production speed, as well as the need for investment in 3D printing technology and expertise. Additionally, companies must consider regulatory requirements, intellectual property rights, and quality assurance considerations when implementing 3D printing in their manufacturing operations. 3D printing offers significant opportunities to enhance supply chain resilience by reducing reliance on centralized production facilities and mitigating risks associated with disruptions in global supply chains. By decentralizing production, enabling rapid response, and implementing inventory-free manufacturing, 3D printing helps companies build more agile, flexible, and resilient supply chains capable of withstanding unforeseen disruptions and uncertainties. As 3D printing technology continues to advance and become more widespread, its role in enhancing supply chain resilience is expected to grow, driving innovation and transformation in the manufacturing industry.

5. **Empowering Small and Medium-Sized Enterprises (SMEs):**

3D printing also empowers small and medium-sized enterprises (SMEs) by providing access to advanced manufacturing technologies and enabling agile, flexible production processes. Traditionally, SMEs may lack the resources and infrastructure to compete with larger companies in global markets. However, 3D printing levels the playing field by enabling SMEs to produce high-quality, customized products without the need for large-scale production facilities. This allows SMEs to respond quickly to changing market demands, innovate new products, and compete effectively with larger competitors. Additionally, 3D printing enables SMEs to engage in collaborative manufacturing networks, sharing design files, and production capabilities with other companies to leverage collective resources and expertise. 3D printing presents vast opportunities for innovation and improvement in supply chain management and distributed manufacturing. By enabling on-demand manufacturing, customization, and localization of production, 3D printing offers companies the ability to reduce inventory costs, enhance product customization, and minimize transportation emissions. Additionally, 3D printing enhances supply chain resilience by decentralizing production and reducing reliance on centralized manufacturing facilities. As technology continues to advance and adoption rates increase, the opportunities presented by 3D printing in supply chain management and distributed manufacturing are boundless, offering endless possibilities for improving efficiency, sustainability, and resilience in the global supply chain.

iv. **Healthcare and Medical Applications**

In the field of healthcare, 3D printing offers transformative opportunities for medical device manufacturing, personalized medicine, and tissue engineering. 3D printing technology enables the production of patient-specific implants, prosthetics, and medical devices tailored to individual anatomical specifications. This personalized approach improves patient outcomes, reduces surgical complications, and enhances quality of life. Additionally, 3D printing is revolutionizing the field of regenerative medicine by enabling the fabrication of complex tissues and organs using bioinks and living cells. This has the potential to revolutionize organ transplantation and address the growing demand for donor organs. The emergence of 3D printing technology has ushered in a new era of innovation and advancement in healthcare and medical applications. With its ability to fabricate intricate and customized objects from digital designs, 3D printing has revolutionized various aspects of patient care, medical device manufacturing, and biomedical research. In this section, we will explore the myriad opportunities presented by 3D printing in healthcare, examining its impact on patient treatment, surgical planning, prosthetics, organ transplantation, and biomedical research. A summary of the key application of 3D printing in the healthcare and medical is shown in the Fig. 4.1

1. **Patient-Specific Treatment and Surgical Planning.**

One of the most significant opportunities offered by 3D printing in healthcare is the ability to create patient-specific anatomical models for treatment planning and surgical simulation. Medical imaging techniques such as CT scans and MRI scans can be used to generate detailed 3D models of patient anatomy, which can then be 3D printed to provide surgeons with a tangible, tactile representation of the patient's anatomy. These models enable surgeons to visualize complex anatomical structures, plan surgical procedures, and practice intricate maneuvers in a risk-free environment. By simulating surgeries on 3D-printed models, surgeons can anticipate potential challenges, optimize surgical approaches, and improve patient outcomes.

Patient-specific treatment and surgical planning have revolutionized the field of medicine, offering tailored interventions that optimize outcomes and minimize risks for

Fig. 4.1 Schematic of 3D
printing application in
healthcare and medical

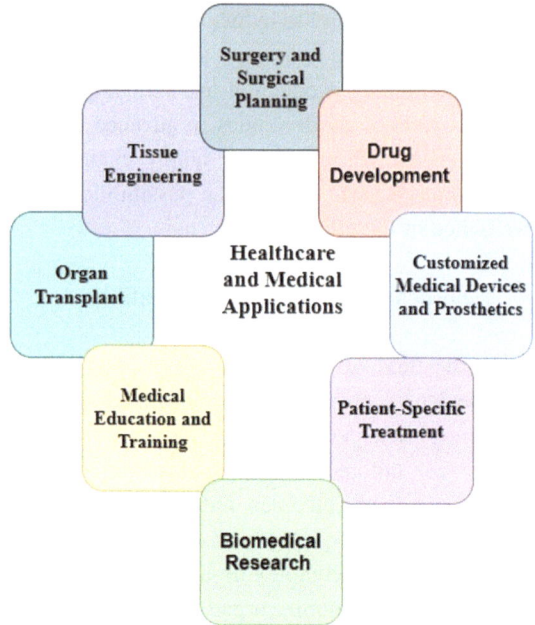

individual patients. This approach integrates advanced imaging, computational modeling, and data analytics to customize treatment strategies based on the unique anatomy, physiology, and pathology of each patient. At the heart of patient-specific treatment is the recognition that medical conditions manifest differently in each individual. While traditional treatment approaches often rely on generalized protocols or population-based guidelines, patient-specific interventions account for the variability and complexity inherent in human biology. By harnessing innovative technologies and methodologies, healthcare providers can personalize treatment plans to address the specific needs and characteristics of each patient. Advanced imaging techniques play a central role in patient-specific treatment and surgical planning. Modalities such as magnetic resonance imaging (MRI), computed tomography (CT), and three-dimensional (3D) reconstruction enable detailed visualization of anatomical structures and pathological abnormalities. High-resolution imaging provides clinicians with comprehensive insights into the spatial relationships, dimensions, and variations of tissues and organs, guiding precise diagnosis and treatment decision-making. Computational modeling and simulation further augment patient-specific treatment strategies by simulating physiological processes, predicting treatment outcomes, and optimizing surgical interventions. Finite element analysis (FEA), computational fluid dynamics (CFD), and biomechanical modeling enable virtual simulations of surgical procedures, implant designs, and treatment responses, allowing clinicians to explore different scenarios and refine treatment plans before actual intervention.

Data analytics and machine learning algorithms enhance the capabilities of patient-specific treatment by leveraging large datasets and clinical insights to inform decision-making and personalize interventions. By analyzing patient-specific data, including medical history, genetic profiles, and biomarkers, machine learning algorithms can identify patterns, predict treatment responses, and optimize therapeutic strategies tailored to individual patients. Patient-specific treatment and surgical planning offer numerous benefits over conventional approaches. By customizing treatment strategies to the unique characteristics of each patient, healthcare providers can optimize treatment outcomes, minimize complications, and improve patient satisfaction. Personalized interventions also enhance efficiency by reducing the need for trial-and-error approaches, optimizing resource utilization, and streamlining healthcare delivery processes. Moreover, patient-specific treatment promotes innovation and advances in medical technology by driving research and development efforts towards personalized medicine solutions. By fostering collaboration between clinicians, engineers, and scientists, patient-specific approaches stimulate interdisciplinary innovation, leading to the development of novel technologies, techniques, and therapies that improve patient care and outcomes. Despite the transformative potential of patient-specific treatment, several challenges must be addressed to realize its full benefits. These include technological limitations, such as the need for advanced imaging equipment ($500,000–$3,000,000), computational resources ($1000–$10,000 per simulation), and expertise in data analytics and modeling. Interoperability issues and data privacy concerns also pose barriers to the seamless integration of patient-specific approaches into clinical practice. Additionally, the adoption of patient-specific treatment may require changes in healthcare policies, reimbursement models, and regulatory frameworks to incentivize innovation, ensure patient access, and safeguard ethical considerations. Moreover, ongoing education and training ($10,000–$50,000 per healthcare provider) are essential to equip healthcare providers with the knowledge and skills necessary to effectively utilize patient-specific technologies and methodologies in their practice. Patient-specific treatment and surgical planning represent a paradigm shift in healthcare, offering personalized interventions that optimize outcomes and improve patient care. By harnessing advanced imaging, computational modeling, and data analytics, healthcare providers can tailor treatment strategies to the unique needs of each patient, driving innovation, efficiency, and excellence in clinical practice. While challenges exist, the continued advancement and adoption of patient-specific approaches hold promise for transforming the future of medicine and enhancing patient outcomes.

2. **Customized Medical Devices and Prosthetics**:

Another significant opportunity presented by 3D printing in healthcare is the ability to create customized medical devices and prosthetics tailored to individual patient needs. Traditional manufacturing methods often involve mass-producing standardized devices that may not fully meet the unique anatomical or functional requirements of every patient.

3D printing enables the production of patient-specific implants, prosthetics, and medical devices customized to fit the patient's anatomy precisely. Whether it's a customized dental implant, a patient-specific orthopedic implant, or a personalized hearing aid, 3D printing allows for the creation of bespoke medical solutions that improve patient comfort, functionality, and quality of life.

Customized medical devices and prosthetics have emerged as transformative solutions in modern healthcare, offering tailored interventions that enhance patient comfort, functionality, and quality of life. These personalized devices are designed to meet the unique anatomical, physiological, and lifestyle needs of individual patients, revolutionizing the treatment of a wide range of medical conditions and disabilities. At the heart of customized medical devices and prosthetics is the principle of patient-centered care, which emphasizes the importance of considering the individual preferences, characteristics, and circumstances of each patient in treatment decision-making. Unlike traditional off-the-shelf devices, which may provide a one-size-fits-all solution, customized medical devices and prosthetics are precisely engineered to match the specific requirements and specifications of the patient, resulting in superior performance and outcomes. Advanced technologies, such as 3D printing, computer-aided design (CAD), and additive manufacturing, play a pivotal role in the fabrication of customized medical devices and prosthetics. These technologies enable healthcare providers to create highly accurate, patient-specific models and prototypes, which can be used to design and manufacture devices with unparalleled precision and customization.

In the field of orthopedics, customized prosthetic limbs and orthoses are revolutionizing mobility and rehabilitation for individuals with limb loss or musculoskeletal disorders. By incorporating patient-specific measurements, anatomical contours, and functional requirements, customized prosthetics can optimize fit, comfort, and performance, enabling users to regain independence and mobility with greater ease and confidence. Similarly, in dentistry, customized dental implants, crowns, and orthodontic appliances are transforming the treatment of dental conditions and disorders. Utilizing digital scanning and CAD/CAM technologies, dental professionals can design and fabricate restorations that seamlessly integrate with the patient's natural dentition, resulting in superior aesthetics, functionality, and longevity compared to conventional dental prosthetics. In the field of ophthalmology, customized intraocular lenses (IOLs) and contact lenses are enhancing visual outcomes for patients with refractive errors, cataracts, and other ocular conditions. By precisely matching the patient's ocular anatomy and visual requirements, customized lenses can correct vision more accurately and effectively than standard lenses, minimizing aberrations and optimizing visual acuity. Furthermore, in the realm of rehabilitation and assistive technology, customized devices such as wheelchairs, braces, and exoskeletons are empowering individuals with disabilities to overcome physical barriers and participate more fully in daily activities. By tailoring these devices to the individual's unique needs, preferences, and functional abilities, customized solutions can improve mobility, posture, and overall quality of life for users. The benefits of customized medical devices and prosthetics extend beyond individual patients to healthcare providers, who can deliver more personalized and effective care, and society as a whole, which benefits from improved health outcomes and reduced healthcare costs. By embracing advanced technologies and patient-centered approaches, healthcare systems can harness the power of customization to address the diverse and evolving needs of patients across a wide range of medical specialties.

Despite the numerous advantages of customized medical devices and prosthetics, challenges remain in terms of accessibility, affordability, and regulatory considerations. Access to customized solutions may be limited by factors such as geographic location, financial resources, and healthcare infrastructure, particularly in underserved and resource-constrained settings. Moreover, the high cost of customized devices and the lack of insurance coverage or reimbursement may pose barriers to adoption for some patients. Additionally, regulatory frameworks governing the design, manufacturing, and distribution of customized medical devices and prosthetics must strike a balance between ensuring safety, efficacy, and quality, while also promoting innovation, accessibility, and affordability. Clear guidelines and standards are needed to govern the use of advanced technologies, protect patient rights and safety, and facilitate the adoption of customized solutions in clinical practice. Customized medical devices and prosthetics represent a paradigm shift in healthcare, offering personalized interventions that improve patient outcomes, enhance quality of life, and promote inclusivity and independence for individuals with medical conditions and disabilities. By leveraging advanced technologies and patient-centered approaches, healthcare providers can deliver more effective, tailored care, while also driving innovation and advancing the field of personalized medicine. Despite challenges, the continued development and adoption of customized solutions hold promise for transforming the future of healthcare and improving the lives of patients worldwide.

3. **Organ Transplantation and Tissue Engineering**:

3D printing has also opened up new possibilities for organ transplantation and tissue engineering, offering hope to patients in need of life-saving organ transplants. Biofabrication techniques such as bioprinting enable the fabrication of complex tissues and organs using bioinks composed of living cells. By layering cells in precise patterns, researchers can create functional tissues and organs that mimic the structure and function of natural tissues. While fully functional organs for transplantation are still in the early stages of development, 3D-printed tissues and organoids are already being used for drug screening, disease modeling, and regenerative medicine applications. In the future, 3D bioprinting holds the potential to revolutionize organ transplantation by providing a limitless supply of patient-specific organs and tissues, eliminating the need for donor organs and reducing transplant waiting times.

3D printing, also known as additive manufacturing, has emerged as a groundbreaking technology with vast potential in the field of medicine. In recent years, it has opened up new possibilities for organ transplantation and tissue engineering, offering hope to patients in need of life-saving organ transplants. Traditionally, organ transplantation has been limited by the scarcity of donor organs and the risk of rejection. Many patients languish on waiting lists for years, hoping for a suitable donor match. However, 3D printing has revolutionized the landscape of organ transplantation by enabling the fabrication of patient-specific organs and tissues, tailored to match the recipient's anatomy and immunological profile. One of the most promising applications of 3D printing in organ transplantation is the fabrication of bioengineered organs using a patient's own cells. This approach, known as bioprinting, involves layer-by-layer deposition of living cells, biomaterials, and bioactive factors to create complex three-dimensional structures that mimic the architecture and function of natural organs. Bioprinting holds immense potential for addressing the critical shortage of donor organs and reducing the risk of rejection in transplant recipients. By using the patient's own cells as building blocks, bioprinted organs can circumvent the need for immunosuppressive drugs and minimize the risk of immune rejection, improving long-term transplant outcomes and quality of life for patients. Researchers have made significant strides in bioprinting a variety of organs and tissues, including skin, bone, cartilage, blood vessels, and even complex organs such as the liver, kidney, and heart. These bioengineered tissues and organs can be used for transplantation, regenerative medicine, drug testing, and disease modeling, offering new avenues for personalized medicine and precision healthcare. In addition to bioprinting, 3D printing technology has facilitated the development of patient-specific implants and prostheses for organ transplantation. By precisely mapping the patient's anatomy using advanced imaging techniques such as magnetic resonance imaging (MRI) or computed tomography (CT), clinicians can design and fabricate custom implants that perfectly fit the recipient's unique anatomy. For example, 3D-printed cranial implants are used to reconstruct cranial defects in patients undergoing neurosurgical procedures, while 3D-printed titanium hip implants are employed in hip replacement surgeries to restore mobility and alleviate pain. These patient-specific implants offer superior fit, stability, and biocompatibility compared to traditional off-the-shelf implants, leading to better clinical outcomes and patient satisfaction.

Moreover, 3D printing technology enables the fabrication of scaffolds and matrices for tissue engineering applications. These scaffolds serve as a framework for the growth and regeneration of new tissue, providing mechanical support and guiding cellular behavior. By precisely controlling the architecture, porosity, and composition of the scaffold, researchers can create biomimetic environments that promote tissue regeneration and repair. In the realm of tissue engineering, 3D-printed scaffolds have been used to regenerate damaged tissues and organs, such as bone, cartilage, and skin. These bioactive scaffolds can be seeded with patient-derived cells and growth factors, facilitating tissue integration and functional recovery. In some cases, 3D-printed scaffolds have been implanted directly into patients to promote tissue regeneration and accelerate healing. Despite the remarkable progress in 3D printing technology, several challenges remain to be addressed before widespread clinical adoption can be realized. These include the need for standardized protocols and regulatory approval processes, the optimization of printing materials and techniques for biomedical applications, and the scalability and cost-effectiveness of bioprinting technologies. Furthermore, ethical considerations surrounding the use of bioprinted organs and tissues, such as consent, equity, and access, must be carefully addressed to ensure equitable distribution and ethical use of these technologies. Additionally, long-term studies are needed to evaluate the safety, efficacy, and durability of 3D-printed organs and tissues in clinical settings. 3D printing technology has revolutionized the field of organ transplantation and tissue engineering, offering new hope to patients in need of life-saving interventions. By enabling the fabrication of patient-specific organs, implants, and scaffolds, 3D printing has the potential to transform the landscape of healthcare, ushering in an era of personalized medicine and regenerative therapies. While challenges remain, the continued advancement of 3D printing technology holds promise for improving patient outcomes and revolutionizing the practice of medicine.

4. **Medical Education and Training**:

3D printing is also transforming medical education and training by providing students and healthcare professionals with hands-on learning experiences and realistic simulation tools. Medical schools and teaching hospitals are integrating 3D printing into their curricula to teach anatomy, surgical procedures, and medical device design. 3D-printed anatomical models allow students to study complex anatomical structures, practice surgical techniques, and develop diagnostic skills in a safe and controlled environment. Additionally, 3D-printed surgical simulators and training phantoms enable surgeons to hone their skills, refine their techniques, and stay current with the latest advancements in medical technology. By providing realistic, high-fidelity training experiences, 3D printing enhances the competency and confidence of healthcare professionals, ultimately improving patient care and safety.

3D printing technology has made significant inroads into medical education and training, revolutionizing the way healthcare professionals learn, practice, and prepare for clinical practice. By leveraging advanced imaging, computer-aided design (CAD), and additive manufacturing techniques, 3D printing has enabled the creation of highly realistic anatomical models, surgical simulators, and educational tools that enhance learning outcomes and competency development across various medical disciplines. One of the most impactful applications of 3D printing in medical education is the creation of anatomical models for teaching and training purposes. These models, generated from patient imaging data such as CT scans or MRIs, provide students and trainees with hands-on experience in visualizing and understanding complex anatomical structures in three dimensions. By replicating the intricacies of human anatomy with unparalleled accuracy and detail, 3D-printed anatomical models facilitate interactive learning, promote spatial understanding, and enhance retention of anatomical knowledge among learners. Moreover, 3D-printed anatomical models serve as valuable tools for surgical planning and rehearsal, allowing surgeons to visualize patient-specific anatomy and practice surgical techniques in a risk-free environment. By incorporating patient-specific pathology, such as tumors, fractures, or congenital anomalies, these models enable surgeons to develop personalized treatment strategies, optimize surgical approaches, and mitigate potential complications before entering the operating room. This enhances patient safety, improves surgical outcomes, and reduces operative time and healthcare costs. In addition to anatomical models, 3D printing technology has enabled the development of patient-specific surgical guides and instruments, which enhance precision and accuracy in surgical procedures. These guides are custom-designed to fit the patient's anatomy and provide real-time feedback to surgeons during procedures, guiding incisions, implant placement, and tissue resection with submillimeter accuracy. By streamlining surgical workflows and reducing intraoperative variability, patient-specific surgical guides improve surgical outcomes and minimize the risk of errors and complications. Furthermore, 3D printing has revolutionized medical training through the creation of high-fidelity surgical simulators and procedural trainers. These simulators replicate the look, feel, and tactile feedback of real surgical procedures, allowing trainees to practice surgical techniques and develop proficiency in a controlled and standardized environment. By simulating a wide range of surgical scenarios, from basic suturing to complex procedures such as laparoscopic surgery or endovascular interventions, these simulators enhance skill acquisition, competency development, and confidence among trainees, ultimately improving patient care and safety. Beyond surgical training, 3D printing technology has expanded into other areas of medical education, including dentistry, radiology, and emergency medicine. In dentistry, 3D-printed dental models and prostheses are used for preoperative planning, orthodontic treatment, and dental implantology, enhancing precision and predictability in dental procedures. In radiology, 3D-printed anatomical models and pathology replicas aid in the interpretation and communication of diagnostic imaging findings, facilitating interdisciplinary collaboration and patient education. In emergency medicine, 3D-printed trauma models and procedural trainers enable trainees to practice critical procedures such as airway management, chest tube insertion, and emergency cricothyroidotomy, improving readiness and preparedness for real-life emergencies.

Despite the numerous benefits of 3D printing in medical education and training, several challenges remain to be addressed. These include the standardization of 3D printing workflows and materials, the integration of 3D printing into existing curricula and accreditation standards, and the scalability and affordability of 3D printing technologies in educational settings. Additionally, ongoing research is needed to evaluate the effectiveness of 3D-printed educational tools in enhancing learning outcomes, clinical skills, and patient care. 3D printing technology has revolutionized medical education and training, offering innovative solutions for teaching, learning, and skill development across various medical specialties. By providing realistic anatomical models, patient-specific surgical guides, and procedural simulators, 3D printing enhances the quality, effectiveness, and efficiency of medical education, ultimately improving patient outcomes and advancing the practice of medicine. As 3D printing continues to evolve and mature, its role in medical education is poised to expand, driving innovation and excellence in healthcare education and training.

5. **Biomedical Research and Drug Development**:

In the realm of biomedical research, 3D printing offers opportunities for innovation and discovery in areas such as drug delivery, tissue engineering, and disease modeling. Researchers are using 3D printing to fabricate complex drug delivery systems, personalized drug formulations, and tissue-engineered constructs for studying disease mechanisms and drug responses. By precisely controlling the composition, structure, and properties of 3D-printed materials, researchers can create sophisticated models of human physiology and pathology, enabling them to better understand disease processes and develop more effective treatments. Additionally, 3D printing enables the rapid prototyping of lab-on-a-chip devices, microfluidic systems, and diagnostic tools for biomedical research and clinical diagnostics. 3D printing presents vast opportunities for innovation and advancement in healthcare and medical applications. From patient-specific treatment planning and surgical simulation to customized medical devices and prosthetics, 3D printing is revolutionizing patient care and improving outcomes. Additionally, 3D printing holds promise for organ transplantation, tissue engineering, medical education, and biomedical research, offering new solutions to longstanding challenges in healthcare. As technology continues to advance and adoption rates increase, the opportunities presented by 3D printing in healthcare are boundless, offering endless possibilities for improving patient care, advancing medical science, and transforming the future of medicine. 3D printing technology has emerged as a powerful tool in biomedical research and drug development, offering innovative solutions for the design, fabrication, and evaluation of pharmaceuticals, medical devices, and tissue-engineered constructs. By leveraging advanced materials, precise control over geometry, and customizable fabrication processes, 3D printing has revolutionized the way researchers and pharmaceutical companies approach drug discovery, personalized medicine, and regenerative therapies. One of the most significant applications of 3D printing in biomedical research is in the fabrication of customized drug delivery systems. By using 3D printing to create intricate drug formulations, researchers can precisely control drug release kinetics, dosage forms, and spatial distribution, optimizing therapeutic efficacy and minimizing side effects. For example, 3D-printed oral tablets can be designed with complex geometries and tailored drug release profiles to achieve controlled release of medications over extended periods, improving patient compliance and treatment outcomes. Furthermore, 3D printing enables the rapid prototyping and production of medical devices and implants for preclinical testing and evaluation. Researchers can use 3D printing to fabricate prototypes of orthopedic implants, prosthetic limbs, and surgical instruments with precise geometries and patient-specific designs. These prototypes can be used to assess device performance, biocompatibility, and mechanical properties in vitro and in vivo, accelerating the development and commercialization of new medical technologies. In addition to drug delivery systems and medical devices, 3D printing technology has revolutionized tissue engineering and regenerative medicine. Researchers can use 3D printing to fabricate scaffolds and matrices with precise control over pore size, porosity, and surface topography, mimicking the architecture and mechanical properties of native tissues. These 3D-printed scaffolds serve as templates for cell seeding and tissue regeneration, enabling the creation of functional tissues and organs for transplantation, disease modeling, and drug screening applications.

Moreover, 3D bioprinting, a specialized form of 3D printing, allows researchers to create complex three-dimensional structures composed of living cells, biomaterials, and bioactive factors. By layering cell-laden bioinks and supporting materials, researchers can fabricate tissue-engineered constructs with spatially defined cell distributions and vascular networks, resembling the organization and function of native tissues. These bioengineered tissues can be used to study disease mechanisms, screen potential therapeutics, and develop patient-specific treatment strategies, advancing the field of personalized medicine and precision healthcare. Furthermore, 3D printing technology facilitates the creation of anatomically accurate models for preclinical testing and education in biomedical research. Researchers can use 3D printing to fabricate patient-specific anatomical models for surgical planning, medical training, and device validation purposes. These models provide a realistic representation of patient anatomy and pathology, allowing researchers to simulate surgical procedures, test medical devices, and train healthcare professionals in a risk-free and controlled environment. Despite the numerous benefits of 3D printing in biomedical research and drug development, several challenges remain to be addressed. These include the optimization of printing materials and processes for biocompatibility, scalability, and regulatory compliance, the standardization of 3D printing workflows and quality control measures, and the integration of 3D printing technologies into existing research and development pipelines. Additionally, ongoing research is needed to explore the long-term safety, efficacy, and cost-effectiveness of 3D-printed pharmaceuticals, medical devices, and tissue-engineered constructs in clinical settings. 3D printing technology has revolutionized biomedical research and drug development, offering innovative solutions for the design, fabrication, and evaluation of pharmaceuticals, medical devices, and tissue-engineered constructs. By enabling precise control over geometry, material properties, and fabrication processes, 3D printing has accelerated the pace of innovation in biomedical research, paving the way for personalized medicine, regenerative therapies, and advanced drug delivery systems. As 3D printing continues to evolve and mature, its impact on biomedical research and drug development is poised to expand, driving innovation and excellence in healthcare and pharmaceutical industries.

v. **Aerospace and Automotive Industries**

The aerospace and automotive industries are also leveraging the opportunities presented by 3D printing to enhance performance, reduce weight, and streamline production processes. In aerospace, 3D printing enables the production of lightweight and complex components, such as turbine blades and aircraft interiors, using advanced materials like titanium and carbon fiber composites. This results in fuel savings, reduced emissions, and improved efficiency. Similarly, in the automotive industry, 3D printing is used to produce lightweight parts, optimize vehicle design, and customize components for high-performance vehicles. Additive manufacturing also enables rapid prototyping of concept cars and production of spare parts on-demand, reducing inventory costs and lead times.

The aerospace and automotive industries are at the forefront of leveraging the opportunities presented by 3D printing to enhance performance, reduce weight, and streamline production processes. 3D printing, also known as additive manufacturing, has revolutionized the way parts and components are designed, prototyped, and manufactured in these industries, offering numerous benefits that traditional manufacturing methods cannot match. One of the key advantages of 3D printing in the aerospace and automotive industries is its ability to enhance performance. Additive manufacturing allows for the creation of complex geometries and intricate designs that are difficult or impossible to achieve using traditional manufacturing methods. For example, in aerospace, 3D printing enables the production of lightweight components with optimized internal structures, resulting in improved fuel efficiency and reduced emissions. Similarly, in the automotive industry, 3D printed parts can be designed to optimize aerodynamics, reduce drag, and enhance overall vehicle performance. By leveraging the design freedom afforded by 3D printing, manufacturers can create innovative solutions that push the boundaries of performance and efficiency. Another significant advantage of 3D printing in the aerospace and automotive industries is its ability to reduce weight. Traditional manufacturing methods often involve machining parts from solid blocks of material, resulting in excess material waste and added weight. In contrast, 3D printing builds parts layer by layer, only using the material necessary to create the final product. This allows for the production of lightweight components without compromising strength or structural integrity. In aerospace, reducing weight is critical for improving fuel efficiency and extending range, while in automotive, it can lead to better handling, acceleration, and fuel economy. By utilizing 3D printing to lightweight components, manufacturers can achieve significant performance gains and cost savings over the life of the product.

3D printing also offers opportunities to streamline production processes in the aerospace and automotive industries. Traditional manufacturing methods often involve multiple steps, including casting, machining, and assembly, which can be time-consuming and labor-intensive. In contrast, 3D printing consolidates these steps into a single, additive process, reducing lead times and production costs. Additionally, 3D printing enables on-demand manufacturing, allowing manufacturers to produce parts and components as needed, rather than maintaining large inventories. This flexibility reduces inventory costs, minimizes waste, and improves supply chain efficiency. Furthermore, additive manufacturing can enable the production of complex assemblies as a single, integrated component, eliminating the need for assembly and reducing the risk of component failure.

Several notable case studies demonstrate the benefits of 3D printing in the aerospace and automotive industries. For example, in aerospace, companies like Boeing and Airbus have embraced 3D printing to produce lightweight, complex components for aircraft engines, interiors, and structural parts. By leveraging advanced materials like titanium and carbon fiber composites, these companies have achieved significant weight savings and performance improvements. Similarly, in automotive, companies like BMW and Ford are using 3D printing to prototype new vehicle designs, produce customized parts, and optimize production processes. Ford, for instance, has used 3D printing to produce prototype brake calipers, resulting in a 75% reduction in development time and significant cost savings.

The aerospace and automotive industries are capitalizing on the opportunities presented by 3D printing to enhance performance, reduce weight, and streamline production processes. By leveraging the design freedom, weight reduction capabilities, and production efficiencies afforded by additive manufacturing, manufacturers in these industries can create innovative solutions that push the boundaries of performance, efficiency, and sustainability. As technology continues to advance and adoption rates increase, 3D printing is poised to play an even more significant role in shaping the future of aerospace and automotive manufacturing, driving progress, and innovation in these critical sectors.

vi. **Education and Research**

3D printing is also transforming education and research by providing students and researchers with access to advanced prototyping and fabrication tools. Educational institutions are integrating 3D printing into their curriculum to teach design thinking, problem-solving skills, and digital manufacturing techniques. Students can experiment with design concepts, create prototypes, and bring their ideas to life using 3D printing technology. Additionally, researchers are using 3D printing to advance scientific discovery and innovation in fields such as materials science, biotechnology, and robotics. This interdisciplinary approach fosters collaboration, creativity, and knowledge sharing, driving progress and innovation across various disciplines.

The advent of 3D printing technology has brought about transformative opportunities in education and research, revolutionizing the way students learn, researchers innovate, and knowledge is disseminated. With its ability to create physical objects from digital designs, 3D printing has become an invaluable tool in classrooms, laboratories, and research facilities, offering hands-on learning experiences, fostering creativity, and enabling breakthroughs across various disciplines. In this section, we will explore the opportunities presented by 3D printing in education and research, examining its impact on teaching, learning, and scientific discovery. One of the most significant opportunities of 3D printing in education is its ability to facilitate hands-on learning and experiential education. Traditional educational models often rely on lectures and textbooks to convey abstract concepts and theoretical knowledge. 3D printing enables educators to supplement traditional teaching methods with tangible, interactive learning experiences that engage students and promote deeper understanding. By allowing students to design, prototype, and fabricate physical objects, 3D printing fosters creativity, problem-solving skills, and critical thinking abilities. Whether it's building scale models of historical landmarks, creating anatomical models for biology classes, or designing prototypes for engineering projects, 3D printing empowers students to explore concepts in a tangible and meaningful way, enhancing their learning experience and retention of knowledge.

3D printing is also transforming STEM (science, technology, engineering, and mathematics) education by providing students with tools to explore real-world applications of theoretical concepts. Project-based learning activities, such as designing and building robots, constructing architectural models, or simulating geological formations, enable students to apply STEM principles in practical contexts. 3D printing allows students to bring their ideas to life, iterate designs, and test hypotheses in a hands-on, collaborative environment. This approach not only reinforces STEM concepts but also develops essential skills such as teamwork, communication, and problem-solving, preparing students for careers in STEM fields and fostering a culture of innovation and entrepreneurship. Another key opportunity presented by 3D printing in education is access to advanced tools and technologies. Historically, access to prototyping and fabrication equipment has been limited to specialized research facilities and industrial settings. However, 3D printing technology has democratized access to these tools, making them accessible to students, educators, and researchers of all backgrounds and skill levels. Educational institutions are increasingly integrating 3D printers into their curricula, providing students with hands-on experience with cutting-edge technology. Whether it's designing custom experiments in physics, fabricating archaeological artifacts in history, or prototyping inventions in entrepreneurship classes, 3D printing empowers students to explore their interests, pursue their passions, and unleash their creativity. In the realm of research, 3D printing offers unprecedented opportunities for innovation and scientific discovery. Researchers across various disciplines, including materials science, biomedical engineering, and robotics, are leveraging 3D printing to prototype, fabricate, and test novel concepts and prototypes. From designing customized medical implants and prosthetics to developing advanced materials and structures for aerospace applications, 3D printing enables researchers to push the boundaries of what's possible and accelerate the pace of innovation. Additionally, 3D printing facilitates collaboration and knowledge sharing among researchers, enabling interdisciplinary teams to tackle complex challenges and address pressing societal needs. 3D printing presents vast opportunities for education and research, empowering students, educators, and researchers to innovate, collaborate, and make meaningful contributions to society. By providing hands-on learning experiences, fostering creativity, and democratizing access to advanced tools and technologies, 3D printing is revolutionizing the way we teach, learn, and conduct research. Whether it's inspiring the next generation of STEM professionals, enabling groundbreaking discoveries in science and technology, or revolutionizing the way we approach education and research, 3D printing is shaping the future of learning and knowledge creation. As technology continues to advance and adoption rates increase, the opportunities presented by 3D printing in education and research are boundless, offering endless possibilities for exploration, discovery, and growth.

3D printing offers a myriad of opportunities for innovation, customization, and efficiency across various industries. From product development and rapid prototyping to healthcare, aerospace, automotive, and education, the potential applications of 3D printing are vast and diverse. By harnessing the power of additive manufacturing, companies can accelerate innovation, reduce costs, and unlock new possibilities for design, manufacturing, and distribution. As technology continues to advance and adoption rates increase, 3D printing is poised to revolutionize the way we design, make, and consume products, shaping the future of manufacturing and beyond.

4.1.2 Regulatory and Ethical Considerations

In the rapidly evolving landscape of the Fourth Industrial Revolution (4IR), thin films and 3D printing stand out as transformative technologies with immense potential across various industries. However, as these technologies continue to advance, it becomes increasingly crucial to address a range of regulatory and ethical considerations to ensure their responsible and sustainable integration into society.

Regulatory Considerations:

Regulatory agencies play a vital role in establishing and enforcing quality and safety standards for thin film deposition processes and 3D printing technologies. These standards ensure that manufactured products meet specific performance criteria and safety requirements, particularly in sectors such as healthcare, aerospace, and automotive, where reliability is paramount.

Intellectual Property Rights (IPR):

Protecting intellectual property rights is essential in fostering innovation and incentivizing investment in research and development. In the context of thin films and 3D printing, regulatory frameworks must address issues related to patent protection, copyright, and licensing agreements to safeguard creators' rights and encourage continued innovation.

Environmental Regulations:

The adoption of thin film deposition processes and 3D printing technologies can have significant environmental implications, including energy consumption, waste generation, and emissions. Regulatory agencies need to establish environmental regulations to mitigate these impacts, promote sustainability, and ensure compliance with waste disposal and emissions standards.

Health and Safety Regulations:

Certain materials and chemicals used in thin film deposition and 3D printing processes may pose health and safety risks to workers and end-users. Regulatory frameworks should address occupational health and safety concerns by implementing measures to minimize exposure to hazardous substances, establish ergonomic standards, and enforce workplace safety protocols.

Ethical Considerations:

Equitable Access:

Ensuring equitable access to thin film technologies and 3D printing capabilities is essential to prevent widening socioeconomic disparities. Ethical considerations should focus on promoting inclusive innovation by addressing barriers to access such as cost, education, and technological literacy, thereby ensuring that the benefits of these technologies are accessible to all segments of society.

Data Privacy and Security:

The digitalization and connectivity inherent in thin film manufacturing processes and 3D printing workflows raise concerns about data privacy and security. Ethical considerations must address issues related to the collection, storage, and use of sensitive information, including design files, manufacturing parameters, and user data, to safeguard privacy and prevent unauthorized access or misuse of data.

Fair Labor Practices:

As automation and digital technologies reshape manufacturing processes, ethical frameworks must prioritize fair labor practices. This includes protecting workers' rights, promoting job training and reskilling initiatives to address potential job displacement, and ensuring fair compensation and working conditions in the transition to advanced manufacturing technologies. Fair labor practices are critical as automation and digital technologies transform manufacturing processes in the Fourth Industrial Revolution (4IR) era. Ethical frameworks must prioritize the well-being of workers and address the challenges and opportunities associated with technological advancements.

1. **Worker Rights Protection:**
   - Fair labor practices entail protecting fundamental worker rights, including the right to safe working conditions, fair wages, freedom from discrimination, and the right to organize and bargain collectively. Regulatory frameworks and corporate policies should uphold these rights to ensure that workers are treated with dignity and respect.
2. **Job Training and Reskilling:**
   - As automation and digital technologies disrupt traditional manufacturing jobs, it's essential to invest in job training and reskilling initiatives to equip workers with the skills needed for new roles. Employers, government agencies, and educational institutions can collaborate to provide training programs that enable workers to transition to emerging fields such as robotics programming, additive manufacturing, and data analytics.
3. **Addressing Job Displacement:**
   - Ethical frameworks should address concerns about job displacement resulting from automation and digital technologies. While these technologies may eliminate some traditional roles, they also create new opportunities for skilled workers. Fair labor practices involve implementing measures to minimize involuntary job loss, such as offering retraining programs, providing transitional support, and facilitating job placement services.
4. **Fair Compensation:**
   - Fair compensation is essential to ensuring that workers receive adequate remuneration for their contributions. Ethical frameworks should advocate for fair wages that reflect the value of labor and enable workers to meet their basic needs and maintain a decent standard of living. Additionally, companies should consider implementing profit-sharing programs or other forms of incentive compensation to align the interests of workers with the success of the organization.
5. **Working Conditions:**
   - Ethical frameworks must prioritize the provision of safe, healthy, and inclusive working conditions for all employees. This includes ensuring compliance with occupational health and safety regulations, addressing workplace hazards, and promoting diversity and inclusion. Companies should foster a culture of safety and respect where workers feel empowered to raise concerns and contribute to continuous improvement initiatives.
6. **Collective Bargaining and Worker Representation:**
   - Fair labor practices support the rights of workers to engage in collective bargaining and representation through trade unions or other worker organizations. Ethical frameworks should recognize the importance of collective action in negotiating fair wages, benefits, and working conditions, and protecting workers' interests in the face of technological change and economic uncertainty.

7. **Worker Well-being and Work-Life Balance:**
   - Beyond traditional compensation and benefits, fair labor practices also encompass promoting worker well-being and achieving a healthy work-life balance. Employers should offer support programs such as employee assistance programs, wellness initiatives, and flexible work arrangements to help employees manage stress, maintain mental and physical health, and achieve greater satisfaction and productivity in their roles. By prioritizing fair labor practices, companies can navigate the challenges of technological disruption while upholding the rights and well-being of their workers. Fair treatment of employees not only fosters a positive organizational culture but also enhances productivity, innovation, and long-term sustainability. In the transition to advanced manufacturing technologies, ethical frameworks that emphasize fairness, equity, and respect for human dignity are essential for creating a more inclusive and equitable future of work.

Responsible Innovation:

Ethical considerations in thin films and 3D printing should emphasize responsible innovation practices that prioritize social and environmental sustainability. This entails ethical sourcing of materials, transparent supply chains, lifecycle assessments, and stakeholder engagement to mitigate potential risks and maximize positive impacts on society and the environment. Responsible innovation in thin films and 3D printing involves adopting a holistic approach that not only focuses on technological advancements but also prioritizes social and environmental sustainability. Here are some key aspects of responsible innovation in these fields:

1. **Ethical Sourcing of Materials**: Responsible innovation begins with the ethical sourcing of materials used in thin films and 3D printing processes. This includes ensuring that raw materials are obtained from suppliers who adhere to ethical labor practices, respect human rights, and minimize environmental impact during extraction and production. Companies should strive to source materials from sustainable and socially responsible suppliers, thereby reducing the risk of supporting unethical practices in the supply chain.

2. **Transparent Supply Chains:** Transparency in the supply chain is essential for responsible innovation. Companies should maintain transparency regarding the origin of materials, manufacturing processes, and distribution channels. This transparency allows stakeholders, including consumers, to make informed decisions about the products they purchase and supports efforts to address social and environmental concerns throughout the supply chain.

3. **Lifecycle Assessments:** Conducting lifecycle assessments (LCAs) is integral to responsible innovation in thin films and 3D printing. LCAs evaluate the environmental impact of products or processes throughout their entire lifecycle, from raw material extraction to disposal. By conducting thorough LCAs, companies can identify areas where

improvements can be made to minimize environmental footprint, reduce resource consumption, and mitigate adverse impacts on ecosystems and communities.

4. **Stakeholder Engagement:** Engaging with stakeholders, including employees, local communities, regulatory agencies, and non-governmental organizations (NGOs), is essential for responsible innovation. By soliciting feedback and incorporating diverse perspectives into decision-making processes, companies can better understand the social and environmental implications of their activities and develop strategies to address concerns and maximize positive impacts. Stakeholder engagement fosters transparency, accountability, and trust, ultimately contributing to more sustainable and socially responsible outcomes.

5. **Technology Assessment:** Responsible innovation requires ongoing assessment of technological developments to anticipate and address potential risks and unintended consequences. Companies should proactively evaluate the ethical, social, and environmental implications of new technologies, processes, and products, and take appropriate measures to minimize negative impacts and maximize benefits. This may involve conducting risk assessments, implementing safeguards, and adhering to ethical principles such as the precautionary principle and the principle of do no harm.

6. **Regulatory Compliance:** Responsible innovation necessitates compliance with relevant laws, regulations, and industry standards. Companies must stay abreast of evolving regulatory requirements related to environmental protection, worker safety, data privacy, and intellectual property rights, among others. Compliance with regulations helps ensure that innovations are developed and deployed in a manner that is ethical, legal, and socially responsible, minimizing the risk of harm to individuals, communities, and the environment.

By embracing responsible innovation practices, companies can leverage the transformative potential of thin films and 3D printing technologies while minimizing negative impacts and maximizing positive contributions to society and the environment. Responsible innovation is not only a moral imperative but also a strategic imperative, as companies that prioritize sustainability and social responsibility are better positioned to create long-term value, build trust with stakeholders, and drive innovation that benefits people and the planet.

Addressing regulatory and ethical considerations is paramount for the responsible adoption and integration of thin films and 3D printing technologies in the Fourth Industrial Revolution. By establishing robust regulatory frameworks and ethical guidelines, stakeholders can navigate the complex challenges and opportunities of these transformative technologies while upholding principles of safety, equity, privacy, and sustainability.

4.2 Summary

Thin films and 3D printing represent two cutting-edge technologies with transformative potential across various industries. While both offer numerous opportunities for innovation and advancement, they also present unique challenges that must be addressed to fully harness their capabilities. Thin films face challenges related to achieving uniformity and consistency in deposition processes, mechanical durability, adhesion to substrates, scalability, and cost-effectiveness. Overcoming these challenges is crucial for realizing the full potential of thin film technology in applications such as electronics, solar energy, and healthcare. However, thin films also offer opportunities for lightweight, flexible, and customizable solutions that can drive innovation in these industries and contribute to sustainability and efficiency. Similarly, 3D printing presents challenges such as limited material availability, inconsistent part quality, scalability, production speed, postprocessing requirements, and intellectual property concerns. Addressing these challenges requires advancements in materials science, process optimization, automation, and legal frameworks. Despite these challenges, 3D printing offers numerous opportunities for innovation in manufacturing, design, and customization across industries such as aerospace, automotive, healthcare, and consumer products.

In conclusion, while thin films and 3D printing face distinct challenges, they also offer significant opportunities for disruptive innovation and advancement. By overcoming technical limitations, improving process efficiency, and addressing regulatory and intellectual property issues, these technologies have the potential to revolutionize industries and drive sustainable growth and development. Continued research, collaboration, and investment are essential for unlocking the full potential of thin films and 3D printing and realizing their transformative impact on society and the economy.

Integration of Thin Films and 3D Printing

5

5.1   Synergies Between Thin Films and 3D Printing Technologies

Thin films and 3D printing technologies, although distinct in their methodologies and applications, possess synergistic potential when integrated and utilized together. Their combined capabilities can unlock new possibilities and enhance the efficiency, functionality, and versatility of manufacturing processes. Here are several ways in which thin films and 3D printing technologies can synergize.

Functional Coatings and Finishes:

Thin films can be applied as functional coatings or finishes on 3D printed objects to enhance their properties and performance. For example, thin films can provide protective coatings to improve durability, corrosion resistance, and weatherability of 3D printed parts. Additionally, functional thin films can be deposited to impart specific properties such as conductivity, optical transparency, or antimicrobial properties to 3D printed components, expanding their range of applications.

Thin films offer a versatile solution to enhance the properties and performance of 3D printed objects by serving as functional coatings or finishes. This integration of thin films with 3D printed structures opens up a wide range of possibilities for improving mechanical, electrical, optical, and surface properties.

1. **Mechanical Reinforcement**: Thin films can be deposited onto the surface of 3D printed objects to improve their mechanical strength, durability, and wear resistance. For example, coatings of hard materials such as diamond-like carbon (DLC), titanium nitride (TiN), or tungsten carbide (WC) can enhance the surface hardness and

scratch resistance of 3D printed components, extending their lifespan and performance in demanding applications.

2. **Corrosion Protection**: Thin film coatings can provide corrosion protection to metallic 3D printed objects by forming a barrier layer that prevents exposure to corrosive environments. Materials such as chromium, nickel, or ceramic coatings can be applied to the surface of 3D printed metal parts to inhibit corrosion, oxidation, and degradation over time, enhancing their longevity and reliability in harsh operating conditions.

3. **Surface Modification**: Thin films can modify the surface properties of 3D printed objects to impart specific functionalities such as hydrophobicity, oleophobicity, or anti-fouling properties. Surface coatings composed of fluoropolymers, silanes, or nanostructured materials can alter the surface energy and wettability of 3D printed surfaces, making them easier to clean, resistant to fouling, and suitable for biomedical, marine, or industrial applications.

4. **Conductive and Dielectric Coatings**: Thin films can be engineered to exhibit electrical conductivity or insulation properties, enabling the integration of electronic components, sensors, or antennas into 3D printed structures. Conductive coatings composed of metals (e.g., gold, silver, copper) or conductive polymers can provide electrical pathways for signal transmission or power distribution, while dielectric coatings can insulate and protect sensitive electronic devices from electromagnetic interference (EMI) or environmental hazards.

5. **Optical Enhancement**: Thin films can enhance the optical properties of 3D printed objects by controlling light transmission, reflection, or absorption. Optical coatings composed of interference filters, anti-reflection coatings, or transparent conductive oxides (TCOs) can improve the clarity, visibility, and aesthetics of 3D printed displays, lenses, or optical components, opening up new applications in automotive, aerospace, and consumer electronics.

6. **Biocompatibility and Bioactivity**: Thin films can impart biocompatibility and bioactivity to 3D printed implants, prosthetics, or medical devices to promote tissue integration and biological functionality. Bioactive coatings composed of hydroxyapatite, titanium oxide, or biodegradable polymers can facilitate osseointegration, cell adhesion, and tissue regeneration, enhancing the performance and safety of 3D printed biomedical products in healthcare applications.

7. **Functional Integration**: Thin films can integrate functional features such as sensors, actuators, or energy harvesting devices directly onto the surface of 3D printed objects, enabling multifunctional capabilities and smart functionalities. Functional coatings composed of piezoelectric materials, conductive polymers, or quantum dots can transform passive 3D printed structures into active components that respond to external stimuli or generate electrical, mechanical, or optical signals for sensing, actuation, or energy conversion purposes.

In summary, thin films offer a versatile and customizable solution to enhance the properties and performance of 3D printed objects across various applications and industries. By integrating thin film coatings or finishes with 3D printed structures, manufacturers can tailor the functionality, durability, and aesthetics of their products to meet specific design requirements and user needs, driving innovation and value creation in additive manufacturing technologies.

Embedded Sensors and Electronics:

Thin film deposition techniques can be used to integrate sensors, conductive traces, and electronic components directly onto 3D printed structures. By embedding thin film electronics during the printing process or post-processing, designers can create fully integrated smart devices and systems with sensing, monitoring, and communication capabilities. This integration enables the development of innovative products such as wearable health monitors, IoT devices, and smart infrastructure. Thin film deposition techniques offer a powerful means to integrate sensors, conductive traces, and electronic components directly onto 3D printed structures, enabling the creation of functionalized and smart devices with enhanced capabilities. This integration combines the design freedom and complexity of 3D printing with the precision and functionality of thin film technologies, opening up new possibilities for diverse applications across industries.

1. **Sensor Integration**: Thin film sensors can be deposited onto the surface of 3D printed objects or embedded within their layers to monitor various physical, chemical, or environmental parameters. For example, thin film temperature sensors composed of materials such as platinum, nickel, or thermoelectric oxides can be integrated onto 3D printed components to measure temperature gradients, thermal stability, or heat dissipation. Similarly, thin film gas sensors based on metal oxides, polymers, or nanomaterials can be incorporated into 3D printed structures to detect gases such as carbon monoxide, methane, or volatile organic compounds (VOCs) for environmental monitoring, industrial safety, or healthcare applications.

2. **Conductive Traces and Interconnects**: Thin film deposition techniques enable the fabrication of conductive traces and interconnects directly onto the surface of 3D printed objects, facilitating electrical connectivity and signal transmission between electronic components. Thin film metallic coatings composed of materials such as gold, silver, or copper can be patterned onto 3D printed substrates using techniques such as sputtering, evaporation, or inkjet printing to create conductive pathways for power distribution, data communication, or sensor interfacing. These conductive traces enable the integration of electronic circuits, antennas, or RFID tags into 3D printed structures, enabling wireless communication, remote monitoring, or IoT functionalities.

3. **Electronic Component Integration**: Thin film deposition techniques can be used to fabricate electronic components directly onto 3D printed structures, eliminating the need for separate assembly and integration steps. For example, thin film transistors,

diodes, resistors, or capacitors can be deposited onto the surface of 3D printed substrates using techniques such as inkjet printing, chemical vapor deposition (CVD), or atomic layer deposition (ALD) to create functional electronic devices with tailored properties and performance characteristics. These integrated electronic components enable the development of smart sensors, actuators, displays, or energy harvesting devices within 3D printed objects, enhancing their functionality, intelligence, and versatility.

4. **Flexible and Stretchable Electronics**: Thin film deposition techniques can be adapted to deposit flexible and stretchable electronic materials onto flexible or elastomeric 3D printed substrates, enabling the fabrication of wearable, conformal, or soft electronics for wearable technology, healthcare monitoring, or human–machine interfaces. Flexible thin film materials such as organic semiconductors, conductive polymers, or graphene can be deposited onto 3D printed textiles, elastomers, or bio-compatible substrates to create flexible electronic circuits, sensors, or electrodes that conform to irregular surfaces, withstand mechanical deformation, and maintain electrical performance under bending, stretching, or twisting conditions.

5. **Functional Coatings and Encapsulation**: Thin film coatings can be applied to protect electronic components, sensors, or conductive traces integrated onto 3D printed structures from environmental hazards such as moisture, temperature fluctuations, or mechanical stress. Thin film encapsulation layers composed of materials such as parylene, silicones, or thin-film ceramics can seal and protect sensitive electronic devices from moisture ingress, chemical exposure, or mechanical damage, prolonging their lifespan and reliability in harsh operating conditions. Additionally, functional coatings with properties such as anti-static, anti-corrosion, or anti-fouling can be applied to 3D printed electronics to enhance performance and durability in specific applications.

In summary, thin film deposition techniques offer a versatile and scalable approach to integrate sensors, conductive traces, and electronic components directly onto 3D printed structures, enabling the development of functionalized, smart, and interconnected devices for diverse applications. By combining the design flexibility of 3D printing with the precision and functionality of thin film technologies, manufacturers can create innovative solutions that enhance the intelligence, functionality, and performance of their products, driving forward the advancement of additive manufacturing and smart technology integration.

Surface Modification and Texturing:
Thin films offer opportunities for precise surface modification and texturing of 3D printed objects. Through techniques such as physical vapor deposition (PVD) or chemical vapor deposition (CVD), thin films can be selectively deposited onto specific regions of 3D printed parts to achieve desired surface properties or textures. This capability

allows for the creation of customized surface functionalities such as hydrophobicity, bio-compatibility, or friction reduction, enhancing the performance and usability of printed components.

Thin films provide significant opportunities for precise surface modification and texturing of 3D printed objects, enabling the enhancement of aesthetic appeal, functionality, and performance across various applications. The integration of thin films with 3D printed structures allows for the creation of tailored surface properties, textures, and functionalities that are difficult or impossible to achieve using traditional manufacturing methods.

1. **Surface Functionalization**: Thin films can be deposited onto the surface of 3D printed objects to modify their chemical, physical, or biological properties. Functional coatings composed of materials such as polymers, ceramics, metals, or biomolecules can impart specific functionalities such as hydrophobicity, hydrophilicity, anti-fouling, or bio-compatibility to 3D printed surfaces. These surface modifications enhance the performance and utility of 3D printed objects in various applications, including biomedical devices, consumer products, and industrial components.
2. **Texture Control**: Thin film deposition techniques enable precise control over surface texture and roughness of 3D printed objects, allowing for the creation of smooth, rough, patterned, or textured surfaces with desired characteristics. Surface texturing can be achieved by controlling deposition parameters such as deposition rate, temperature, pressure, or substrate orientation during thin film deposition processes. Techniques such as physical vapor deposition (PVD), chemical vapor deposition (CVD), or plasma-enhanced chemical vapor deposition (PECVD) can be used to deposit thin films with controlled surface morphologies, nanostructures, or microtextures onto 3D printed substrates, enhancing their tactile feel, visual appearance, or functional properties.
3. **Optical Effects**: Thin films can create optical effects such as coloration, iridescence, or light diffraction on the surface of 3D printed objects, enhancing their aesthetic appeal and visual impact. Thin film coatings composed of interference filters, photonic crystals, or nanostructured materials can manipulate light transmission, reflection, or absorption properties to create vibrant colors, metallic finishes, or iridescent patterns on 3D printed surfaces. These optical effects can be customized to meet specific design requirements and aesthetic preferences, making 3D printed objects more visually appealing and engaging to end-users.
4. **Surface Engineering**: Thin films enable surface engineering of 3D printed objects to improve adhesion, lubrication, or wear resistance properties. Functional coatings such as self-assembled monolayers (SAMs), diamond-like carbon (DLC), or anti-friction coatings can be deposited onto 3D printed surfaces to reduce friction, minimize wear, or enhance lubrication properties in mechanical components, bearings, or moving parts. These surface modifications extend the lifespan, reliability, and performance of 3D printed objects in demanding applications, reducing maintenance costs and downtime.

5. **Biomedical Applications**: Thin films offer opportunities for surface modification of 3D printed biomedical implants, prosthetics, or medical devices to promote tissue integration, biocompatibility, or antibacterial properties. Functional coatings composed of bioactive ceramics, hydrogels, or antimicrobial agents can be deposited onto 3D printed substrates to enhance osseointegration, minimize inflammation, or prevent bacterial colonization in implantable devices. These surface modifications improve the biocompatibility, safety, and efficacy of 3D printed biomedical products, facilitating better patient outcomes and healthcare delivery.

6. **Functional Gradients**: Thin films enable the creation of functional gradients or multi-layered coatings on 3D printed surfaces, allowing for the spatial control of surface properties and functionalities. By depositing thin films with varying compositions, thicknesses, or properties onto 3D printed substrates, manufacturers can create surfaces with tailored gradients in mechanical, chemical, or optical properties. Functional gradients enable precise control over surface interactions, adhesion, or performance characteristics, enhancing the functionality and versatility of 3D printed objects in diverse applications.

In summary, thin films offer unique opportunities for precise surface modification and texturing of 3D printed objects, enabling the creation of functionalized, aesthetically appealing, and high-performance products across industries. By leveraging thin film technologies, manufacturers can customize the surface properties, textures, and functionalities of 3D printed objects to meet specific design requirements, enhance user experience, and drive innovation in additive manufacturing and surface engineering.

Conformal Coatings and Encapsulation:
Thin film deposition technologies enable the application of conformal coatings and encapsulation layers onto complex geometries produced by 3D printing. Conformal coatings provide uniform protection against moisture, chemicals, and mechanical damage, ensuring the longevity and reliability of 3D printed products. Encapsulation with thin films can also enhance the stability and performance of sensitive electronic or photonic devices fabricated using additive manufacturing techniques. Thin film deposition technologies play a crucial role in applying conformal coatings and encapsulation layers onto complex geometries produced by 3D printing, offering precise and uniform coverage that conforms to the intricate surfaces of printed objects. This capability is essential for protecting sensitive components, enhancing functionality, and ensuring the reliability of 3D printed products in various applications.

1. **Conformal Coatings**: Thin film deposition techniques such as atomic layer deposition (ALD), chemical vapor deposition (CVD), or plasma-enhanced chemical vapor deposition (PECVD) enable the deposition of conformal coatings onto complex 3D printed surfaces. Conformal coatings adhere tightly to the irregular contours, recesses, and

features of 3D printed objects, providing uniform coverage and protection against environmental factors such as moisture, chemicals, or abrasion. These coatings enhance the durability, reliability, and performance of 3D printed components in harsh operating conditions, extending their lifespan and functionality.

2. **Encapsulation Layers**: Thin film encapsulation layers provide a protective barrier that seals and shields internal components of 3D printed objects from external hazards such as moisture, humidity, dust, or contaminants. Encapsulation layers are deposited onto the surface of 3D printed substrates using techniques such as physical vapor deposition (PVD), chemical vapor deposition (CVD), or spin coating to create a hermetic seal that encapsulates and protects sensitive electronic components, sensors, or circuitry. These encapsulation layers prevent corrosion, oxidation, or degradation of internal components, ensuring the long-term reliability and performance of 3D printed devices in challenging environments.

3. **Complex Geometries**: Thin film deposition technologies are well-suited for coating and encapsulating complex 3D printed geometries with intricate features, overhangs, or internal cavities that are difficult to access using traditional coating methods. Techniques such as ALD, CVD, or spray coating can deposit thin films onto the surface of 3D printed objects from multiple directions, ensuring uniform coverage and penetration into tight spaces without compromising the integrity or functionality of printed structures. This capability enables the application of conformal coatings and encapsulation layers onto complex 3D printed components in aerospace, automotive, electronics, and medical device industries, where precise protection and reliability are paramount.

4. **Material Compatibility**: Thin film deposition technologies offer versatility in material selection, allowing manufacturers to choose coatings and encapsulation materials that are compatible with the substrate material and the intended application. Thin film coatings composed of metals, ceramics, polymers, or hybrid materials can be deposited onto various 3D printed substrates, including plastics, metals, ceramics, or composites, to provide tailored protection and functionality. Material compatibility ensures optimal adhesion, adhesion, and performance of conformal coatings and encapsulation layers, minimizing the risk of delamination, cracking, or failure over time.

5. **Customization and Tailoring**: Thin film deposition technologies enable customization and tailoring of conformal coatings and encapsulation layers to meet specific design requirements, performance criteria, or regulatory standards. Manufacturers can optimize deposition parameters such as thickness, composition, or morphology of thin films to achieve desired properties such as hardness, transparency, conductivity, or barrier properties. Customized coatings and encapsulation layers can be tailored to address specific challenges or requirements in diverse applications, ensuring the reliability, functionality, and compliance of 3D printed products with industry standards and regulations.

In summary, thin film deposition technologies provide a versatile and precise solution for applying conformal coatings and encapsulation layers onto complex geometries produced by 3D printing. By leveraging thin film deposition techniques, manufacturers can enhance the durability, reliability, and functionality of 3D printed components in various industries, ensuring optimal performance and protection in demanding applications.

Hybrid Manufacturing Processes:
Integrating thin film deposition and 3D printing technologies in hybrid manufacturing processes offers flexibility and efficiency in producing complex, multi-material components. Hybrid approaches combine the design freedom of 3D printing with the precision and functionality of thin films, enabling the fabrication of integrated assemblies with diverse material properties and functionalities. This synergy opens up new opportunities in sectors such as aerospace, automotive, and electronics, where lightweight, multifunctional components are in demand. Integrating thin film deposition and 3D printing technologies in hybrid manufacturing processes offers significant advantages, including enhanced flexibility and efficiency in producing complex, multi-material components. This synergistic approach combines the design freedom and material versatility of 3D printing with the precision, functionality, and surface modification capabilities of thin film deposition techniques.

1. **Material Diversity and Compatibility**: Hybrid manufacturing processes enable the integration of multiple materials with diverse properties and functionalities into a single component. 3D printing allows for the fabrication of complex geometries using a wide range of materials, including polymers, metals, ceramics, and composites, while thin film deposition techniques offer additional material options for surface modification, coating, or functionalization. By combining these technologies, manufacturers can create multi-material components with tailored properties, such as enhanced mechanical strength, conductivity, corrosion resistance, or biocompatibility, to meet specific application requirements.

2. **Functional Gradients and Surface Textures**: Hybrid manufacturing enables the creation of functional gradients and surface textures by selectively depositing thin films onto 3D printed substrates. Thin film deposition techniques such as atomic layer deposition (ALD), chemical vapor deposition (CVD), or physical vapor deposition (PVD) can be used to apply coatings with controlled thicknesses, compositions, or properties onto complex 3D printed geometries, allowing for precise control over surface characteristics. Functional gradients and surface textures enhance the performance, aesthetics, and functionality of components in applications such as optics, electronics, biomedical devices, and consumer products.

3. **Design Flexibility and Complexity**: Hybrid manufacturing processes offer unparalleled design flexibility and complexity by combining the capabilities of 3D printing

with thin film deposition techniques. 3D printing allows for the creation of complex geometries, internal structures, and lattice designs that are difficult or impossible to achieve using traditional manufacturing methods. Thin film deposition techniques enable the modification and customization of surface properties, interfaces, and functionalities to enhance the performance and functionality of 3D printed components. This synergy enables the fabrication of innovative products with optimized designs, improved functionality, and enhanced aesthetics.

4. **Process Integration and Automation**: Hybrid manufacturing facilitates seamless process integration and automation by combining 3D printing and thin film deposition into a single manufacturing platform. Integrated systems can automate material handling, deposition, curing, and post-processing steps to streamline production workflows, reduce lead times, and increase productivity. Automated quality control and inspection systems ensure the consistency, accuracy, and reliability of manufactured components, minimizing defects and rework. Process integration enables manufacturers to optimize resource utilization, minimize waste, and improve overall efficiency in producing complex, multi-material components.

5. **Customization and Personalization**: Hybrid manufacturing processes enable customization and personalization of components to meet individual customer preferences and requirements. By integrating 3D printing and thin film deposition technologies, manufacturers can create bespoke products with unique designs, features, and functionalities tailored to specific user needs. Customized coatings, surface textures, or functional layers can be applied to 3D printed substrates to enhance performance, aesthetics, or usability, offering added value and differentiation in competitive markets.

In summary, integrating thin film deposition and 3D printing technologies in hybrid manufacturing processes offers flexibility, efficiency, and innovation in producing complex, multi-material components. By leveraging the complementary capabilities of these technologies, manufacturers can create customized products with enhanced performance, functionality, and aesthetics, driving forward the advancement of additive manufacturing and surface engineering.

Prototyping and Iterative Design:
Thin films can facilitate rapid prototyping and iterative design processes in conjunction with 3D printing. By coating or modifying 3D printed prototypes with thin films, designers can quickly evaluate different material combinations, surface treatments, or functional enhancements before finalizing the design for production. This iterative approach accelerates the development cycle, reduces time-to-market, and enables rapid innovation in product design and development. The integration of thin films and 3D printing technologies offers synergistic benefits that enhance the capabilities and performance of manufactured products. By leveraging the complementary strengths of these technologies, designers and manufacturers can unlock new opportunities for innovation, customization,

and efficiency across various industries. Thin films play a crucial role in facilitating rapid prototyping and iterative design processes when used in conjunction with 3D printing technologies. This combination offers several advantages, including increased flexibility, accelerated development cycles, and cost-effective prototyping.

1. **Functional Coatings and Surface Modification**: Thin films can be applied as functional coatings or surface modifications to 3D printed prototypes to enhance their properties or simulate specific functionalities. For example, coatings composed of conductive materials such as metals or conductive polymers can enable the integration of electronic components or sensors into 3D printed prototypes, allowing for testing and validation of electrical functionality. Similarly, thin films with specific optical, mechanical, or chemical properties can be deposited onto 3D printed surfaces to replicate real-world conditions or requirements, facilitating more accurate performance evaluations and design iterations.

2. **Customized Surface Textures and Finishes**: Thin films offer the ability to customize surface textures, finishes, and aesthetics of 3D printed prototypes, enhancing their visual appeal and tactile feel. By applying thin film coatings with controlled thicknesses and morphologies, manufacturers can create smooth, textured, matte, or glossy surfaces on 3D printed prototypes, simulating different materials, textures, or surface treatments. These customized surface finishes allow designers and engineers to evaluate the look and feel of prototypes in real-world conditions, gather feedback from stakeholders, and iterate on design concepts more effectively.

3. **Functional Integration and Testing**: Thin films enable the integration of functional features, such as sensors, actuators, or electronic components, directly onto 3D printed prototypes, enabling rapid testing and validation of design concepts. By depositing thin film coatings with specific properties onto 3D printed substrates, manufacturers can create prototypes with embedded functionalities, such as temperature sensing, pressure sensing, or chemical sensing capabilities. These functional prototypes allow engineers to evaluate performance, reliability, and usability under simulated operating conditions, identifying design flaws, optimizing parameters, and iterating on design iterations in a timely manner.

4. **Prototyping of Multi-material Components**: Thin films offer opportunities for prototyping multi-material components with complex geometries and heterogeneous properties. By combining 3D printing with thin film deposition techniques, manufacturers can create prototypes with integrated structures, interfaces, or layers composed of different materials, such as polymers, metals, ceramics, or composites. This integration enables the rapid fabrication of prototypes with customized material properties, mechanical characteristics, or functional attributes, facilitating design validation, performance testing, and optimization of multi-material components for various applications.

5. **Cost-Effective and Time-Efficient Prototyping**: Thin films enable cost-effective and time-efficient prototyping by reducing material waste, production lead times, and iteration cycles. Compared to traditional manufacturing methods, 3D printing and thin film deposition technologies offer rapid turnaround times, allowing designers and engineers to quickly produce and test multiple iterations of prototypes within a short timeframe. This accelerated prototyping process enables faster innovation, design refinement, and product development, reducing time-to-market and increasing competitiveness in dynamic industries.

In summary, thin films play a vital role in facilitating rapid prototyping and iterative design processes when integrated with 3D printing technologies. By leveraging the complementary capabilities of these technologies, manufacturers can create functional prototypes with enhanced properties, functionalities, and aesthetics, enabling faster innovation, design validation, and product development in diverse industries.

5.2   Examples of Combined Applications and Innovations

Smart Wearable Devices:

Thin films and 3D printing technologies are combined to create smart wearable devices with integrated sensors and electronics. For instance, flexible sensors and conductive traces made using thin film deposition techniques are embedded within 3D printed wearable accessories such as fitness trackers, smart clothing, and medical monitoring devices. This integration allows for seamless monitoring of vital signs, movement tracking, and biometric data collection, enabling personalized health monitoring and lifestyle management.

The combination of thin films and 3D printing technologies has revolutionized the development of smart wearable devices, leading to the creation of innovative solutions that integrate sensors, electronics, and flexible materials into wearable form factors. These smart wearable devices offer a wide range of functionalities, including health monitoring, fitness tracking, biometric sensing, and real-time data analysis, enhancing the user's overall well-being, performance, and connectivity.

Flexible Sensor Integration: Thin films enable the integration of various sensors onto flexible substrates, allowing for the development of lightweight, conformable sensor arrays for wearable applications. Sensors such as accelerometers, gyroscopes, heart rate monitors, temperature sensors, and electrochemical sensors can be printed or deposited onto flexible materials using thin film deposition techniques. These sensors provide real-time data on physical activity, vital signs, environmental conditions, and biochemical parameters, enabling continuous health monitoring and performance tracking for users.

1. **Biocompatible Materials and Comfortable Wearables**: 3D printing enables the fabrication of wearable devices with customized form factors and ergonomic designs that ensure comfort and ease of use for the wearer. Thin films can be applied to 3D printed components to enhance biocompatibility, reduce skin irritation, and improve comfort during prolonged wear. Additionally, thin film coatings can provide moisture-wicking properties, antimicrobial protection, and UV resistance, enhancing the durability and hygiene of wearable devices in various environmental conditions.

2. **Integration of Thin Film Electronics**: Thin film electronics, such as organic semiconductors, conductive polymers, and flexible printed circuits, are integrated into wearable devices to process, transmit, and analyze sensor data in real-time. These thin film electronic components can be deposited onto flexible substrates using printing or deposition techniques, enabling the fabrication of lightweight, low-power electronic systems for wearable applications. Thin film electronics offer advantages such as flexibility, stretchability, and energy efficiency, making them ideal for integration into smart wearable devices.

3. **Wireless Connectivity and Data Transmission**: Smart wearable devices incorporate wireless connectivity features such as Bluetooth, Wi-Fi, or NFC to communicate with smartphones, tablets, or other IoT devices. Thin film antennas and communication modules can be integrated into wearable devices to enable seamless wireless data transmission, remote monitoring, and synchronization with cloud-based platforms. These wireless connectivity features enhance the functionality and usability of smart wearable devices, enabling users to access and share data anytime, anywhere.

4. **Energy Harvesting and Power Management**: Thin films and 3D printing technologies enable the integration of energy harvesting and power management systems into smart wearable devices, reducing the reliance on traditional batteries and extending device runtime. Thin film photovoltaic materials can be deposited onto wearable surfaces to harvest solar energy, while energy storage devices such as thin film batteries or supercapacitors can be integrated into the device to store and manage harvested energy. These energy harvesting and power management systems provide sustainable, self-powered operation for smart wearable devices, reducing the need for frequent battery replacements and enhancing device longevity.

5. **Customized Design and Personalization**: 3D printing allows for the customization and personalization of smart wearable devices to match individual preferences, styles, and functional requirements. Users can choose from a variety of design options, colors, and materials to create personalized wearable devices that reflect their personality and lifestyle. Thin films can be applied to 3D printed components to add customized finishes, textures, or branding elements, ensuring that each wearable device is unique and tailored to the user's needs.

In summary, the convergence of thin films and 3D printing technologies enables the development of smart wearable devices with integrated sensors, electronics, and flexible materials, offering advanced functionalities, comfort, and personalization for users. These smart wearable devices have applications in various fields, including healthcare, fitness, sports, entertainment, and fashion, contributing to the growth of the wearable technology market and improving the quality of life for users around the world.

Advanced Automotive Components:

In the automotive industry, thin films and 3D printing technologies are utilized to produce advanced components with enhanced functionality and performance. For example, 3D printed automotive parts such as lightweight structures or complex geometries can be coated with thin films to improve surface properties, reduce friction, or provide corrosion resistance. Additionally, thin film sensors and electronic circuits integrated into 3D printed components enable real-time monitoring of vehicle performance, environmental conditions, and driver safety. In the automotive industry, the integration of thin films and 3D printing technologies has revolutionized the production of advanced components, leading to enhanced functionality, performance, and efficiency in vehicles. This convergence of technologies enables automotive manufacturers to overcome traditional manufacturing constraints, accelerate product development cycles, and create innovative solutions that meet the evolving demands of consumers and regulatory requirements.

1. **Lightweight Structural Components**: 3D printing allows for the fabrication of lightweight structural components with intricate geometries and complex designs using advanced materials such as polymers, metals, and composites. Thin films can be applied to 3D printed components to enhance surface properties, reduce weight, and improve performance. For example, thin film coatings can provide corrosion resistance, improve wear resistance, or increase surface hardness, ensuring the durability and longevity of lightweight components such as chassis parts, brackets, and body panels.

2. **Customized Interior Features**: 3D printing enables the customization of interior features and accessories in vehicles, allowing for personalized design elements tailored to individual preferences and tastes. Consumers can choose from a variety of customizable options for dashboard trim, center console accents, door handles, and other interior components. Thin films can be applied to 3D printed interior parts to add decorative finishes, textures, or branding elements, enhancing the aesthetic appeal and exclusivity of the vehicle's interior design.

3. **Functional Prototyping and Iterative Design**: 3D printing facilitates rapid prototyping and iterative design processes in the automotive industry, enabling engineers to quickly test and validate new concepts before moving into full-scale production. Thin films can be used to simulate different surface finishes, coatings, or material properties on 3D printed prototypes, allowing designers to evaluate aesthetic options, functional requirements, and performance characteristics. This iterative approach accelerates the

development cycle, reduces development costs, and improves the overall quality of automotive components and systems.

4. **Advanced Sensor Integration**: Thin films enable the integration of advanced sensors and electronics into automotive components to enhance safety, performance, and driver experience. For example, thin film sensors embedded within 3D printed components can monitor structural integrity, detect mechanical stress, or measure environmental parameters such as temperature, humidity, and air quality. These sensors provide real-time data feedback to vehicle systems, enabling predictive maintenance, condition monitoring, and proactive safety measures.

5. **Optimized Aerodynamics and Fuel Efficiency**: 3D printing technologies enable the fabrication of aerodynamic components and streamlined shapes that optimize airflow and reduce drag in vehicles. Thin films can be applied to 3D printed surfaces to further improve aerodynamic performance by reducing surface roughness, minimizing frictional losses, and enhancing laminar flow characteristics. These aerodynamic enhancements contribute to improved fuel efficiency, reduced emissions, and enhanced overall vehicle performance, particularly in electric and hybrid vehicles where energy conservation is paramount.

6. **Integrated Electronic Systems and Connectivity**: Thin films and 3D printing technologies are utilized to integrate electronic systems and connectivity features into automotive components, enhancing vehicle functionality and connectivity. For example, 3D printed antennas embedded with thin film conductive traces enable wireless communication, GPS navigation, and vehicle-to-vehicle (V2V) or vehicle-to-infrastructure (V2I) connectivity. Additionally, thin film coatings can provide electromagnetic shielding, EMI/RFI protection, or thermal management for electronic components, ensuring reliable operation in harsh automotive environments.

7. **Customized Exterior Styling and Accessories**: 3D printing allows for the customization of exterior styling elements and accessories in vehicles, enabling consumers to personalize their vehicles with unique design features. From customized grilles and emblems to aerodynamic spoilers and side skirts, 3D printed exterior accessories can be tailored to match individual preferences and design aesthetics. Thin films can be applied to 3D printed exterior parts to enhance surface finishes, weather resistance, and color durability, ensuring that customized styling elements maintain their appearance and performance over time.

In summary, the integration of thin films and 3D printing technologies is driving innovation and advancement in the automotive industry, enabling the production of advanced components with enhanced functionality, performance, and customization capabilities. This convergence of technologies not only improves vehicle efficiency, safety, and reliability but also enhances the overall driving experience for consumers, ushering in a new era of automotive design and manufacturing.

Biomedical Implants and Prosthetics:

Thin films and 3D printing are combined to fabricate biomedical implants and prosthetics with customized features and enhanced biocompatibility. In orthopedics and dentistry, 3D printed implants are coated with thin films to promote osseointegration, reduce bacterial adhesion, and improve long-term implant stability. Furthermore, thin film electronics and sensors can be integrated into prosthetic devices to provide sensory feedback, adjust mechanical properties, and enhance user comfort and mobility.

The combination of thin films and 3D printing technologies has revolutionized the field of biomedical implants and prosthetics, allowing for the fabrication of customized devices with enhanced biocompatibility, functionality, and performance. This integration offers numerous advantages in terms of design flexibility, patient-specific customization, and improved patient outcomes.

1. **Customized Design and Patient-Specific Adaptation**: 3D printing enables the creation of patient-specific implants and prosthetics tailored to individual anatomical characteristics and medical needs. Using advanced imaging techniques such as MRI or CT scans, clinicians can generate digital models of patient anatomy, which serve as the basis for designing customized implants or prosthetic components. Thin films can then be applied or integrated onto the 3D printed structures to enhance their surface properties, biocompatibility, and functionality, ensuring a precise fit and optimal performance.

2. **Enhanced Biocompatibility and Tissue Integration**: Thin film coatings can be applied to 3D printed implants and prosthetics to improve their biocompatibility and promote tissue integration. For example, bioactive thin films containing growth factors, peptides, or other bioactive molecules can stimulate cell adhesion, proliferation, and differentiation, facilitating tissue regeneration and healing at the implant interface. Additionally, thin film coatings with anti-inflammatory or antimicrobial properties help mitigate immune responses and reduce the risk of infection, enhancing the long-term stability and success of implanted devices.

3. **Functionalization and Drug Delivery**: Thin films deposited onto 3D printed implants and prosthetics can serve as carriers for therapeutic agents, enabling localized drug delivery to the surrounding tissues. By incorporating drugs, growth factors, or other bioactive compounds into thin film coatings, clinicians can precisely control the release kinetics and dosage, providing targeted treatment for conditions such as bone fractures, osteoarthritis, or tissue defects. This functionalization enhances the therapeutic efficacy of implants and prosthetics while minimizing systemic side effects and improving patient compliance.

4. **Mechanical Properties and Performance Optimization**: Thin films can be used to modify the mechanical properties and performance of 3D printed implants and prosthetics to better match the physiological demands of the recipient. For instance, coatings with wear-resistant or lubricating properties can reduce friction and enhance the longevity of joint implants, while thin film reinforcement layers can improve the

strength, stiffness, and fatigue resistance of load-bearing components. This optimization of mechanical properties ensures the durability, reliability, and functionality of implanted devices under physiological loading conditions.

5. **Biofunctional Coatings and Surface Modifications**: Thin films enable the application of biofunctional coatings and surface modifications to 3D printed implants and prosthetics, imparting specific functionalities such as hydrophilicity, cell adhesion, or tissue patterning. By precisely controlling surface topography, chemistry, and wettability, clinicians can create biomimetic interfaces that mimic the natural extracellular matrix and facilitate cell-material interactions. These biofunctional coatings promote tissue ingrowth, vascularization, and remodeling, improving the biointegration and performance of implanted devices in vivo.

6. **Personalized Rehabilitation and Functional Restoration**: 3D printing technologies enable the fabrication of personalized prosthetic devices with intricate designs and functional features tailored to the individual needs of patients. Thin films can be used to enhance the functionality and aesthetics of prosthetic limbs, orthoses, and other assistive devices by incorporating sensors, actuators, or electronic components. This integration enables personalized rehabilitation programs, adaptive assistive technologies, and functional restoration solutions that improve mobility, dexterity, and quality of life for individuals with limb loss or musculoskeletal impairments.

In summary, the combination of thin films and 3D printing technologies holds tremendous potential for advancing the field of biomedical implants and prosthetics. By integrating these technologies, clinicians can design and fabricate customized devices with enhanced biocompatibility, functionality, and patient-specific adaptation, leading to improved treatment outcomes, enhanced patient comfort, and better quality of life for individuals with medical conditions or disabilities.

Flexible Electronics and Displays:

Thin films and 3D printing technologies converge to create flexible electronics and displays for various applications. Flexible substrates produced using 3D printing techniques are coated with thin films to fabricate bendable, stretchable, and lightweight electronic devices such as flexible displays, electronic skins, and wearable sensors. This integration enables the development of next-generation electronics with conformal form factors, enabling seamless integration into curved surfaces and unconventional shapes.

The convergence of thin films and 3D printing technologies has led to significant advancements in the development of flexible electronics and displays, revolutionizing industries such as consumer electronics, healthcare, wearables, and IoT devices. By combining the benefits of both technologies, manufacturers can create innovative solutions that offer flexibility, lightweight design, and improved performance.

1. **Flexible Substrates and Structural Components**: 3D printing allows for the fabrication of flexible substrates and structural components for electronic devices using a variety of flexible materials such as thermoplastics, elastomers, and flexible resins. These substrates can be customized to conform to curved surfaces, irregular shapes, or wearable form factors, enabling the development of flexible electronic devices with enhanced ergonomics and aesthetics. Thin films can then be deposited onto these flexible substrates to create functional electronic circuits, sensors, and displays.

2. **Printed Electronics and Conductive Inks**: 3D printing techniques such as inkjet printing, aerosol jet printing, and screen printing enable the deposition of conductive inks and functional materials onto flexible substrates to create electronic circuits and components. Thin films of conductive materials such as silver nanoparticles, graphene, or conductive polymers can be precisely printed onto flexible substrates to form conductive traces, electrodes, and interconnects. This approach allows for the rapid prototyping and customization of flexible electronic devices with complex geometries and intricate designs.

3. **Integrated Sensors and Actuators**: Thin film sensors and actuators can be integrated into flexible electronic devices using 3D printing technologies to enable sensing, actuation, and feedback functionalities. For example, piezoelectric thin films deposited onto flexible substrates can convert mechanical deformation into electrical signals, enabling the development of flexible pressure sensors, touch sensors, and energy harvesters. Similarly, 3D printed microfluidic channels can be combined with thin film sensors to create flexible biosensors for medical diagnostics and environmental monitoring applications.

4. **Flexible Displays and Lighting**: Thin film deposition techniques enable the fabrication of flexible displays and lighting devices with high-resolution, low-power consumption, and bendable form factors. Organic light-emitting diodes (OLEDs), quantum dot displays, and electrophoretic displays can be printed onto flexible substrates to create lightweight, portable, and energy-efficient display panels for smartphones, tablets, e-readers, and wearable devices. Additionally, thin film light-emitting materials such as phosphorescent or quantum dot films can be integrated into flexible lighting solutions for architectural lighting, automotive lighting, and signage applications.

5. **Energy Harvesting and Storage**: Flexible electronics and energy storage devices can be combined to create self-powered systems for energy harvesting and storage applications. Thin film photovoltaic materials deposited onto flexible substrates can capture solar energy to power electronic devices and sensors in remote or off-grid environments. Similarly, flexible supercapacitors or batteries can be fabricated using 3D printing techniques to store harvested energy and provide on-demand power for wearable electronics, IoT devices, and portable electronics.

6. **Wearable Health Monitoring and Biomedical Devices**: The convergence of thin films and 3D printing enables the development of wearable health monitoring and biomedical devices with flexible form factors and biocompatible materials. Flexible sensors,

electrodes, and biosensors printed onto skin-like substrates can monitor vital signs, detect biomarkers, and deliver therapeutic interventions in real-time. These wearable devices offer non-invasive, continuous monitoring of physiological parameters and enable personalized healthcare solutions for remote patient monitoring, chronic disease management, and health and wellness tracking.

In summary, the convergence of thin films and 3D printing technologies opens up exciting possibilities for the development of flexible electronics and displays with diverse applications across industries. By leveraging the complementary strengths of these technologies, manufacturers can create innovative solutions that combine flexibility, functionality, and customization to meet the evolving needs of consumers and industries in the digital age.

Hybrid Energy Systems:

Thin films and 3D printing technologies are employed in the development of hybrid energy systems for renewable energy generation and storage. For instance, 3D printed components such as structural supports or housing enclosures are coated with thin film photovoltaic materials to create integrated solar panels with enhanced efficiency and durability. Similarly, thin film batteries or supercapacitors can be integrated into 3D printed energy storage devices for compact and lightweight energy solutions in portable electronics, unmanned aerial vehicles (UAVs), and off-grid applications.

Thin films and 3D printing technologies play a significant role in advancing the development of hybrid energy systems for renewable energy generation and storage. These technologies offer innovative approaches to address challenges associated with conventional energy sources and storage solutions, paving the way for more sustainable and efficient energy systems.

1. **Solar Energy Harvesting**: Thin film photovoltaic (PV) materials, such as amorphous silicon, cadmium telluride, and organic photovoltaics, are lightweight, flexible, and can be manufactured using cost-effective deposition techniques. These thin film solar cells can be integrated into various substrates or printed onto surfaces, enabling the fabrication of solar panels with custom shapes and sizes. Additionally, 3D printing technologies allow for the rapid prototyping and customization of solar panel components, such as mounting structures, trackers, and concentrators, to optimize energy capture and conversion efficiency.

2. **Energy Storage Devices**: 3D printing enables the fabrication of complex, multi-material structures for energy storage devices such as batteries, supercapacitors, and fuel cells. Additive manufacturing techniques allow for the precise control of electrode geometries, porosity, and surface area, optimizing electrochemical performance and energy density. Thin film coatings can further enhance the performance and longevity of energy storage devices by providing protective layers, improving conductivity, and preventing degradation due to corrosion or electrolyte leakage.

3. **Hybrid Photovoltaic-Thermal Systems**: Hybrid photovoltaic-thermal (PV-T) systems combine solar electricity generation with heat recovery to maximize energy output and efficiency. Thin film solar collectors integrated with heat exchangers can capture solar radiation for both electricity and thermal energy production. 3D printing enables the fabrication of customized heat exchanger designs with intricate geometries and enhanced heat transfer properties, allowing for efficient heat extraction and utilization. This integration enhances the overall energy yield and versatility of PV-T systems, making them suitable for various applications such as residential, commercial, and industrial heating and cooling.

4. **Flexible and Portable Energy Solutions**: Thin film solar cells printed onto flexible substrates offer lightweight, portable, and adaptable energy solutions for off-grid and mobile applications. Flexible solar panels can be integrated into backpacks, tents, clothing, and portable electronic devices, providing on-the-go power generation for outdoor activities, emergency situations, and remote locations. 3D printing enables the fabrication of lightweight, durable enclosures and mounting accessories for portable energy systems, enhancing their ruggedness, weather resistance, and user-friendliness.

5. **Hydrogen Production and Storage**: Thin film catalysts and membranes are employed in electrochemical and photoelectrochemical processes for hydrogen production from renewable sources such as water and biomass. 3D printing techniques enable the fabrication of custom-designed reactors, electrodes, and electrolyte compartments for hydrogen generation and storage systems. By combining thin film catalysts with 3D printed reactor architectures, researchers can optimize reaction kinetics, improve mass transport, and enhance overall system efficiency for sustainable hydrogen production.

6. **Grid Integration and Microgrids**: Hybrid energy systems incorporating thin films and 3D printing technologies support grid integration and the development of resilient microgrid networks. Thin film solar panels and energy storage devices can be integrated into building facades, roofs, and infrastructure components, enabling distributed generation and storage of renewable energy. 3D printing facilitates the customization and deployment of energy management systems, smart meters, and grid-connected devices for monitoring, control, and optimization of energy flows within microgrids, enhancing their reliability, flexibility, and sustainability.

In summary, the integration of thin films and 3D printing technologies in the development of hybrid energy systems represents a promising approach to advancing renewable energy generation and storage. These technologies offer versatility, scalability, and customization capabilities that enable the efficient harnessing and utilization of renewable resources, contributing to a more sustainable and resilient energy infrastructure for the future.

Customized Consumer Products:

Thin films and 3D printing technologies enable the customization and personalization of consumer products to meet individual preferences and requirements. For example, 3D printed product prototypes are coated with thin films to simulate different surface finishes, textures, or colors before finalizing the design for mass production. Additionally, thin film electronics and sensors integrated into 3D printed consumer goods enable interactive features, smart functionalities, and IoT connectivity, enhancing user experience and product utility.

Thin films and 3D printing technologies are at the forefront of enabling the customization and personalization of consumer products, allowing for the creation of tailored solutions that meet individual preferences, requirements, and tastes. This convergence of technologies offers unprecedented flexibility, versatility, and creativity in product design and manufacturing, empowering consumers to participate in the co-creation process and express their unique identity.

1. **Customized Product Design**: 3D printing technologies enable the fabrication of custom-designed products with intricate geometries, complex structures, and personalized features. Consumers can collaborate with designers or use online design platforms to create custom product designs that reflect their preferences, aesthetic preferences, and functional requirements. Thin films can then be applied or integrated into 3D printed components to enhance surface properties, add decorative finishes, or provide additional functionalities, allowing for truly unique and personalized consumer products.

2. **On-Demand Manufacturing and Mass Customization**: The on-demand manufacturing capabilities of 3D printing enable mass customization of consumer products, where individual items are produced according to specific customer orders or preferences. This eliminates the need for large-scale production runs and inventory stockpiling, reducing waste, and minimizing environmental impact. Thin film coatings can be applied during or after the 3D printing process to add customized finishes, colors, textures, or branding elements to consumer products, ensuring that each item is tailored to the customer's preferences and specifications.

3. s**Personalized Fit and Functionality**: 3D printing technologies enable the creation of consumer products with personalized fit and functionality, particularly in sectors such as footwear, eyewear, and orthotics. By scanning or digitizing individual body measurements, designers can create custom-fit products that conform to the unique contours and proportions of each user. Thin films can be incorporated into 3D printed products to enhance comfort, support, and performance, such as adding cushioning layers, arch support structures, or moisture-wicking coatings to footwear or orthopedic devices.

4. **Customized Branding and Brand Engagement**: Thin films offer opportunities for customized branding and brand engagement in consumer products. Logos, graphics, or text can be printed onto 3D printed surfaces using thin film transfer techniques or direct printing methods, allowing brands to create personalized branding elements that resonate with their target audience. Customized branding enhances brand recognition, strengthens brand loyalty, and fosters deeper connections with consumers, who appreciate the ability to personalize their products with unique design elements or personal messages.

5. **Design Iteration and Rapid Prototyping**: 3D printing enables rapid iteration and prototyping of consumer products, allowing designers to quickly test and refine product concepts before finalizing the design for production. Thin films can be used to simulate different surface finishes, textures, or color variations on 3D printed prototypes, enabling designers to evaluate aesthetic options and make informed design decisions. This iterative approach accelerates the product development cycle, reduces time-to-market, and facilitates continuous innovation in consumer product design.

6. **Personalized Consumer Electronics and Accessories**: The convergence of thin films and 3D printing enables the customization of consumer electronics and accessories such as smartphone cases, headphones, and wearable devices. Consumers can personalize their devices with custom-designed covers, grips, or decorative elements using 3D printing, while thin films can be applied to enhance durability, add anti-scratch coatings, or incorporate functional features such as wireless charging capabilities or NFC tags. This customization enhances user experience, fosters brand loyalty, and differentiates products in a competitive market landscape.

In summary, the combination of thin films and 3D printing technologies empowers consumers to participate in the design and customization of consumer products, resulting in personalized solutions that reflect individual preferences, lifestyle choices, and aesthetic sensibilities. This convergence of technologies not only enhances user satisfaction and brand loyalty but also drives innovation and creativity in product design and manufacturing, shaping the future of consumer-centric industries in the digital age. These examples demonstrate the diverse range of applications and innovations that can be achieved through the synergistic combination of thin films and 3D printing technologies. By leveraging the unique capabilities of each technology, designers and manufacturers can create innovative solutions that address complex challenges and deliver enhanced performance, functionality, and customization across various industries and domains.

5.3   Future Directions and Emerging Trends

Future Directions and Emerging Trends of Thin Films and 3D Printing in the Fourth Industrial Revolution (4IR):

1. **Nanotechnology Integration**:

   In the era of the Fourth Industrial Revolution (4IR), there is a noticeable shift towards the integration of nanotechnology with thin films and 3D printing, marking a significant convergence that unlocks novel opportunities for fabricating nanoscale structures, devices, and materials with unparalleled precision and control. This integration leverages the strengths of each technology to create nanomaterials with unique properties, thus expanding the horizons of various fields including nanoelectronics, nanomedicine, and nanophotonics. The integration of nanotechnology with thin films and 3D printing enables the fabrication of nanoscale structures and devices with enhanced properties.

   Nanomaterials synthesized using thin film deposition techniques or incorporated into 3D printing resins offer a myriad of advantages. Thin film deposition methods such as physical vapor deposition (PVD) and chemical vapor deposition (CVD) enable precise control over the thickness and composition of thin films, allowing for the creation of nanoscale layers with tailored properties. By incorporating nanomaterials into thin films, researchers can enhance their mechanical, electrical, and optical properties, paving the way for advancements in nanoelectronics, photonics, and sensors. Moreover, the integration of nanotechnology with 3D printing revolutionizes the fabrication of complex three-dimensional structures with nanoscale features. By incorporating nanomaterials into 3D printing resins, researchers can enhance the mechanical strength, conductivity, and catalytic activity of printed objects. This opens up new possibilities in areas such as nanomedicine, where 3D-printed implants and drug delivery systems can be engineered to release therapeutic agents with precise control over dosage and release kinetics. One of the key advantages of integrating nanotechnology with thin films and 3D printing is the ability to create multifunctional materials with tailored properties. For example, nanocomposites comprising nanoparticles dispersed within a polymer matrix exhibit synergistic properties derived from the combination of the individual components. By controlling the size, shape, and composition of nanoparticles, researchers can engineer nanocomposites with enhanced mechanical strength, thermal stability, and electrical conductivity, making them ideal for a wide range of applications including structural materials, energy storage devices, and sensors. In the field of nanomedicine, the integration of nanotechnology with thin films and 3D printing holds promises for developing advanced drug delivery systems, diagnostic devices, and tissue engineering scaffolds. Nanoparticles functionalized with targeting ligands can be encapsulated within thin film coatings or incorporated into 3D-printed constructs to enable targeted drug delivery to specific tissues or cells.

Similarly, 3D-printed tissue scaffolds incorporating nanomaterials can mimic the extra-cellular matrix and provide cues for cell growth, differentiation, and tissue regeneration. The integration of nanotechnology with thin films and 3D printing represents a transfor-mative approach for fabricating nanoscale structures, devices, and materials with enhanced properties and functionalities. This convergence enables researchers to exploit the unique properties of nanomaterials and the versatility of thin films and 3D printing to address challenges across various fields including nanoelectronics, nanomedicine, and nanopho-tonics. As technology continues to advance, the synergy between nanotechnology, thin films, and 3D printing will drive innovation and unlock new possibilities for addressing complex societal and environmental challenges.

2. **Multi-material Printing**:

Future advancements in 3D printing technology are poised to revolutionize manufac-turing processes by enabling multi-material printing capabilities. This evolution involves incorporating multiple materials into the printing process, such as polymers, metals, ceramics, and composites, to produce complex, multi-functional components with diverse material properties and functionalities. This capability holds immense potential for driv-ing innovation across various industries, including aerospace, automotive, healthcare, and consumer goods. The ability to print with multiple materials opens up a wide range of possibilities for fabricating hybrid structures with tailored properties. For exam-ple, in aerospace applications, manufacturers can combine lightweight polymers with high-strength metals to create components that are both durable and lightweight. Sim-ilarly, in automotive manufacturing, multi-material printing allows for the production of parts with optimized mechanical properties, such as stiffness, strength, and impact resistance, to enhance vehicle performance and safety. Moreover, multi-material print-ing enables the fabrication of graded materials with varying compositions and properties along specific axes. This capability is particularly advantageous in healthcare, where customized implants and prosthetics can be printed with bioresorbable materials that grad-ually degrade over time, promoting tissue regeneration and integration. Additionally, in electronics manufacturing, multi-material printing enables the integration of conductive and insulating materials to create complex circuitry and embedded sensors in 3D-printed devices. Furthermore, the ability to print with multiple materials facilitates the customiza-tion of products tailored to specific applications and user requirements. Consumer goods manufacturers can leverage multi-material printing to create personalized products with unique designs, colors, and functionalities. This customization capability enhances con-sumer satisfaction and brand loyalty by offering products that meet individual preferences and needs. In addition to these applications, multi-material printing has the potential to revolutionize the field of material science by enabling the fabrication of novel materi-als with customizable properties and functionalities. By combining different materials at the molecular level, researchers can create advanced materials with enhanced mechanical, electrical, and thermal properties, opening up new opportunities for innovation in materials design and engineering. Future advancements in 3D printing technology will focus on enabling multi-material printing capabilities, allowing manufacturers to produce complex, multi-functional components with diverse material properties and functionalities. This capability holds tremendous potential for driving innovation across various industries, including aerospace, automotive, healthcare, and consumer goods, by facilitating the fabrication of hybrid structures, graded materials, and customized products tailored to specific applications and user requirements. As technology continues to evolve, multi-material printing will play an increasingly important role in shaping the future of manufacturing and materials science, unlocking new possibilities for innovation and advancement.

3. **Additive Manufacturing at Scale**:

   As 3D printing technology continues to mature, a significant shift towards additive manufacturing at scale is anticipated, where large-scale production of end-use parts and components becomes economically viable. This shift is driven by advances in printing speed, process automation, and material efficiency, enabling manufacturers to achieve high throughput and cost-effectiveness. As a result, additive manufacturing is poised to become a mainstream manufacturing technique for mass production, with far-reaching implications for supply chain logistics, customization, and distributed manufacturing, fundamentally transforming the way products are designed, produced, and distributed globally. Advancements in printing speed play a pivotal role in facilitating additive manufacturing at scale. As printing speeds increase, manufacturers can produce parts and components more quickly, reducing production lead times and increasing overall throughput. Additionally, improvements in process automation streamline the printing process, reducing manual intervention and minimizing the risk of errors or defects. These advancements in speed and automation contribute to the cost-effectiveness of additive manufacturing, making it a viable option for large-scale production. Furthermore, advancements in material efficiency enable manufacturers to optimize material usage and minimize waste during the printing process. By optimizing print parameters, such as infill density and support structures, manufacturers can reduce material consumption while maintaining part quality and integrity. Additionally, advancements in material science enable the development of new materials with improved properties, expanding the range of applications for additive manufacturing. The shift towards additive manufacturing at scale has profound implications for supply chain logistics. Traditional manufacturing methods often involve complex supply chains with multiple suppliers and distribution channels, leading to longer lead times and increased logistics costs. In contrast, additive manufacturing enables on-demand production of parts and components, reducing the need for extensive warehousing and inventory management. This just-in-time manufacturing approach enhances supply chain agility and responsiveness, enabling manufacturers to adapt quickly to changing market demands and customer preferences. Moreover, additive manufacturing facilitates customization and personalized production on a mass scale. By leveraging digital design and printing technologies, manufacturers can tailor products to individual customer preferences and requirements without the need for costly tooling or setup. This capability enables mass customization at a fraction of the cost and time required by traditional manufacturing methods, opening up new opportunities for product differentiation and customer engagement. Additionally, additive manufacturing enables distributed manufacturing models, where production facilities are decentralized and located closer to the point of use. This decentralization reduces transportation costs and carbon emissions associated with traditional manufacturing and distribution networks, contributing to environmental sustainability and resilience. The shift towards additive manufacturing at scale represents a paradigm shift in manufacturing, with significant implications for supply chain logistics, customization, and distributed manufacturing. As 3D printing technology continues to mature and evolve, additive manufacturing is poised to become a mainstream manufacturing technique for mass production, revolutionizing the way products are designed, produced, and distributed globally.

4. **Smart Materials and Structures**:

The integration of smart materials and structures with thin films and 3D printing technologies represents a significant opportunity to drive innovation in smart manufacturing and Internet of Things (IoT) applications. Smart materials, characterized by their ability to sense, respond, and adapt to external stimuli, can be seamlessly integrated into 3D printed structures or coated with thin films to create adaptive, responsive, and self-monitoring systems. These smart structures have the potential to revolutionize various industries including robotics, aerospace, infrastructure, and wearable technology by enabling enhanced functionality, improved performance, and greater efficiency. Smart materials encompass a diverse range of materials with unique properties and functionalities. These materials include shape memory alloys (SMAs), piezoelectric polymers, electroactive polymers, and self-healing composites, among others. By incorporating these smart materials into 3D printed structures or coating them with thin films, researchers can create innovative systems capable of sensing, responding, and adapting to changes in their environment. One of the key advantages of integrating smart materials with thin films and 3D printing technologies is the ability to create adaptive structures that can adjust their shape, stiffness, or mechanical properties in response to external stimuli. For example, shape memory alloys can undergo reversible phase transitions in response to changes in temperature, enabling the creation of self-deploying structures or morphing aerospace components. Similarly, piezoelectric polymers can generate electrical signals in response to mechanical stress, allowing for the development of self-powered sensors or actuators for robotics and wearable technology. Moreover, the integration of smart materials with thin films and 3D printing technologies enables the creation of self-monitoring systems capable of detecting and repairing damage autonomously. Self-healing composites, embedded within 3D printed structures or coated with thin films, can repair cracks or fractures when subjected to mechanical or thermal stress, prolonging the lifespan of components and reducing maintenance costs. This self-healing capability is particularly beneficial for applications in infrastructure, automotive, and aerospace industries where durability and reliability are paramount. Furthermore, smart structures integrated with thin films and 3D printing technologies have the potential to enhance the performance and efficiency of IoT devices and systems. By embedding sensors and actuators directly into 3D printed components or coating them with thin films, researchers can create intelligent systems capable of real-time monitoring, control, and optimization. These smart structures can optimize energy consumption, improve safety, and enhance user experience in various IoT applications such as smart homes, smart cities, and industrial automation. The integration of smart materials and structures with thin films and 3D printing technologies holds tremendous potential to drive innovation in smart manufacturing and IoT applications. By creating adaptive, responsive, and self-monitoring systems, researchers can unlock new possibilities for enhancing performance, improving efficiency, and advancing functionality across various industries. As technology continues to evolve, the synergy between smart materials, thin films, and 3D printing technologies will pave the way for the development of next-generation smart devices and systems that can address complex challenges and transform the way we interact with the world.

5. **Bioprinting and Biofabrication**:

In the field of biomedicine, there is a burgeoning interest in leveraging bioprinting and biofabrication techniques that incorporate thin films and 3D printing to engineer intricate biological constructs and tissues. These innovative approaches involve depositing bioinks, which consist of living cells and biomaterials, layer-by-layer to create functional tissues, organoids, and organ-on-a-chip devices. Such advancements hold promise for various applications in drug screening, disease modeling, and regenerative medicine. The integration of thin film coatings with bioprinted constructs plays a crucial role in enhancing cell adhesion, viability, and functionality. Thin film coatings can provide a supportive matrix for cell growth and proliferation, mimicking the extracellular environment and promoting tissue development. Additionally, thin films can serve as reservoirs for growth factors, cytokines, and other bioactive molecules, regulating cellular behavior and promoting tissue regeneration. Furthermore, thin film coatings can impart mechanical properties to bioprinted constructs, enhancing their structural integrity and stability. By controlling the composition and thickness of thin films, researchers can tailor the mechanical properties of bioprinted tissues to mimic native tissues and organs more closely. This enables the development of biohybrid systems with improved biological and mechanical properties, facilitating their integration into the body for therapeutic purposes. Moreover, the integration of thin films with bioprinting techniques enables the fabrication of organ-on-a-chip devices, which mimic the structure and function of human organs on a microscale. These microfluidic devices incorporate bioprinted tissues or organoids onto thin film-coated substrates, allowing for precise control over cell culture conditions, nutrient supply, and drug exposure. Organ-on-a-chip devices offer a versatile platform for studying physiological processes, modeling disease states, and screening potential drug candidates in a more physiologically relevant context. The integration of thin films and 3D printing techniques with bioprinting holds tremendous potential for advancing the field of biomedicine. By enabling the fabrication of complex biological constructs and tissues, these innovative approaches offer new opportunities for drug discovery, disease modeling, and regenerative medicine. The development of biohybrid systems and organ-on-a-chip devices represents a significant step towards more personalized and effective medical treatments, ultimately improving patient outcomes and quality of life.

6. **Digital Twin Technology**:

The concept of digital twins, virtual replicas of physical objects or systems, will play a significant role in the future of thin films and 3D printing. By creating digital twins of manufacturing processes, materials, and products, designers can simulate and optimize performance, predict behavior, and iterate designs in a virtual environment before physical realization. Digital twins enable real-time monitoring, analysis, and control of manufacturing processes, leading to improved efficiency, quality, and sustainability across the product lifecycle. The emergence of digital twins, which are virtual replicas of physical objects or systems, is poised to revolutionize the future of thin films and 3D printing. Digital twins offer a powerful tool for designers and manufacturers to simulate, optimize, and predict the behavior of materials, processes, and products in a virtual environment before physical realization. By creating digital twins of manufacturing processes, designers can simulate various scenarios and parameters to identify optimal settings for thin film deposition and 3D printing. This enables manufacturers to optimize process parameters such as temperature, pressure, and material flow, leading to improved efficiency and quality in the production of thin films and 3D-printed components. Moreover, digital twins allow for the creation of virtual prototypes of products and components, enabling designers to iterate designs and assess performance in a virtual environment. This iterative design process reduces the need for physical prototypes, saving time and resources while accelerating the product development cycle. Designers can simulate the behavior of thin film coatings and 3D-printed parts under different operating conditions, optimizing their performance and reliability before physical manufacturing. Furthermore, digital twins enable real-time monitoring, analysis, and control of manufacturing processes, leading to improved efficiency and sustainability across the product lifecycle. By integrating sensors and data analytics into manufacturing equipment, manufacturers can collect real-time data on process parameters and performance metrics. This data can be used to create digital twins that accurately represent the physical manufacturing environment, allowing for predictive maintenance, quality control, and optimization of resource utilization. The concept of digital twins holds significant promise for the future of thin films and 3D printing. By enabling virtual simulation, optimization, and real-time monitoring of manufacturing processes, digital twins offer opportunities to improve efficiency, quality, and sustainability across the product lifecycle. As digital twin technology continues to evolve, it will play an increasingly important role in driving innovation and advancing the capabilities of thin film deposition and 3D printing technologies.

7. **Sustainability and Circular Economy**:

As the world progresses into the Fourth Industrial Revolution (4IR), sustainable practices and circular economy principles are set to significantly influence the future of thin films and 3D printing. The integration of eco-friendly materials, energy-efficient processes, and waste reduction strategies will shape the development of more sustainable manufacturing solutions in these domains. Furthermore, advancements in recycling and remanufacturing technologies will foster the reuse and repurposing of materials and products, contributing to the circular economy and minimizing environmental impact throughout the product lifecycle. Innovations in eco-friendly materials will play a crucial role in advancing sustainability in thin films and 3D printing. Researchers are actively developing biodegradable and renewable materials that can be used as feedstocks for thin film coatings and 3D printing filaments. These materials offer comparable performance to traditional counterparts while reducing reliance on finite resources and minimizing environmental pollution. Additionally, bio-based materials derived from sustainable sources such as plant-based polymers and bioplastics are gaining traction for their potential to reduce carbon emissions and mitigate climate change. Energy-efficient processes will also contribute to sustainability in thin film deposition and 3D printing. Manufacturers are adopting technologies such as additive manufacturing with reduced energy consumption, optimized process parameters, and renewable energy sources. By minimizing energy usage during production, manufacturers can lower operational costs and reduce their carbon footprint, contributing to environmental sustainability. Furthermore, waste reduction strategies are essential for promoting sustainability in thin films and 3D printing. Manufacturers are implementing closed-loop systems that minimize material waste and enable the recycling of unused materials and byproducts. For example, excess material from thin film deposition processes can be collected and recycled for future use, reducing raw material consumption and landfill waste. Similarly, 3D printing technologies are being developed with built-in recycling capabilities, allowing for the reuse of printed parts and support structures. These waste reduction strategies not only conserve resources but also reduce the environmental burden associated with disposal and waste management. Advancements in recycling and remanufacturing technologies will further enhance sustainability in thin films and 3D printing. Researchers are exploring innovative methods for reclaiming and repurposing materials from end-of-life products and manufacturing waste streams. By reintroducing recycled materials into the production process, manufacturers can close the loop and minimize the need for virgin materials, thereby conserving natural resources and reducing environmental impact. sustainable practices and circular economy principles are poised to shape the future of thin films and 3D printing in the 4IR. Innovations in eco-friendly materials, energy-efficient processes, and waste reduction strategies will drive the development of more sustainable manufacturing solutions. Additionally, advancements in recycling and remanufacturing technologies will enable the reuse and repurposing of materials and products, contributing to a more sustainable and environmentally conscious approach to manufacturing in these domains.

The future of thin films and 3D printing in the Fourth Industrial Revolution is characterized by convergence, innovation, and sustainability. By embracing emerging trends and leveraging synergies between these technologies, manufacturers can unlock new possibilities for customization, efficiency, and performance across diverse industries and applications, driving forward the era of digital manufacturing and smart production.

5.4   Summary

The integration of thin films and 3D printing represents a convergence of two cutting-edge technologies, offering synergistic advantages and unlocking new possibilities for advanced manufacturing applications. Thin films, characterized by their nanoscale thickness and tailored properties, and 3D printing, known for its additive manufacturing capabilities and design freedom, complement each other in various ways to enhance functionality, performance, and customization across diverse industries. One of the key benefits of integrating thin films with 3D printing lies in enhancing the surface properties of printed objects. Thin films can be deposited onto 3D-printed surfaces to impart specific functionalities such as improved mechanical strength, electrical conductivity, optical transparency, or chemical resistance. For example, by coating 3D-printed components with thin film layers of materials like graphene, carbon nanotubes, or metal oxides, manufacturers can enhance their structural integrity, conductivity, or corrosion resistance, expanding their range of applications in electronics, aerospace, and automotive industries. Moreover, the integration of thin films and 3D printing enables the fabrication of multi-material structures with graded or functionalized interfaces. By selectively depositing thin film coatings onto specific regions of a 3D-printed object, designers can create complex material gradients, surface textures, or functional patterns that optimize performance and functionality. This capability is particularly valuable in fields such as biomedical engineering, where customized implants, drug delivery systems, and tissue scaffolds with tailored surface properties are in high demand. Another advantage of integrating thin films and 3D printing lies in the potential for embedding functional elements directly into 3D-printed objects. Thin film sensors, actuators, or electronic components can be integrated into the layers of a 3D-printed structure during the printing process, enabling the fabrication of smart, responsive, and interactive devices. For instance, by embedding thin film sensors for temperature, pressure, or strain monitoring into 3D-printed parts, manufacturers can create intelligent components for predictive maintenance, structural health monitoring, or wearable technology applications. Furthermore, the combination of thin films and 3D printing facilitates the development of novel material systems and hybrid manufacturing processes. Researchers are exploring innovative techniques such as in-situ deposition of thin films during 3D printing, hybrid additive-subtractive manufacturing, or multi-material jetting with embedded thin film layers to achieve unprecedented levels of complexity, functionality, and performance in printed objects. These hybrid approaches offer new avenues for customized product design, rapid prototyping, and on-demand manufacturing across a wide range of industries.

In summary, the integration of thin films and 3D printing holds immense potential for advancing additive manufacturing capabilities and expanding the scope of applications in fields ranging from electronics and healthcare to aerospace and consumer goods. By combining the strengths of thin film technology and 3D printing, manufacturers can create highly customized, functional, and intelligent products with enhanced performance, efficiency, and sustainability. As research and development efforts continue to push the boundaries of both technologies, the integration of thin films and 3D printing is poised to drive innovation and disruption in the manufacturing landscape for years to come.